AF902505

VOLUME ONE HUNDRED AND ELEVEN

ADVANCES IN PROTEIN CHEMISTRY AND STRUCTURAL BIOLOGY

Protein-Protein Interactions in Human Disease, Part B

VOLUME ONE HUNDRED AND ELEVEN

ADVANCES IN PROTEIN CHEMISTRY AND STRUCTURAL BIOLOGY

Protein-Protein Interactions in Human Disease, Part B

Edited by

ROSSEN DONEV

Biomed Consult Ltd
Swansea, United Kingdom

ELSEVIER

ACADEMIC PRESS
An imprint of Elsevier

Academic Press is an imprint of Elsevier
50 Hampshire Street, 5th Floor, Cambridge, MA 02139, United States
525 B Street, Suite 1800, San Diego, CA 92101-4495, United States
The Boulevard, Langford Lane, Kidlington, Oxford OX5 1GB, United Kingdom
125 London Wall, London, EC2Y 5AS, United Kingdom

First edition 2018

Copyright © 2018 Elsevier Inc. All rights reserved.

No part of this publication may be reproduced or transmitted in any form or by any means, electronic or mechanical, including photocopying, recording, or any information storage and retrieval system, without permission in writing from the publisher. Details on how to seek permission, further information about the Publisher's permissions policies and our arrangements with organizations such as the Copyright Clearance Center and the Copyright Licensing Agency, can be found at our website: www.elsevier.com/permissions.

This book and the individual contributions contained in it are protected under copyright by the Publisher (other than as may be noted herein).

Notices
Knowledge and best practice in this field are constantly changing. As new research and experience broaden our understanding, changes in research methods, professional practices, or medical treatment may become necessary.

Practitioners and researchers must always rely on their own experience and knowledge in evaluating and using any information, methods, compounds, or experiments described herein. In using such information or methods they should be mindful of their own safety and the safety of others, including parties for whom they have a professional responsibility.

To the fullest extent of the law, neither the Publisher nor the authors, contributors, or editors, assume any liability for any injury and/or damage to persons or property as a matter of products liability, negligence or otherwise, or from any use or operation of any methods, products, instructions, or ideas contained in the material herein.

ISBN: 978-0-12-814342-1
ISSN: 1876-1623

For information on all Academic Press publications
visit our website at https://www.elsevier.com/books-and-journals

ELSEVIER Book Aid International Working together to grow libraries in developing countries

www.elsevier.com • www.bookaid.org

Publisher: Zoe Kruze
Acquisition Editor: Ashley M. Jackman
Editorial Project Manager: Alina Cleju
Production Project Manager: James Selvam
Cover Designer: Victoria Pearson

Typeset by SPi Global, India

CONTENTS

CONTRIBUTORS

Clara Aicart-Ramos
Universidad Complutense de Madrid, Madrid, Spain

Diego Alonso-López
Bioinformatics and Functional Genomics Group, Cancer Research Center (CiC-IMBCC, CSIC/USAL/IBSAL), Consejo Superior de Investigaciones Científicas (CSIC) and University of Salamanca (USAL), Salamanca, Spain

Mónica M. Arroyo
Bioinformatics and Functional Genomics Group, Cancer Research Center (CiC-IMBCC, CSIC/USAL/IBSAL), Consejo Superior de Investigaciones Científicas (CSIC) and University of Salamanca (USAL), Salamanca, Spain; Pontifical Catholic University of Puerto Rico (PCUPR), Ponce, Puerto Rico

Klaus Bister
Institute of Biochemistry and Center for Molecular Biosciences Innsbruck, University of Innsbruck, Innsbruck, Austria

Philippe Bouvet
Université de Lyon, Centre de Recherche en Cancérologie de Lyon, Centre Léon Bérard; Ecole Normale Supérieure de Lyon, Lyon, France

Mounira Chalabi-Dchar
Université de Lyon, Centre de Recherche en Cancérologie de Lyon, Centre Léon Bérard, Lyon, France

Carlos Costas-Insua
Universidad Complutense de Madrid, Madrid, Spain

Javier De Las Rivas
Bioinformatics and Functional Genomics Group, Cancer Research Center (CiC-IMBCC, CSIC/USAL/IBSAL), Consejo Superior de Investigaciones Científicas (CSIC) and University of Salamanca (USAL), Salamanca, Spain

C. George Priya Doss
School of Biosciences and Technology, VIT University, Vellore, Tamil Nadu, India

Seetharama D. Jois
Basic Pharmaceutical Sciences, School of Pharmacy, University of Louisiana at Monroe, Monroe, LA, United States

James L. Keck
University of Wisconsin School of Medicine and Public Health, Madison, WI, United States

Manuel Llano
University of Texas at El Paso, El Paso, TX, United States

Javier Merino-Gracia
Universidad Complutense de Madrid, Madrid, Spain

Mario A. Peña-Hernandez
University of Texas at El Paso, El Paso, TX, United States; Laboratorio de Inmunología y Virología, Facultad de Ciencias Biológicas, Universidad Autónoma de Nuevo León, San Nicolas de los Garza, Mexico

Maria Petrova
"Roumen Tsanev" Institute of Molecular Biology, Bulgarian Academy of Sciences, Sofia, Bulgaria

Ignacio Rodríguez-Crespo
Universidad Complutense de Madrid, Madrid, Spain

Alexander B. Sigalov
SignaBlok, Inc., Shrewsbury, MA, USA

Sitanshu S. Singh
Basic Pharmaceutical Sciences, School of Pharmacy, University of Louisiana at Monroe, Monroe, LA, United States

P. Sneha
School of Biosciences and Technology, VIT University, Vellore, Tamil Nadu, India

Eduard Stefan
Institute of Biochemistry and Center for Molecular Biosciences Innsbruck, University of Innsbruck, Innsbruck, Austria

D. Thirumal Kumar
School of Biosciences and Technology, VIT University, Vellore, Tamil Nadu, India

Jakob Troppmair
Daniel Swarovski Research Laboratory, Medical University of Innsbruck, Innsbruck, Austria

Iva Ugrinova
"Roumen Tsanev" Institute of Molecular Biology, Bulgarian Academy of Sciences, Sofia, Bulgaria

Jennifer Uppin
School of Biosciences and Technology, VIT University, Vellore, Tamil Nadu, India

S. Usha
School of Social Sciences and Languages, VIT University, Vellore, Tamil Nadu, India

Andrew F. Voter
University of Wisconsin School of Medicine and Public Health, Madison, WI, United States

PREFACE

Proteins interact with one another and with other biological macromolecules such as nucleic acids (DNA and RNA) and lipids. These interactions form a functional network that mediates signaling and metabolic pathways in individual cells, tissues, and organisms. While nucleic acids are primarily involved in coding and transcription of genetic information, proteins play a central role in biological function controlling the mechanisms that keep organisms in a healthy state. However, due to mutations/polymorphisms in proteins or imbalance in their expression, organisms develop metabolic and signaling dysfunction that leads to a state of disease. Thus, the molecular basis of diseases can be elucidated by in-depth study of protein interaction networks. The knowledge about the functioning of these interaction networks can lead to the design of new methods for diagnosis, prevention, and treatment of diseases that would lack or have minimal adverse effects.

This volume of the *Advances in Protein Chemistry and Structural Biology* is a second part dedicated to protein–protein interactions in human disease. While the previous volume (Volume 110) was focused on multidisciplinary approaches in applying cutting-edge techniques in detection and analysis of protein–protein interactions and the application of these techniques in identifying molecular mechanisms underlying different human diseases, the second part is centered on the role of protein complexes in different types of cancer, neurological disorders, and virus–host protein interactions. The design and development of therapeutic protein interaction target inhibitors is also discussed in several chapters. Thus, we hope that articles in this volume would inspire new ideas in researchers and specialists from a wide number of fields such as protein chemists, pharmacologists, medical doctors, cell biologists, biomedical researchers, immunologists, structural biologists, computational biochemists, and other researchers working in this field. This volume is also intended as a scholarly material for use by students with interest in the above topics.

Dr. Rossen Donev
MicroPharm Ltd and Biomed Consult,
United Kingdom

CHAPTER ONE

Homo- and Heterodimerization of Proteins in Cell Signaling: Inhibition and Drug Design

Sitanshu S. Singh, Seetharama D. Jois[1]
Basic Pharmaceutical Sciences, School of Pharmacy, University of Louisiana at Monroe, Monroe, LA, United States
[1]Corresponding author: e-mail address: jois@ulm.edu

Contents

Advances in Protein Chemistry and Structural Biology, Volume 111
ISSN 1876-1623
https://doi.org/10.1016/bs.apcsb.2017.08.003
© 2018 Elsevier Inc.
All rights reserved.

Abstract

Protein dimerization controls many physiological processes in the body. Proteins form homo-, hetero-, or oligomerization in the cellular environment to regulate the cellular processes. Any deregulation of these processes may result in a disease state. Protein–protein interactions (PPIs) can be inhibited by antibodies, small molecules, or peptides, and inhibition of PPI has therapeutic value. PPI drug discovery research has steadily increased in the last decade, and a few PPI inhibitors have already reached the pharmaceutical market. Several PPI inhibitors are in clinical trials. With advancements in structural and molecular biology methods, several methods are now available to study protein homo- and heterodimerization and their inhibition by drug-like molecules. Recently developed methods to study PPI such as proximity ligation assay and enzyme-fragment complementation assay that detect the PPI in the cellular environment are described with examples. At present, the methods used to design PPI inhibitors can be classified into three major groups: (1) structure-based drug design, (2) high-throughput screening, and (3) fragment-based drug design. In this chapter, we have described some of the experimental methods to study PPIs and their inhibition. Examples of homo- and heterodimers of proteins, their structural and functional aspects, and some of the inhibitors that have clinical importance are discussed. The design of PPI inhibitors of epidermal growth factor receptor heterodimers and CD2–CD58 is discussed in detail.

1. INTRODUCTION

Most of the physiological processes in the body are controlled by cellular interactions that, in turn, are controlled by interacting biomolecules. Among the biomolecules, proteins are responsible for most of the biochemical pathways that control the physiological processes. Proteins seldom act alone; they bind to other proteins or biomolecules, eliciting a physiological response. It is estimated that there are nearly 650,000 such interactions that control actions that enable the human body to function normally (Stumpf et al., 2008). The complex network of protein–protein interactions (PPIs) that carries out the biological process in an organism is termed "interactome" (Bogan & Thorn, 1998). A number of proteins self-associate to form dimers or oligomers and also form heterodimers. Homodimerization and heterodimerization of proteins regulate several of the biochemical pathways, and any deregulation of this process leads to disease states.

The association of two proteins could result in homodimers (complexation of identical monomers) or heterodimers (complexation of nonidentical monomers) in the cellular environment. Apart from these complexes, proteins can also form oligomers either to perform functions in cells or, in some cases, to control the functions of these proteins. Protein homo–hetero- and oligomerizations can be classified as stable or transient, depending on the timescale used, and the method of detection used. Based on their affinity, a lifetime of the complex, and composition, PPIs are classified as (i) homo- and hetero-oligomeric complexes, (ii) nonobligate and obligate complexes, and (iii) transient and permanent complexes (Acuner Ozbabacan, Engin, Gursoy, & Keskin, 2011). As described earlier, homo- and hetero complexation depends on identical or nonidentical monomers that form the complex. Whether they are classified as obligate or nonobligate is based on whether the monomers of the complex exist in the stable form in vivo on their own or not. An example of obligate homodimers is DNA-binding homodimer Ku proteins (Krishna & Aravind, 2010). Nonobligate proteins dissociate after they carry out a biochemical process. Protein complexes that participate in the signaling process form transient dimers, and after signaling, they dissociate and hence are examples of nonobligate interacting proteins. Whether they are transient or permanent PPI depends on the lifetime of the complex. Permanent interactions have K_d values $<\mu$M, whereas strong transient interactions have K_d values in the nM range, and weak transient interactions have K_d values in the μM range. Enzymes are extensively studied in terms of structure and multimerization; among the reported 452 human enzymes, only one-third are monomers, and the remaining enzymes are known to form homomultimers (Marianayagam, Sunde, & Matthews, 2004). Heterodimeric interactions are commonly found in enzyme inhibitors, enzyme complexes, antibody–antigen, signal proteins, and cell cycle proteins (Sowmya, Breen, & Ranganathan, 2015). The well-known G-protein-coupled receptors (GPCRs) are known to form dimers and oligomers. For proteins that undergo oligomerization, the equilibration between monomer–dimer and oligomer kinetics seems to control the physiological activity. In this chapter, we have provided some of the methods used for detecting PPI and its inhibition with examples. Most of the PPI inhibitors we described here are used for the purpose of modulation of biochemical pathways and for therapeutic purposes. It is not our intention to exhaustively cover the PPI inhibitors. We have highlighted some well-known examples of PPIs that are described in the literature and have covered some recent examples of PPIs and their inhibition. Readers can refer to reviews described in the literature for more extensive coverage (Arkin, Tang, & Wells, 2014; Bakail & Ochsenbein, 2016; Fry et al., 2013;

Guo, Wisniewski, & Ji, 2014; Iyer, 2016; Jin, Wang, & Fang, 2014; London, Raveh, & Schueler-Furman, 2013; Morelli, Bourgeas, & Roche, 2011; Petta, Lievens, Libert, Tavernier, & De Bosscher, 2016; Sable & Jois, 2015; Skwarczynska & Ottmann, 2015; Zhang, Ben-David, & Sidhu, 2016). Examples of PPI inhibition related to epidermal growth factor receptor (EGFR) and CD2–CD58 that we have been working on in our laboratory are provided in detail.

2. METHODS TO STUDY PPIs AND THEIR INHIBITION

Most cellular assays used to screen compounds for biological activity or inhibitors of proteins employ either enzymatic assays or drug-like molecules binding to one of the proteins of interest, or measure the IC_{50} values of inhibition. These screening methods do not provide information about whether the designed compounds inhibit PPI. When PPI inhibitors are designed, pharmacological assays do not provide direct evidence of PPI inhibition and hence experimental methods that provide evidence of PPI inhibition have to be provided. We have highlighted some of the most commonly used methods of PPI and its inhibition. Coimmunoprecipitation and pull-down assays are used in the early stages of PPI inhibitor design, whereas assays such as surface plasmon resonance (SPR), proximity ligation assay (PLA), and enzyme fragment complementation assays are used in the later stages.

2.1 Coimmunoprecipitation

Coimmunoprecipitation (Hall, 2005) is an assay that is used to analyze PPIs from cells and provide physiologically relevant information concerning PPI since proteins are extracted from their native environment in most cases. In principle, the assay is similar to immunoprecipitation (IP), with modifications for studying PPI. In an IP assay, an antibody forms an immune complex with the protein of interest. This complex is captured using an immobilized protein (on a bead) ligand that binds to the antibody; hence the immune complex is captured on the bead. Washing the beads will wash away any proteins not precipitated on the beads. The antigen–antibody complex is eluted from the support, analyzed by sodium dodecyl sulfate-polyacrylamide gel electrophoresis (SDS-PAGE), and detected by Western blotting. In co-IP assays, the antibody–antigen complex will have a binding partner of the antigen; this means that the target antigen precipitated by the antibody "coprecipitates" with a binding partner/protein complex from a lysate. The detection of the coprecipitates provides information about the interacting partner protein. The two proteins can be identified by their respective antibodies in a Western blot assay, which can be performed in any laboratory equipped to perform biochemistry or molecular biology techniques. Although the assay is quite straightforward and sounds simple, there are some limitations. Antibody contamination and nonspecific binding are the major factors related to negative results of the assay. Apart from this, the stability of the complex of two proteins poses a challenge during the performance of the experiments. Since the assay involves several washing steps, the condition of the washing buffers and the way the precipitate is handled can cause the complex to dissociate, resulting in only antibody–one protein complex detection. In addition, if the two proteins interact with low affinity or if the interactions are transient, the proteins that take part in PPI may not be detected. Many protein complexes will remain intact after lysis using standard nondenaturing lysis buffers such as buffers with low ionic strength (i.e., <120 mM NaCl) that contain nonionic detergents (NP-40 and Triton X-100). To prevent the disruption of the interaction between the proteins and loss of PPI in the sample, harsh conditions such as cell lysing by sonication or vortexing should be avoided. The samples should be handled gently to prevent the loss of bound complex proteins during centrifugation and washing. A different version of the same assay using a cross-linking technique can be used to stabilize the PPI. Depending on the size of the cross-linking agent, proteins that are interacting at a particular distance will be cross-linked, and these can be immunoprecipitated to detect a particular protein pair. More specific methods using highly specific association between streptavidin and biotin can be used. A wide range of affinity resins, magnetic beads, and coated plates based on immobilized avidin, streptavidin, are available commercially to be used in co-IP (Adams, Seeholzer, & Ohh, 2002; Hall, 2005).

2.2 Pull-Down Assay

This assay can be used as an initial screening assay to identify previously unknown PPI. It can be used along with co-IP to prove that two proteins interact with one another in a hypothesized biochemical pathway. In the pull-down assay, to capture the proteins that interact with one another, a "bait" protein is used (Nguyen & Goodrich, 2006). The bait protein is tagged with affinity tags such as histidine, glutathione, or biotin. These tags will bind to an immobilized affinity ligand in a column. The bait protein and another protein that interacts with the bait protein that forms PPI will both be captured on the affinity column and purified, and the proteins that interact with one another can be characterized. Tags can be attached during protein expression or after protein purification, depending on the type of affinity tag used.

The complex formed is eluted from the column and detected to confirm the presence of proteins that interact with one another. Stable protein complex formed can be washed with high ionic strength to eliminate any nonspecific interactions. If the complex has weak interactions, assay conditions can be modified by pH and salt concentration. SDS-PAGE is a harsh treatment that denatures all proteins in the sample and thus restricts analysis. A nondenaturing method called competitive analyte elution can elute a biologically functional protein complex. In all pull-down assays, control experiments are necessary to generate biologically significant results.

2.3 Proximity Ligation Assay

PLA is an assay that can be used to study endogenous PPIs (Fredriksson et al., 2002; Trifilieff et al., 2011). It can be used to study PPI in in vitro or in vivo tissue samples. There are different versions of the PLA; here, we describe antibody-based and fluorescent visualization assays. This assay is based on the principle that, if two proteins interact and the distance between the proteins is within 16 nm, the interaction between the proteins can be detected by specific antibodies against these proteins. The primary antibodies used should be from different species so that secondary antibodies with probes can be detected. Furthermore, using this assay, one can quantify PPI in cells, and PPI inhibition can be studied based on the decrease in the number of PPI probes detected. The important consideration in this assay is the availability of antibodies to proteins of interest. The overall assay principle is depicted in Fig. 1.

The two proteins of interest are targeted with primary antibodies, one from mouse and another from rabbit. These antibodies bind to two proteins of interest, and secondary antibodies (probes) to the primary antibodies are added. The secondary antibodies that bind to primary antibodies have DNA probes. Two oligonucleotides are added that bind to DNA probes on the primary antibodies, and then ligase is added. The DNA forms a circle if the two primary antibodies are within a certain distance (Fig. 1). Once polymerase and nucleotides are added, a rolling circle amplification of probe DNA takes place. The amplified probe DNA is detected with a fluorescent

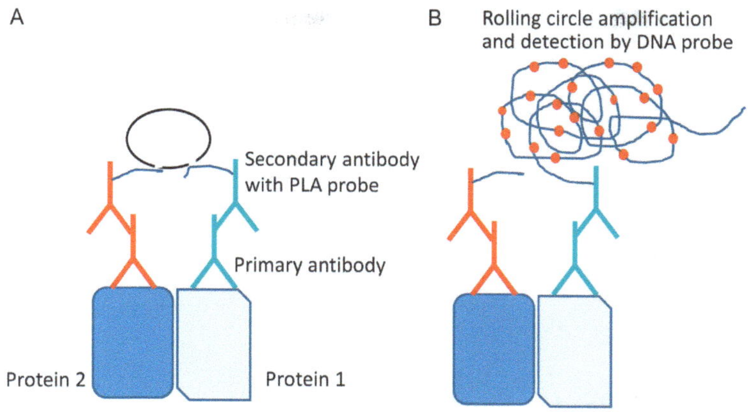

Fig. 1 The principle of proximity ligation assay (PLA). (A) Two proteins of interest are targeted by primary antibody from different species. Corresponding secondary antibodies with DNA probes are added. If the two proteins are in proximity the hybridized DNA will be used for rolling circle amplification. (B) Amplified DNA will be detected by a DNA probe. For visualization, a DNA probe with red fluorescence is used. Each dimer of protein in cells is viewed as a *red dot* with a high-resolution microscope. *Adapted from Trifilieff, P., Rives, M. L., Urizar, E., Piskorowski, R. A., Vishwasrao, H. D., Castrillon, J., et al. (2011). Detection of antigen interactions ex vivo by proximity ligation assay: Endogenous dopamine D2-adenosine A2A receptor complexes in the striatum.* BioTechniques, 51*(2), 111–118. https://doi.org/10.2144/000113719.*

probe. When a high-resolution microscope is used, each PPI is visible as red dot due to rolling circle amplification of the probe DNA. Thus, each red dot/fluorescent dot corresponds to a dimer pair of proteins. The assay can be carried out on fixed cells or tissues. As an example, we provide the HER2:HER3 interaction. PLA was used to demonstrate that HER2: HER3 PPI is present in SKBR-3 breast cancer cells. Addition of a peptidomimetic to cells such as SKBR-3 results in inhibition of PPI, as shown by PLA (Fig. 2). The PPI is shown as red fluorescent dots in the cells without any treatment; upon addition of a PPI inhibitory compound at different concentrations, the number of red fluorescent dots decreased, indicating inhibition of a particular set of PPI (in the figure, the PPI is HER2: HER3). The results can also be quantified for dose–response curves.

2.4 Enzyme Fragment Complementation Assay (PathHunter Assay)

PathHunter™ assay (DiscoveRx, Fremont, CA) or enzyme fragment complementation assay can be used to confirm the interaction between two proteins and to evaluate whether the ligand designed inhibits the PPIs. These

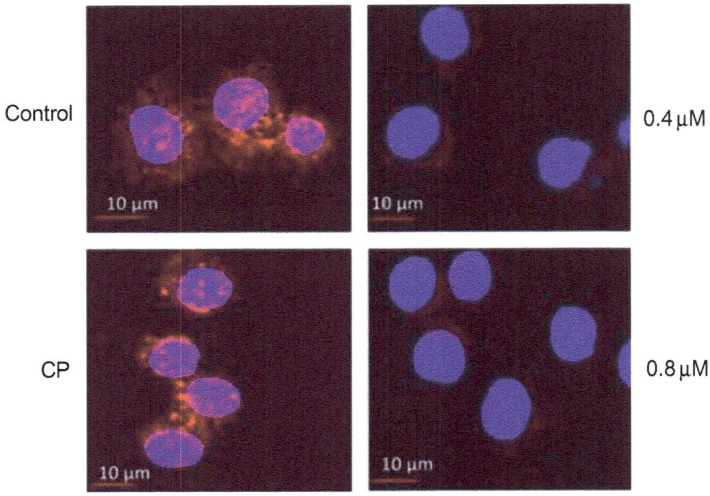

Fig. 2 HER2:HER3 heterodimerization and its inhibition by compound **9** observed by PLA. Concentration-dependent inhibition of HER2:HER3 heterodimers was observed. Control-SKBR-3 cells showing only HER2:HER3 heterodimerization as *red spots*. CP-A control compound did not inhibit HER2:HER3 heterodimerization. At a sub-optimum dose of compound **9** (0.4 μM), HER2:HER3 heterodimerization was inhibited to a lesser extent. At an optimum dose of compound **9** (0.8 μM), HER2:HER3 heterodimerization was significantly inhibited. *Reproduced with permission from Kanthala, S., Banappagari, S., Gokhale, A., Liu, Y. Y., Xin, G., Zhao, Y., & Jois, S. (2015). Novel peptidomimetics for inhibition of HER2:HER3 heterodimerization in HER2-positive breast cancer.* Chemical Biology & Drug Design, 85, *702–714. https://doi.org/10.1111/cbdd.12453. Copyright (2014) John Wiley and Sons.*

assays may only be applicable to certain proteins since the method requires the expression of proteins of interest in model cell lines, and the two proteins are attached with tags of fragments of beta-galactosidase. The assay is widely used for dimerization of receptor tyrosine kinase (RTK) proteins. The enzyme beta-galactosidase is split into two inactive fragments. The two fragments are expressed as tags in proteins of interest using engineered cells. When the two proteins interact with one another, the tags from the two proteins which were inactive form an active complex. The activity of the complex can be detected by an enzymatic assay using luminescence. The relative luminescence intensity can be quantified as the formation of PPI (Yin et al., 2009). In cells, RTK activation results in dimerization of receptors, and this leads to phosphorylation and binding of the SH2 domain to the kinase domain. When RTK and SH2 domains come into proximity, the inactive galactosidase fragments come together to form the active enzyme. The active enzyme produces a chemiluminescent signal that can be detected and directly correlated to kinase activation, phosphorylation, and interaction with SH2 domain-containing proteins. This assay is specific for RTK dimerization. The limitation of this method in the present form is that the assay is carried out on engineered cells and not on native forms. However, the assay is used to evaluate the inhibition of dimerization also. The application of this method is used in studying G-protein-coupled receptor (GPCR) binding by different proteins. Since GPCR is the target for most of the drugs, the assay is useful in designing drug-like molecules (Yin et al., 2009). In our laboratory, we have used the enzyme fragment complementation assay to evaluate the ability of peptidomimetics to inhibit HER2–HER3 dimerization in U2OS cell lines (Fig. 3).

2.5 Surface Plasmon Resonance

While many methods such as IP and PLA as well as enzyme fragment complementation assays provide information about PPI, the detection methods are indirect, and methods such as IP involve several steps. A technique that directly detects the interaction of two purified proteins is SPR (de Mol & Fischer, 2010; Wilson, 2002). In this method, one of the proteins is immobilized on a sensor chip surface, the other is made to flow over this surface, and the binding kinetics are detected by a change in the refractive

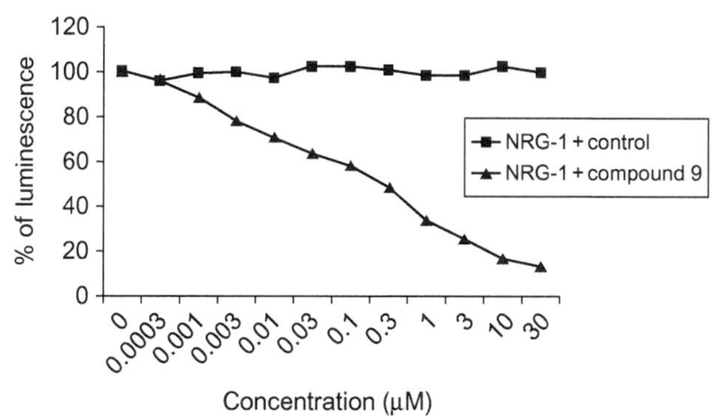

Fig. 3 Inhibition of heterodimerization of HER2:HER3 in HER2, HER3 transfected U2OS cells by compound **9** at different concentrations using enzyme fragment complementation assay (DiscoveRx). Dose–response curve for inhibition of HER2:HER3 heterodimerization by compound **9** in the presence of 0.3 μM NRG1 (*triangles*). Control compound (CP) in the presence of 0.3 μM NRG-1 (*filled squares*). *Reproduced with permission from Kanthala, S., Banappagari, S., Gokhale, A., Liu, Y. Y., Xin, G., Zhao, Y., & Jois, S. (2015). Novel peptidomimetics for inhibition of HER2:HER3 heterodimerization in HER2-positive breast cancer.* Chemical Biology & Drug Design, 85, *702–714. https://doi.org/10.1111/cbdd.12453. Copyright (2014) John Wiley and Sons.*

index. The sensor chip is created by applying a thin layer of gold to a glass surface. A dextran matrix is applied to the gold to create an environment for biomolecular interactions, and this surface can be combined with or modified by different chemicals to provide linkers for different types of chemical reactions for immobilization. Most widely used sensor chips are carboxymethyl-dextran linked to a gold surface. Such layers are useful for immobilization of proteins and peptides that have free amine groups. SPR chips that are ready to use are available from commercial sources (GE Healthcare, Pittsburgh, PA). A source of light passes through a prism and strikes the surface of the flow cell at an angle such that the beam is totally reflected. This light beam creates surface plasmon wave of excited electrons on the gold surface (Fig. 4A). Although the light undergoes total reflection, an electromagnetic field component called an evanescent wave penetrates into the medium <500 nm of the lower refractive index, in this case, into the dextran layer where PPIs can be studied. Depending on the material on the gold surface and dextran layer, the reflective beam will have a reduced intensity. The angle at which the light beam strikes the detector from a reference surface point is called the SPR angle. The SPR angle is sensitive to the composition of the layer at the surface of the gold. As two molecules interact with one another, there will be an accumulation of molecules at the dextran surface because of the binding of one protein to another; hence, the refractive index of the surface changes, causing a change in the SPR angle.

The SPR angle is directly proportional to the amount of bound molecules at the surface with respect to a reference surface. If the two molecules do not interact, there is no binding, and the SPR angle will not change; therefore, there will be no SPR signal. Thus, PPIs can be directly detected. A schematic diagram of SPR sensorgrams of the interaction of two proteins

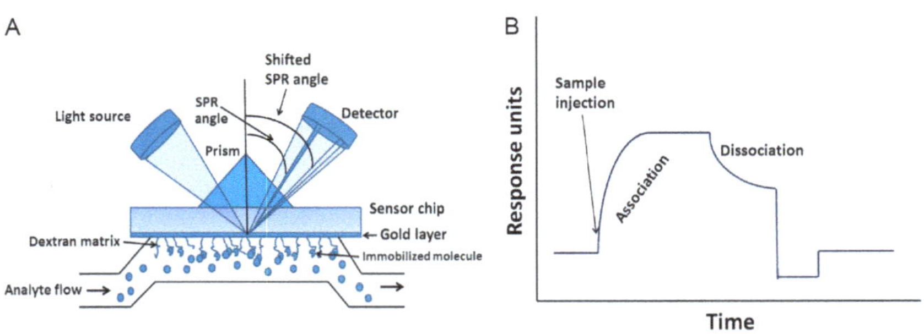

Fig. 4 The principle of SPR analysis. (A) SPR chip with analyte flow and SPR angle. (B) SPR sensorgram indicating association and dissociation phases.

is shown in Fig. 4B. The advantages of this method are that no labeling is required for molecules and the interaction between the molecules is detected in real time. This means that the kinetics of association and dissociation can be measured, which is difficult in many other experiments that determine the binding affinity of two molecules. SPR technology requires a very small amount of sample. With BIACORE technology, 1000 resonance units (RU) corresponds to an SPR angle of 0.1 degrees. For most proteins, binding of 1 ng/mm^2 of protein at a dextran surface causes a change of 1000 RU. This technique also determines interaction affinity of molecules over a wide range of binding affinity (Table 1) and molecular weights. The area of detection of this interaction extends up to 300 nm from the gold surface and, therefore, the interaction of an immobilized protein with a protein on a cell surface can be detected using the proper flow rate and buffer.

SPR chips have two- or three-channel flow cells. The volume of each flow cell is around 0.02 μL and, hence, the amount of sample needed is very small. After the protein has been immobilized on the chip surface, the chip can be reused for only a certain period of time since the bound protein or ligand can be washed away; however, regeneration buffer can be used to make the chip surface available for a second analyte or another experiment. An SPR sensorgram consists of different phases—a baseline to start with and, after injection of the analyte if there is binding, an association phase and steady state where the binding equilibrium reaches saturation, and a dissociation phase. After dissociation, the chip can be regenerated, which is referred to as the regeneration phase. Binding kinetics can be obtained by titrating the analyte over a wide variety of concentration ranges against the protein immobilized on the chip surface. A typical binding experiment is shown in Fig. 4B. Apart from protein–protein or protein–ligand binding, PPI inhibition can be studied using the coinjection mode (Banappagari, Corti, Pincus, & Satyanarayanajois, 2012).

Table 1 Range of Binding and Kinetics Parameters That Can Be Measured Using SPR Analysis

Concentration Range	Kinetics	Affinity
1 mM–10 pM	Association rate constant $k_a = 10^3$–$10^7\ M^{-1} s^{-1}$	K_d 100 μM to 200 pM
	Dissociation rate constant $k_d = 10^{-1}$–$5 \times 10^{-6}\ s^{-1}$	

2.6 Mutational Studies to Identify Hot Spots in PPI

Most of the PPI inhibitors target the interface of proteins in small regions called hot spots. A region of protein surface is called a hot spot when replacement of an amino acid residue by alanine in that spot lowers the free energy of binding by at least 2 kcal/mol (Clackson & Wells, 1995). Computational as well as experimental methods are available to identify hot spots on the surface of interacting proteins. Experimental study involves very tedious work where the two proteins of interest are expressed and purified, and binding of the two proteins is studied by isothermal calorimetry or any other method that can detect protein binding. Then, particular amino acids on each protein are mutated with alanine using site-directed mutagenesis, and binding studies on the mutated proteins are carried out. The change in the affinity of binding upon mutation is evaluated. One such example is the interaction between CD2 and CD58 proteins that are involved in cell adhesion and immune response. As an example, mutation studies carried out on CD2–CD58 protein pairs are illustrated in Fig. 5A and B. Point mutation was carried out on the proposed binding of the CD2 protein to CD58 protein (Kim et al., 2001). The crystal structure of CD2 complexed with CD58 has been reported (Wang et al., 1999). Detailed analysis of amino acid residues involved in PPI of CD2 and CD58 indicated that the interface

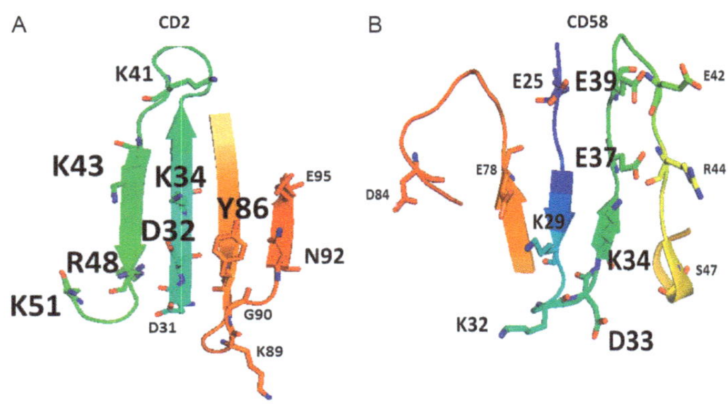

Fig. 5 The structure of PPI domains of (A) CD2 and (B) CD58 with amino acids that form hydrogen bonding, salt bridges, and hydrophobic interaction (PDB ID 1QA9). Mutation studies suggested that the affinity of interaction between the two proteins is reduced when alanine is replaced by some of the residues in the proteins (Kim et al., 2001; Wang et al., 1999). Residues that have the most effect on affinity of binding are represented by large-sized letters for amino acid labels. Residues that partially affect the affinity are represented with *medium-sized letters*. Residues that do not affect the affinity by replacement with alanine are represented with *small-sized letters*.

interaction is mainly electrostatic in nature with 10 salt bridges and 5 hydrogen bonds (Wang et al., 1999) and that the interface area is around 1200 $Å^2$. When the amino acid residues that form salt bridges and hydrogen bonds in the two proteins were replaced with alanine by point mutation, the interaction between CD2 and CD58 was not significantly altered, suggesting that electrostatic interactions did not contribute significantly to form the heterodimer. When a hydrophobic amino acid Tyr86 in the interface region was subjected to point mutation by replacement of alanine, the binding affinity between the two proteins reduced nearly by 1000-fold, suggesting that the hot spot is a hydrophobic region. Based on this mutation study, the researchers proposed that hot spots on these proteins are at Tyr86 (Fig. 5A). Such hot spots can be used for the design of PPI inhibitors.

Other experimental techniques such as fluorescence resonance energy transfer (FRET) (Karpova & McNally, 2006), bioluminescence resonance energy transfer (Kocan & Pfleger, 2011), mass spectrometry (Kaake et al., 2014), and isothermal calorimetry (Velazquez-Campoy, Leavitt, & Freire, 2004) are also available for studying PPI (Berg, 2005). In terms of investigating PPI in the pure form of the protein, NMR and X-ray crystallography are used. These methods provide detailed knowledge about the binding surface of the two proteins involved. However, protein purification, crystallization, and data collection are time consuming. Before proceeding with NMR or X-ray crystallography to determine details of PPI, relatively faster and physiologically more important methods such as Co-IP and pull-down assay are used.

3. INHIBITION OF DIMERIZATION OF PROTEIN AND PPI

The first step in the design of PPI inhibitors is to evaluate whether the proteins that interact with one another and the biochemical pathway that results have any significance in any disease state. If they have importance in a disease state, will the inhibition of PPI lead to alleviation of this disease state? Most of the data related to the PPI importance in a disease state come from basic and clinical studies or genomics studies related to the disease. Once the proteins of interest that have significance in a disease state have been identified, the next question to ask is whether the inhibition of this particular PPI leads to any significant adverse effects on the general physiological processes in the body. If the PPI inhibition might result in significant effects on normal physiology, then that PPI should not be targeted for the drug discovery process. Once the target PPI partners are identified, is there a model system such as cell lines that is relevant for the disease state? If the cell lines chosen express the proteins of interest, they would serve as a good model, and molecular biology methods such as the coimmunoprecipitation assay and pull-down assay (described earlier in methods to study PPI) can be used to show that the two proteins of interest form heterodimers. On the other hand, if the crystal structures of the complex of proteins of interest have already been elucidated and are available in the literature, one can jump directly to the design of inhibitors. There are several steps in the design of PPI inhibitors. The overall process of the design of PPI inhibitors is schematically represented in Fig. 6.

As more and more crystal structures of the physiologically important proteins became available, and interactions of proteins were deciphered, it was realized that PPI surfaces are important in modulating the physiological processes for alleviating many disease states. PPIs control most of the physiological processes in the body, and any deregulation in PPI leads to disease states (Chene, 2006). This understanding led to a new area in drug design and discovery in which the molecules designed have to bind to PPI surfaces. In the early 1980s and 1990s, it was presumed that PPI surfaces were rather flat, and that inhibiting such surfaces could be done only by large molecules such as antibodies or fusion proteins (Wells & McClendon, 2007). However, this concept was overridden when peptides, peptidomimetics, and small molecules were designed to inhibit PPI. Typically, PPI surfaces have a span of 700–3000 $Å^2$ and have shallow grooves and channels or small ridges that can accommodate functional hydrophobic groups to interact with the receptor protein. For such large surface interactions, designing a drug seems difficult compared to designing traditional small molecules having drug-like properties (Lo Conte, Chothia, & Janin, 1999; Moreira, Fernandes, & Ramos, 2007; Reichmann, Rahat, Cohen, Neuvirth, & Schreiber, 2007; Wells & McClendon, 2007).

Detailed 3D-structure analyses of PPI revealed that PPIs are made up of secondary structure epitopes that are derived from a continuous or a discontinuous epitope. It is well documented in the literature that PPI surfaces are generally hydrophobic in nature and all of the interfaces between the two proteins do not necessarily contributes to binding energy. Small hydrophobic spots contribute to the free energy of binding and help to hold the two proteins together. Such regions on PPI interfaces are called hot spots (Clackson & Wells, 1995). These hot spots have a core region and a rim region; the amino acid composition of the core region contains aromatic residues, whereas, in the rim region, the amino acid composition is similar to

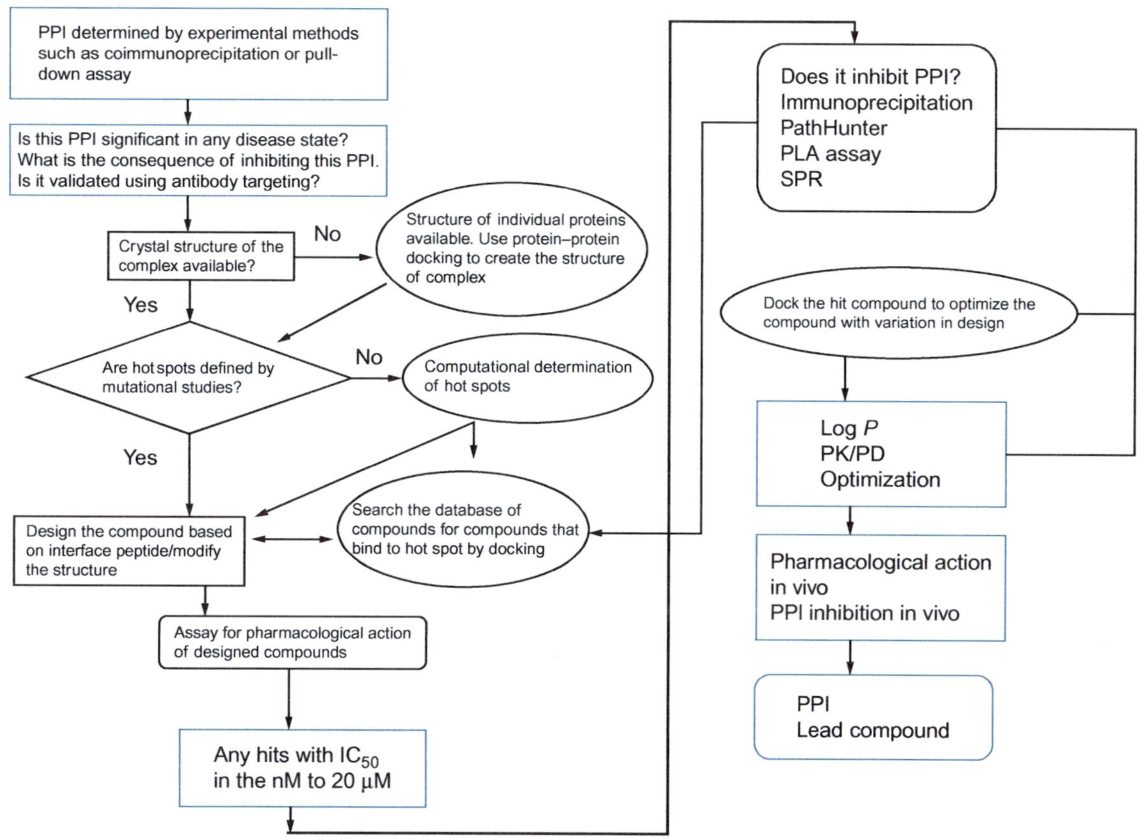

Fig. 6 A schematic diagram of a decision tree for the design of PPI inhibitors. *Adapted from Sable, R., & Jois, S. (2015). Surfing the protein-protein interaction surface using docking methods: Application to the design of PPI inhibitors.* Molecules, 20(6), 11569–11603. *https://doi.org/10.3390/molecules200611569. MDPI.*

that of the rest of the protein surface (Chakrabarti & Janin, 2002; Chene, 2006; DeLano, 2002). Amino acids that were frequently found in PPI hot spots were Trp, Tyr, Leu, Ile, Phe, and Arg (Bogan & Thorn, 1998). Amino acid Trp can form hydrophobic as well as hydrogen bonds with ligand molecules without the introduction of water at the PPI site. Among hydrophobic amino acids with similar structural properties, such as Leu and Ile, Ile seems to be preferred at PPI (Moreira et al., 2007). Based on the knowledge of PPI surface, compounds that can bind to PPI surface can be designed.

4. DESIGN OF COMPOUNDS BASED ON PROTEIN INTERFACE

At present the methods used to design the PPI inhibitors can be classified into three major groups: (1) structure-based drug design (SBDD), (2) high-throughput screening (HTS), and (3) fragment-based drug design (FBDD). Each of these methods is described briefly here.

4.1 Structure-Based Drug Design

SBDD uses the fact that peptide fragments that are involved in PPI surfaces can be mimicked by peptides, peptidomimetics, or small organic molecules. These molecules are called "proteomimics" (Jubb, Higueruelo, Winter, & Blundell, 2012). Since PPIs are concentrated in small regions, the interface contact is held by a few key residues that are arranged in a particular three-dimensional arrangement. The functional groups and geometric arrangement of these functional groups that exist in the interface of proteins can form a template for pharmacophores. These protein recognition sites usually consist of particular secondary structures that can be used in the design of PPI inhibitors. The most common are α-helical structures, extended or β-strand structures, loops, β-turns, and proline-rich motifs (Gokhale & Satyanarayanajois, 2014; Wilson, 2009). However, these small fragments of peptides acquire particular stable secondary structures in a globular protein because of different interactions within the protein structure. When such fragments of peptide structures are isolated, they do not acquire stable secondary structures that mimic those of the proteins. Hence, these secondary structures have to be constrained using different functional groups and different strategies. One of the best examples of constrained secondary structure is that of an α-helix mimic for p53–MDM2 PPI is important in the regulation of cancer development (Vassilev et al., 2004).

4.2 Fragment-Based Discovery in PPI Inhibition

FBDD concept was initiated three decades ago (Jencks, 1981; Verlinde, Rudenko, & Hol, 1992); however, the method was restricted to enzyme-based drug discovery and, in particular, applied to kinase inhibitors. The first success of this method was achieved in 2011 after the approval by the Food and Drug Administration (FDA) of the drug Zelboraf (PLX4032) (Bollag et al., 2010), a B-Raf inhibitor. The same principle can be applied to the design of PPI inhibitors. In the first step in this method, a ligand is constructed from fragments of chemical building blocks that are optimized for binding to small subsites on protein surfaces. A library of low molecular weight fragments are screened for binding to proteins of interest (Scott, Coyne, Hudson, & Abell, 2012). Screening can be performed by different experimental techniques such as fluorescence assay or NMR. Using a method called structure–activity relationship (SAR) by NMR (Shuker, Hajduk, Meadows, & Fesik, 1996), a large number of small molecular fragments can be screened to bind to one subsite of a protein surface. In the second step, another set of compounds is screened to bind to another site (near the first site). Lead compounds that show relatively high-affinity binding are evaluated further to find the exact binding site of these two fragments on the protein surface. Then, using the medicinal chemistry knowledge, these two fragments are linked chemically to arrive at the new lead compound (Erlanson, 2006). Usually, the linked compound will have higher affinity to bind to protein compared to individual fragments that bind to different sites (Fig. 7A and B). Note that the fragments of compounds designed in the first and second steps bind with low affinity to the target protein (Fig. 8). When the fragments are linked, the affinity of the new molecule

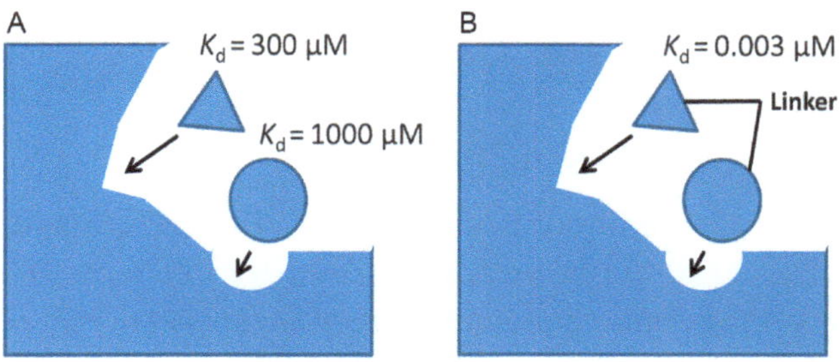

Fig. 7 A schematic representation of the fragment-based drug design principle. (A) Fragments are screened from a database or are built based on PPI hot spots on the protein. Each fragment has a relatively weak dissociation constant. (B) The screened fragments are linked chemically to obtain a high-affinity binding ligand.

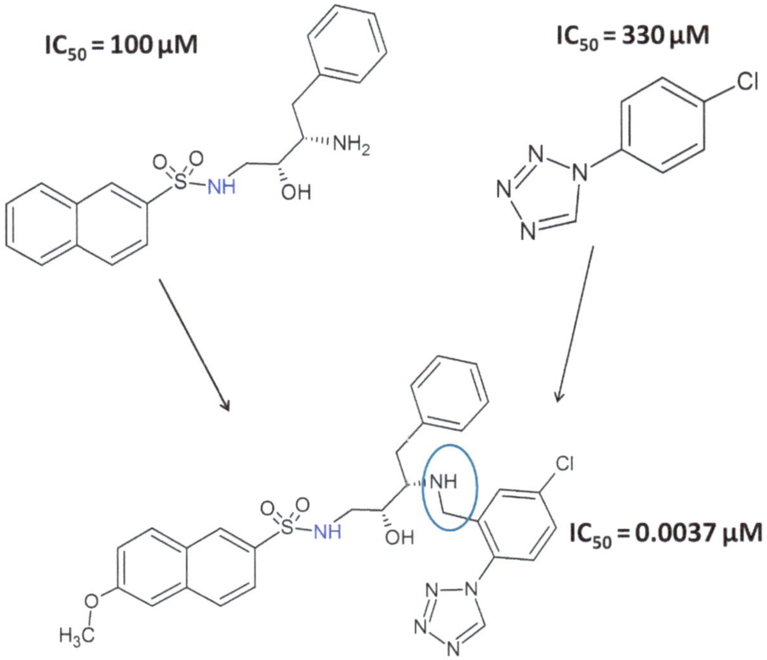

Fig. 8 Example of fragment-based approach. Two molecules with IC_{50} values of 100 and 330 μM, respectively, were linked to obtain a high-affinity ligand that has an IC_{50} value of 37 nM. *Adapted from Erlanson, D. A. (2006). Fragment-based lead discovery: A chemical update.* Current Opinion in Biotechnology, 17*(6), 643–652. https://doi.org/10.1016/j.copbio.2006.10.007. Copyright (2006) Elsevier.*

is significantly increased. The overall idea is to build a molecule that can fit into the shallow groove of the PPI using fragments of chemicals and knowledge of the binding site. An example of the SAR by the NMR method is the discovery of BH3-Bcl-X_L inhibitor (Oltersdorf et al., 2005). Apart from these methods, natural product screening is also used for PPI inhibition (Sperl, Seifert, & Berg, 2009).

4.3 High-Throughput Screening

Conventional tools used in drug discovery such as HTS (Fox et al., 2006; Mayr & Fuerst, 2008) can be used to discover PPI inhibitors. The first step in such cases is the creation of the structure of a complex of two proteins and identification of hot spots on one of the protein interfaces. If the hot spots have a cleft or a relatively deep pocket, the databases of the molecules are screened using experimental as well as computational analysis to find out whether any of the organic molecules bind to the cleft. A compound that binds to the groove in the lower nanomolar range will be used for lead compound generation. The HTS method involves screening large numbers of compounds in a chemical library to find a lead compound that binds to the target protein with high affinity ($K_d < 1\,\mu M$). One important aspect of HTS is the development of a fast, reliable cell-based or enzyme-based assay to screen large numbers of compounds. Usually, these assays are carried out in microwell plates that have 1536 wells per plate and use 2.5–10 μL of the medium. With such microplates, one can screen 200,000 compounds per day (Mayr & Fuerst, 2008). In the pharmaceutical and biotechnology industries, these systems are completely automated once the assay is established. For PPI inhibitor studies, such assays are modified to evaluate the binding of these compounds to one of the protein partners. Identification of a small molecule that inhibits the interaction of p53 with MDM2 provides an example of the use of HTS (Vassilev et al., 2004).

5. HOMODIMERS

It is reported that in eukaryotic organisms that contain a large number of self-interacting proteins and the ability to self-interact have several structural and functional advantages over proteins (Ispolatov, Yuryev, Mazo, & Maslov, 2005). Self-association of proteins is known to provide improved stability (Dunbar et al., 2004) and control over the accessibility and specificity of active sites in proteins (Marianayagam et al., 2004). Furthermore, self-association can help to minimize genome size. Apart from self-association, structurally similar proteins have statistically significant interaction propensity compared to structurally nonsimilar proteins (Ispolatov et al., 2005). Structural analysis of interface residues in homodimers using the available crystal structures has been reported, and it was found that hydrophobic residues such as Ala, Val, Leu, Met, Ile, and Phe are predominant at the homodimer interface (Zhanhua, Gan, Lei, Sakharkar, & Kangueane, 2005). Analysis of structures of homodimers and oligomers indicated that most of the homodimers and oligomers are symmetric (Dayhoff, Shoemaker, Bryant, & Panchenko, 2010; Levy, Pereira-Leal, Chothia, & Teichmann, 2006). Assymetric homodimers are also formed, but are known to perform certain specialized functions (Goodsell & Olson, 2000). Swapna, Srikeerthana, and Srinivasan (2012) have analyzed a nonredundant set of high-resolution crystal structures for symmetric and asymmetric homodimers and found that some functional proteins do form asymmetric structures. For the inhibition of drug design, symmetry in the dimerization is also important. Here we discuss some of the homodimers and their inhibition for drug discovery.

5.1 G-Protein-Coupled Receptors

GPCRs are a major component of cell surface receptors, and there are more than 800 known GPCRs. These receptors are involved in the signal transduction process. The structure of GPCRs consists of an N-terminal region that is typically outside the cell membrane and seven transmembrane helices (TM-I–TM-VII). The C-terminus extends into the cytoplasm. GPCRs are involved in a variety of physiological processes, including visual sense, taste, smell, behavior regulation, autonomic nervous system, immune system and inflammation, homeostasis, and tumor formation (Luttrell, 2006). GPCRs are classified into glutamate, rhodopsin, adhesion, frizzled/taste2, and secretin families (Fredriksson, Lagerstrom, Lundin, & Schioth, 2003). Nearly half of the drugs used in clinical practice directly or indirectly target GPCRs (Jacoby, Bouhelal, Gerspacher, & Seuwen, 2006; Tyndall & Sandilya, 2005), clearly indicating the importance of these receptors in controlling physiological processes and drug targeting. GPCRs were initially considered to be monomers on the cell surface of neuronal and nonneuronal cells. However, because GPCR can form homodimers, heterodimers, and higher order oligomers on the plasma membrane, the concept of GPCRs being only monomers can no longer be substantiated (Bouvier, 2001; Franco et al., 2007; Szidonya, Cserzo, & Hunyady, 2008). In the past decade, several experimental reports that were available suggested that GPCR oligomerization may be important for receptor function, including agonist binding, potency, efficacy, and G-protein selectivity. One of the limitations of the biophysical experimental methods that were used to detect GPCR oligomers is that these methods could not distinguish between homodimers and oligomers (Chabre, Cone, & Saibil, 2003; James, Oliveira, Carmo, Iaboni, & Davis, 2006). For the past few years, several groups have designed "bivalent ligands" that bind to GPCR dimeric structure to prove the importance of dimerization as well as the possibility of drug design using the bivalent ligand concept (Fig. 9). These molecules consist of two agonist/antagonist moieties separated by spacers of variable lengths. By bridging the pharmacophores at the two sites, a bivalent ligand can be designed that can bind to two sites, one on each homodimer or heterodimer. Such bivalent ligands can also be used to detect protein dimers (Brogi, Tafi, Desaubry, & Nebigil, 2014; Franco, Martinez-Pinilla, Lanciego, & Navarro, 2016; Hiller, Kuhhorn, & Gmeiner, 2013; Xu et al., 2012).

5.2 Receptor Tyrosine Kinase-Like Orphan Receptor 2

Receptor tyrosine kinase-like orphan receptors 1 and 2 (Ror1 and Ror2) are two members of Ror, which is a neurotrophic tyrosine kinase receptor

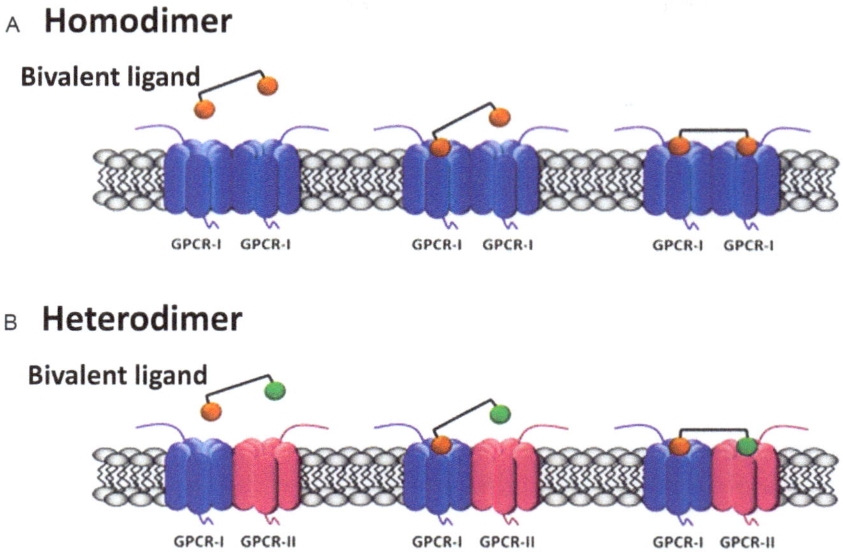

Fig. 9 Scheme of the simultaneous binding of bivalent ligands with linkers of appropriate length to two receptors in a GPCR dimer. The agonist/antagonist moieties of the bivalent ligand are selective for their respective receptors (*orange* for GPCR-I; *green* for GPCR-II) and are linked by a spacer. (A) Binding to two equal GPCRs forming a homodimer. (B) Binding to two different GPCRs forming a heterodimer. *Reproduced with permission from Franco, R., Martinez-Pinilla, E., Lanciego, J. L., & Navarro, G. (2016). Basic pharmacological and structural evidence for class A g-protein-coupled receptor heteromerization.* Frontiers in Pharmacology, *7, 76. https://doi.org/10.3389/fphar.2016.00076. Copyright (2016).*

within the RTK family. Ror receptors are very closely related to Trk neurotrophin (NT) receptors and muscle-specific kinase. Ror2 plays an important role in developmental morphogenesis, specifically of the cartilage-derived skeleton (Roarty, Shore, Creighton, & Rosen, 2015). It has been found that disruption of mouse Ror2 corresponds to extensive skeletal abnormalities in which all endochondrally derived bones are foreshortened or misshapen, while in humans, the mutation in Ror2 gene accounts for short height, limb bone shortening, and segmental defects of the spine (Aglan et al., 2015). Receptor dimerization is induced by ligand binding to Ror2. Elucidation of molecular mechanism indicated that Ror2 binds to Wnt family glycoproteins and modulates the Wnt signaling. Ror2 is also known to interact with a bone morphogenetic protein receptor type Ib (BRI-b) that modulates cartilage development. The coexpression of Ror2 and casein kinase I is known to result in tyrosine phosphorylation of GPCR kinase (Liu, Ross, Bodine, & Billiard, 2007). Inhibition of Ror2 homodimerization can be useful in different types of cancer also via Wnt pathway. Recently, it has been shown that Ror2 is upregulated in renal cell carcinoma (RCC) tissues and cell lines. Knockdown of Ror2 also inhibits proliferation, migration, and invasion and induces G1 phase cell cycle arrest and apoptosis of RCC cell lines. Knockdown of Ror2 is also known to inhibit tumor growth in vivo. RCC represents one of the most resistant tumors to radiation and chemotherapy (Yang et al., 2017). Hence, the design of molecules to modulate Ror2 dimerization may lead to useful therapeutic agents. At present, there are no reports of known inhibitors of Ror2 homodimerization.

5.3 Glucocorticoid Receptor

Human glucocorticoid receptor (GR), a nuclear receptor superfamily receptor, is associated with many physiological processes such as immune regulation and metabolism. Homodimerization of GR is important for control of GR transcriptional activity (Oasa, Sasaki, Yamamoto, Mikuni, & Kinjo, 2015). GR generally binds to glucocorticoid response elements that modulate the transcription process as homodimers. GR consists of an N-terminal transactivation domain, a central DNA-binding domain (DBD), a C-terminal ligand-binding domain (LBD), and a flexible "hinge region" that separates the DBD and the LBD. Of several members of the nuclear receptor superfamily, the DBD is the most critical region. The two zinc-finger motifs present in the DBD recognize and bind specific DNA sequences on target response elements (Kadmiel & Cidlowski, 2013). GR is also known to participate in nongenomic signaling that does not require nuclear-GR-mediated transcription or translation. Nongenomic signaling effects of GR are rapid and have implications in various systems, including the cardiovascular, immune, and neuroendocrine systems. GR isoforms are expressed in nearly all tissue types, and glucocorticoid signaling is almost ubiquitously prevalent in the various organ systems (Oakley & Cidlowski, 2011). Due to their antiinflammatory, antiproliferative, proapoptotic, and antiangiogenic roles, glucocorticoids have been remarkably effective in treating various diseases. Although the details of dimerization of structural aspects of GR with ligand are relatively recent (Bledsoe et al., 2002; Madauss et al., 2008), GR has been a successful drug target for nearly 60 years. The first report of the clinical use of GR targeting was for rheumatoid arthritis (RA) in 1940, and since that time GR has been targeted for many conditions such as chronic inflammatory conditions, including asthma, skin infections, and ocular infections, as well as for immunosuppression in patients undergoing an organ transplant. In addition to their antiinflammatory properties, the antiproliferative and antiangiogenic actions of corticosteroids have been exploited for the treatment of cancers (Vilasco et al., 2011).

5.4 Amyloid Precursor Protein Dimers

Amyloid precursor protein (APP) is a type I transmembrane protein expressed in many cell types, including neurons. APP is a 695 amino acid protein with a large ectodomain and relatively short intracellular region. APP has been shown to form homodimers (Khalifa et al., 2010). In APP, dimerization is known to be induced by the N-terminal region of APP, referred to as the E1 region (Fig. 10), which contains a growth factor-like domain and a copper-binding domain (CuBD) (Soba et al., 2005). A loop formed by disulfide bridges is required for the stabilization of the homodimeric state. Further, juxtamembrane (JM) and transmembrane (TM) regions also participate in homodimerization. APP is processed into smaller fragments, and there are two known catabolic pathways, namely, nonamyloidogenic and amyloidogenic pathways (Khalifa et al., 2010). APP processing seems to be a critical event in the onset and progression of Alzheimer's disease (AD) (De Strooper, Vassar, & Golde, 2010) and hence homodimerization of APP and the details of domains and amino acid residues involved in particular domains are studied in detail. In AD, the main component of plaques is the amyloid beta (Aβ) peptides with 40–42 amino acids (Masters et al., 1985). These peptides are released from a precursor protein APP by sequential cleavage by beta-site APP-cleaving enzyme 1 (BACE1) and by the γ-secretase complex. Cleavage of APP that consists of 695 amino acids by BACE1 releases the large ectodomain of APP and membrane-anchored C-terminal APP fragment (CTF) of 99 amino acids. The 99 amino acid polypeptide will undergo further cleavage by γ-secretase resulting in Aβ peptides of various lengths. APP intracellular domain (ICD) is released into the cytosol (Eggert, Midthune, Cottrell, & Koo, 2009; Jung et al., 2014; Vassar et al., 1999). It was proposed that dimerization of TM domain and amino acids in the TM domain is important in this cleavage process. APP contains three glycine-xxx-glycine (GxxxG) motifs at the extracellular JM/TM boundary. It is reported that the GxxxG motifs in the APP TM domain participate in dimerization and this domain is located in the region where cleavage occurs. Structural studies on the APP JM/TM region in isolation showed that the GxxxG motifs mediate TM helix

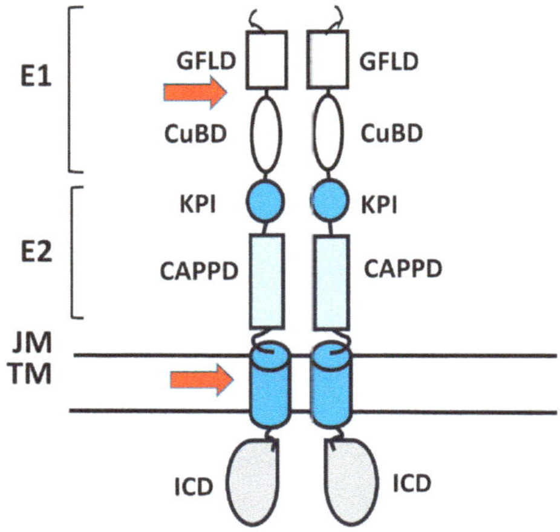

Fig. 10 Schematic representation of different domains of APP that form homodimers. *Arrows* indicate the proposed dimerization regions. *CAPP*, central APP domain; *CuBD*, copper-binding region; *GFLD*, growth factor-like domain; *ICD*, intracellular domain; *JM*, juxtamembrane region; *KPI*, Kunitz protease inhibitor domain; *TM*, transmembrane region. *Schematic diagram is drawn based on Khalifa, N. B., Van Hees, J., Tasiaux, B., Huysseune, S., Smith, S. O., Constantinescu, S. N., et al. (2010). What is the role of amyloid precursor protein dimerization?* Cell Adhesion & Migration, 4*(2), 268–272; Eggert, S., Midthune, B., Cottrell, B., & Koo, E. H. (2009). Induced dimerization of the amyloid precursor protein leads to decreased amyloid-beta protein production.* The Journal of Biological Chemistry, 284*(42), 28943–28952. https://doi.org/10.1074/jbc.M109.038646.*

homodimerization of the protein in the lipid bilayer (Sato et al., 2009; Fig. 10). Mutational studies by the introduction of a cysteine residue at the junction of the JM/TM region were shown to form stable dimers linked by disulfide bridges. The stabilization of dimerization leads to increased Aβ production (Scheuermann et al., 2001). Aβ is produced as a stable dimer, indicating that the amyloidogenic secretases (β and γ) are able to process APP under its dimeric form. Thus, dimerization seems to help Aβ production. The motifs involved in dimerization of C-terminal APP fragments (CTFs) are also responsible for the packing of Aβ peptides into protofibrillar structures (Sato et al., 2006). The glycines present in GxxxG motifs are important in the PPI of TM helices as well as in the formation of the cross β-sheet structures found in the Aβ fibrils. The GxFxGxF framework seems to be the hot spot for designing drug-like molecules for AD. Peptides can be designed to disrupt sheet-to-sheet packing and inhibit the formation of mature toxic Aβ fibrils. Antibodies mapping to an epitope in this Aβ region are also able to significantly reduce the accumulation of intracellular Aβ, which is known to be highly neurotoxic (Tampellini et al., 2007). Thus, the dimerization process, the GxxxG motifs, the details of structure in the dimerization region, and the cleavage of this region by secretase are important in designing drugs for AD. Richter et al. (2010) have studied the molecular mechanism of γ-secretase modulators such as sulindac sulfide and indomethacin and, using molecular docking studies, have suggested that these compounds bind at the smooth surface provided by glycines arranged in GxxxG motifs (Richter et al., 2010).

Munter et al. (2007) have shown that γ-secretase processivity is reduced when CTFβ forms dimers, because of interactions of TM domain GxxxG motifs. This leads to the formation of fragments of Aβ isoforms which are larger in size compared to 40 amino acid Aβ. There are reports indicating that APP CTFβ dimers are not γ-secretase substrates. Jung et al. (2014) studied the importance of residues at the interface of APP ectodomain and TMD by mutating the lysine residues at the interface of the APP ectodomain and transmembrane domain (TMD) and evaluated the Aβ production. Based on their studies, they concluded that the monomeric form of the mutant increased long Aβ production without altering the initial ε-cleavage utilization, whereas dimeric forms of APP are not efficient γ-secretase substrates and primary sequence determinants within APP substrates alter γ-secretase processivity. Thus, there is controversy regarding the dimerization of APP and its link to cleavage of APP by γ-secretase. The design of inhibitors of APP has to be carefully considered when targeting a particular region of APP that helps for homodimerization.

5.5 EGFR Homodimers

EGFR (also known as ErbB1 or HER1) is a well-known tyrosine kinase receptor involved in the signal transduction process. EGFR has importance in key stages of the development of organisms, such as cell proliferation, motility, differentiation, and tissue homeostasis. Overexpression of EGFR or enhancement of the receptor activity results in tumorigenesis. EGFR has an extracellular domain (ECD) consisting of 621 amino acids, a single TMD with 25 amino acids, and a cytoplasmic kinase domain with nearly 800 amino acids, forming the complete structure (Baselga & Swain, 2009; Ferguson, 2008). Peptide growth factor ligands such as epidermal growth factor (EGF), transforming growth factor α (TGFα), amphiregulin, betacellulin, epigen, epiregulin, and heparin-binding EGF-like growth factor are known to bind to EGFR. Binding of these ligands is known to induce change in the conformation of the ECD of EGFR. Among these, only EGF,

TGFα, amphiregulin, and epigen associate specifically with the EGFR homodimer (Roskoski, 2014). The homodimer of EGFR ECD structure has been elucidated by X-ray crystallography (Lu et al., 2010) and electron microscopy (Mi et al., 2008, 2011). The ECD of EGFR consists of four domains, namely domains I–IV (domain I residues 1–165, domain II residues 166–309, domain III residues 310–481, and domain IV residues 482–621). In the homodimer, domains II and IV interact with one another, forming a PPI interface (Fig. 11A). Domain II and domain IV are composed of eight and seven disulfide modules, respectively. The homodimer crystal structure has a twofold symmetry around the dimerization arm of domain II. The ligand is known to bind within a cleft formed by domains I and III (Ogiso et al., 2002). Crystal structures of the monomeric EGFR with and without the ligand suggested that there is a substantial change in the conformation of the ECD between monomer and dimer. In the ligand unbound state, EGFR exists in so-called closed conformation (Fig. 11D) with domains II and IV interacting with one another. Domains III and IV undergo significant movement in their structure upon binding of the ligand (Ferguson, 2004, 2008; Fig. 11E). How this change in conformation results in the transmission of signaling from outside the cell into the cytoplasmic domain is not explained because the complete structure of EGFR molecule including ECD, TM, and kinase domain is difficult to elucidate. However, each domain structure is available as a fragment (Ferguson et al., 2003; Lu et al., 2010; Mineev et al., 2010; Stamos, Sliwkowski, & Eigenbrot, 2002). There have been attempts at modeling the complete 3D structures of EGFR and its homodimer. Molecular dynamics simulations have been carried out to explain the transmission of signaling from outside of the cell to inside the cell in terms of EGFR structure (Endres et al., 2013; Poger & Mark, 2014). In terms of PPI, domain II of EGFR has β-hairpins that interact with one another in hand-shaking fashion (Fig. 11B). It has been shown that deletions or mutations in domain II completely prevent ligand-induced EGFR activation (Garrett et al., 2002; Ogiso et al., 2002). Domain IV of EGFR extends out from domains I to II and seem to form PPI at the C-terminal part (Fig. 11C). Crystal structures revealed that domain IV is flexible and that the electron density around the C-terminal portion is not well defined. However, based on experimental data, the mode of interaction of domain IV was proposed (Lu et al., 2010).

EGFR homodimer formation and its inhibition can be detected by PLA assay as described by Fichter et al. (2014). Inhibition of dimerization of EGFR homodimers by small molecules and antibodies is reported. Based

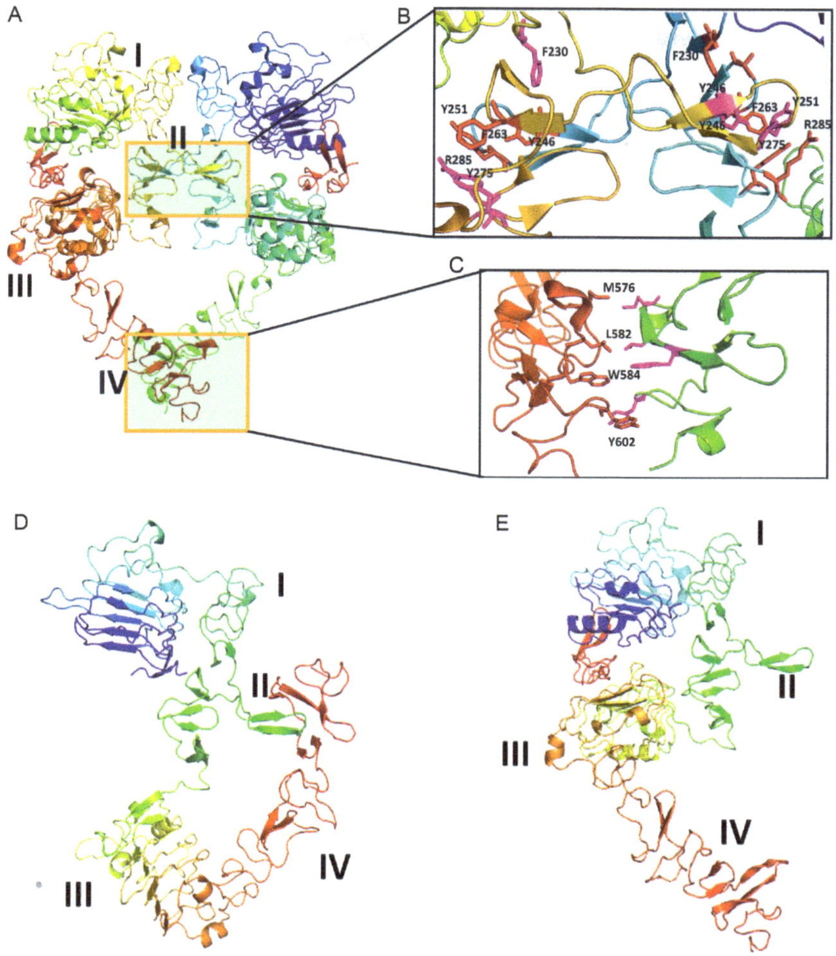

Fig. 11 EGFR extracellular domain (ECD). (A) Homodimer (PDB ID 1NJP). Domains II and IV participate in dimerization. (B) and (C) Expanded regions of PPI of domains II and IV. Domain II region has a β-turn structure in the PPI region, forming a hydrophobic core at the interaction site. Molecules that can mimic this region are designed to inhibit PPI. Domain IV also has hydrophobic interactions that can be used to target inhibitors. (D) and (E) EGFR open (PDB ID 1NJP) and closed conformations (1NQL). Domains I–IV are labeled. In the closed conformation, domains II and IV interact, blocking the dimerization arm of domain II and forming a dimer. Notice that in open conformation, the domain IV is moved away from domain II, allowing domain II to interact with other EGFR molecules. Figure was generated using PyMol software (Schrodinger LLC, OR).

on the structure of dimerization arm β-loop peptide-based molecules were designed to inhibit the domain II of EGFR. These peptides were modified to peptidomimetic using a triazolyl bridge between the peptide strands to constrain the EGFR dimerization arm β-loop (Fig. 11B). The designed peptides have significantly improved proteolytic stability over the nonmodified peptide sequence, and their inhibitory effects are dependent on the number of the methylene units and orientation of the introduced triazolyl bridge (Hanold et al., 2015). Yang, Yang, Pike, and Marshall (2010) have reported a small molecule that targets the β-hairpin structure of domain II to inhibit EGFR homodimerization. Using chemical cross-linking methods, they have shown that the small molecule designed using a computational approach inhibits EGFR homodimer. The antibody cetuximab used for the treatment of colorectal cancer does not directly inhibit the dimerization of EGFR. It binds to an EGF-binding site and blocks the ligand-binding site, thus indirectly inhibiting the dimerization and downstream signaling process of EGFR (Graham, Muhsin, & Kirkpatrick, 2004). EGFR domain IV can also be targeted to design small molecules or peptides (Fig. 11C).

6. HETERODIMERIZATION OF PROTEINS AND INHIBITION

When PPI occurs between nonidentical chains, heterodimerization results. The stability of heterodimers can vary. For example, α/β tubulins form a stable dimer, and these dimers form long protofilaments, which are constituents of microtubules (Lowe, Li, Downing, & Nogales, 2001). The number of crystal structure complexes of heterodimers available from the Protein Data Bank is relatively small compared to those of homodimers. Sowmya et al. (2015) have analyzed a nonredundant set of 278 heterodimer complexes for interfacial structural features and found that there was a correlation between the interfacial surface area of PPI and the possible function of the protein. Computational methods are used to predict the heterodimeric complexes using the crystal structures of individual proteins based on binding studies and mutational data. However, a recent critical assessment of predicted interactions (CAPRI) report indicates that prediction of homodimers is easier and superior to the prediction of heterodimers of protein complexes (Lensink et al., 2016). Thus, there is still a long way to go in terms of the availability of structural biology information for PPI of heterodimers.

The most widely covered topics concerning PPI are p53–MDM2 (Vassilev et al., 2004), Bcl-Xl (Hikita et al., 2010; Oltersdorf et al., 2005), and IL-2–IL-2Rα interactions. In the case of the p53 and HDM2 interaction, the p53-binding site on HDM2 is a cleft rather than a flat surface. Identification of a small molecule that inhibits the interaction of p53 with MDM2 provides an example of the use of HTS (Vassilev et al., 2004). An example of modulation of PPI of proteins that interact via flat surfaces is seen in the case of IL-2 and its receptor IL-2Rα (Braisted et al., 2003; Tilley et al., 1997). Here we describe some of the heterodimers and their importance in physiological function and possible inhibition of dimerization for clinical applications.

6.1 p45–p75 Heterodimers

Injury to the brain and spinal cord results in major loss of physical and other functions. In many cases, these injuries are permanent because injured nerves cannot regrow to perform their function. The NTs are a family of neurotrophic factors that control multiple aspects of nervous system development and function. p75 is a member of the tumor necrosis factor receptor superfamily. The structure of p75 consists of four extracellular cysteine-rich domains, a single TM domain, and an ICD that consists of a JM and a death domain (DD) (Dechant & Barde, 2002; Lin et al., 2015). The p75 receptor has different effects, depending on its interactions with different partners and copartner proteins. For example, p75 interacts with Trk receptors (tropomyosin receptor kinase) to promote NT-dependent nerve growth. On the other hand, p75 inhibits nerve growth mediated by myelin-associated inhibitors via functioning in part as a coreceptor for the glycophosphatidylinositol-linked neuronal Nogo-66 receptor (NgR) or another non-NgR molecule (Gentry, Rutkoski, Burke, & Carter, 2004). The binding of p75 to proneurotrophins and with the coreceptor sortilin was shown to play a role in apoptosis (Nykjaer, Willnow, & Petersen, 2005). p75 is known to form homodimers in solution, and homodimerization (Nadezhdin et al., 2016) seems to be important for complexation with NgR that leads to inhibition of nerve growth. p45, an NT receptor homolog 2 (NRH2), NT receptor-like DD protein (NRADD). p45 exhibits vast sequence similarity to p75 in the TM, JM, and DD regions. p45 contains a short and truncated ECD with no NT-binding domain. p75 plays a role during injury to the brain and spinal cord. At the site of the injury in the brain and spinal cord, there are proteins that are released from the damaged myelin that binds to Nogo receptor (NgR) on the nerve and inhibits nerve growth. NgR has to form a complex with the p75 neurotropin receptor to inhibit the signaling. p45 can bind to p75 and impedes the formation of p75 homodimer that is required for p75/NgR complex formation and its downstream activation of RhoAGTPase. The complex formation of p75/NgR requires the binding of p75 through its TM and ICDs. Vilar et al. (2014) have shown that p45 binds specifically to conserved regions in the p75 TM and the ICD and that this blocks p75 dimerization along with its downstream signaling. Thus, modulation of oligomerization of p75 is a good strategy to overcome the effect of p75's inhibitory effects on nerve regeneration, and hence the design of p75 inhibitors will have therapeutic applications for brain and spinal cord injury. In addition, p45 itself can be used as a therapeutic agent to injured neurons and can prevent the blocking of nerve growth by inhibiting p75 interactions in paralysis or spinal cord damage injuries (Vilar et al., 2014). At present, there are no known inhibitors of p75/NgR complex.

6.2 IL-6–IL-6Rα

Interleukin 6 receptor, a cytokine receptor also known as CD126, interacts with IL-6 a cytokine and regulates cell growth, apoptosis, proliferation, and immune responses. IL-6 interacts with IL-6Rα and forms a binary complex and then guides glycoprotein GP130 to form the IL-6/IL-6Rα/GP130 heterotrimer. The IL-6/IL-6Rα/GP130 heterotrimers occur by the interaction between IL-6 of one trimer and the D1 domain of GP130 of the other trimer to form a hexamer. These IL-6/IL-6Rα/GP130 trimers trigger a signaling cascade of phosphorylation of Janus kinases (JAKs) and a downstream effector signal transducer and activator of transcription 3 (STAT3) that is followed by reciprocal dimerization of the Tyr705-phosphorylated STAT3, resulting in STAT3 nucleus translocation, DNA binding, and multiple oncogene transcriptions (Li et al., 2014). These JAKs and STAT3 pathways are very crucial for progression of various types of cancer. Madindoline A (MDL-A), a natural product, is known to bind to the ECD of GP130 and inhibit IL-6-dependent STAT3 tyrosine phosphorylation in hepatocellular carcinoma (HepS2) cells (Saleh, Greenman, Billings, Van Vranken, & Krolewski, 2005; Fig. 12). However, there are limitations in using MDL-A for therapeutic purposes because of several steps involved in the synthesis, weak-binding affinity to the receptor, and extraction of the natural product has very low yield. Li et al. (2014) used multiple ligand simultaneous

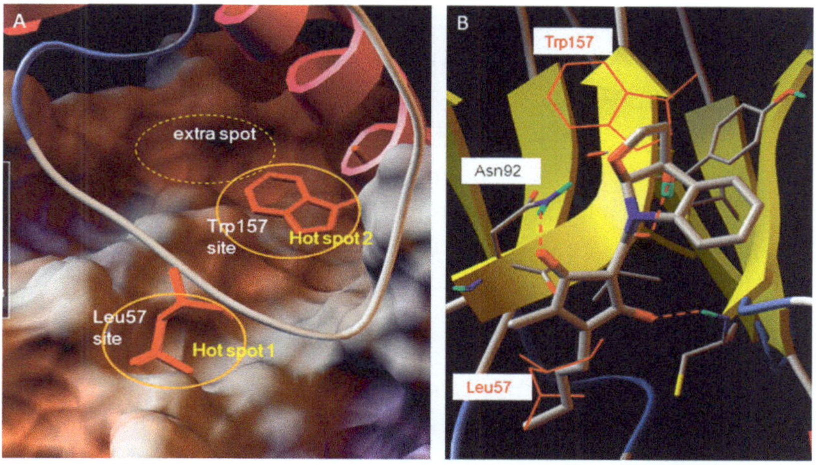

Fig. 12 Hot spots at the interface of PPI in the IL-6/GP130 D1 domain. Modeling of binding hot spots at the IL-6/GP130 D1 domain interface. (A) D1 domain is represented as electrostatic potential surface (*red*, negatively charged; *blue*, positively charged; *white*, hydrophobic). IL-6 is in ribbon representation. The two larger *yellow* eclipses indicate the two main binding "hot spots," Leu57-binding site and Trp157-binding site, between IL-6 and GP130. (B) Binding modeling of MDL-A to the GP130 D1 domain. D1 domain is in ribbon representation, and MDL-A is in thick ball-and-stick rendering. Hydrogen bonds are shown as *red dotted lines*. MDL-A disrupts both binding spots of the GP130 D1 domain. MDL-A forms three hydrogen bonds with Asn92, Cys6, and the carbonyl backbone of Val93 residues of GP130. The modeling indicates that the long butyl tail of MDL-A displaces Leu57 (*thin red line*), and the indoline moiety partially disrupts Trp157 (*thin red line*) of the helix D of IL-6. *Reproduced with permission from Li, H., Xiao, H., Lin, L., Jou, D., Kumari, V., Lin, J., & Li, C. (2014). Drug design targeting protein-protein interactions (PPIs) using multiple ligand simultaneous docking (MLSD) and drug repositioning: Discovery of raloxifene and bazedoxifene as novel inhibitors of IL-6/GP130 interface.* Journal of Medicinal Chemistry, *57(3), 632–641. https://doi.org/10.1021/jm401144z. Copyright (2014) American Chemical Society.*

docking (MLSD) and drug-repositioning approaches to identify compounds that inhibit PPI of IL-6 and GP130. Using this computational approach, Li et al. (2014) found two compounds, raloxifene and bazedoxifene, that were able to inhibit PPI. Raloxifene is a well-known oral selective estrogen receptor modulator (SERM) that has estrogenic actions on bone and antiestrogenic actions on the uterus and breast (Jones et al., 1984). Bazedoxifene, an analog of raloxifene, is also an SERM that is under development for the treatment of osteoporosis (Biskobing, 2007). Such drug molecules that are already approved for therapeutic purposes that are also PPI inhibitors have a high chance of becoming successful PPI inhibitor drugs as the drug is repurposed.

6.3 Epidermal Growth Factor Receptors

Here we provide an example of the design of a PPI inhibitor starting from the structure of a protein complex. The procedure described is a rational drug design or an SBDD method, which does not include any database screening or HTS. The description starts with the importance of proteins and PPI in normal and disease states and the significance of the biochemical pathway. The design information and the experimental results that indicate that the designed molecule inhibits PPI are described in detail to give the reader extensive knowledge of the way that PPI inhibitors are designed.

As mentioned earlier, the human epidermal growth factor receptor (HER) system of RTKs plays an important role in cell growth and differentiation in normal physiology (Ferguson, 2008). The receptor system consists of four members: HER1 or EGFR and HER2–4 (also called ErbB1–4). Among the EGFR, HER2 is known to always exist in the open conformation and is a preferred dimerization partner for other EGFR (Baselga & Swain, 2009). Deregulation of homo- and heterodimerization processes of these receptors or overexpression of receptors leads to different types of cancer and plays a key role in tumor progression (Lee-Hoeflich et al., 2008). Possible dimers such as EGFR–HER2, HER2–HER3, and HER2–HER4 have been proposed in the literature (Shankaran, Wiley, & Resat, 2006; Tao & Maruyama, 2008). Among these, EGFR–HER2 and HER2–HER3 are well known in different types of cancer. Mutation in EGFR seems to play a major role in breast and lung cancers. Because of the limitations of chemotherapy for cancer, EGFR-targeted therapy has attracted attention. Studies related to breast and non-small-cell lung cancer (NSCLC) have shown a link between HER2 expression and poor prognosis in patients with cancer (Hirsch et al., 2014; Hirsch, Varella-Garcia, & Cappuzzo, 2009). Approximately 18%–33% of breast and NSCLC tumors show a positive result for HER2 overexpression, suggesting the importance of HER2 in these types of cancers. The coexpression of EGFR and HER2 was associated with a significantly shortened overall survival rate in patients whose tumors expressed high levels of EGFR or HER2 (Brabender et al., 2001). Since HER2 protein is overexpressed in different cancer types, targeting the HER2 pathway will most likely target only cancer cells, and possible side effects on normal cells will be minimal. The kinase domain of EGFR has been targeted for cancer therapy using a tyrosine kinase inhibitor. However, most of these develop resistance within 5 years and, hence, ECDs are viable targets for cancer therapy (Oxnard et al., 2011).

6.3.1 Structure of ECDs of Proteins

Based on the biochemical pathway, one can target the different dimerization and PPI sites on HER2 protein for developing therapeutic effects on cancer. Possible dimerization inhibition sites are domain II of ECD, domain IV of ECD, and a TMD. Detailed 3D structures of ECD of EGFR, HER2, and HER3 are all known. Structures of homodimers of EGFR ECD have been elucidated by X-ray crystallography (Lu et al., 2010; Fig. 11A). However, the structures of heterodimers of EGFR:HER2 or HER2:HER3 are not known. Since EGFRs have nearly 50% homology and similar domains, one can model the HER2:HER3 ECD using EGFR as a template structure. In the ECD of EGFR, domains II and IV are involved in PPI. The importance of domain II of the EGFR dimerization arm is well known (Burgess et al., 2003; Cho et al., 2003; Lu et al., 2010; Ogiso et al., 2002). The structure of HER2 monomer as well as HER2 complexed with antibodies trastuzumab and pertuzumab has been elucidated (Fig. 13A and B). HER2 domain IV is a clinically validated target since trastuzumab, an antibody, binds to domain IV of HER2 and has therapeutic value against HER2-positive breast cancer (Piccart-Gebhart et al., 2005). However, domain IV has not been well studied because of its flexibility. A homodimer of EGFR domain IV indicates the PPI and possible hot spots. Based on this, a heterodimer of HER2:HER3 was built (template-based modeling/docking), and possible hot spots were identified by FTMAP (Kozakov et al., 2015).

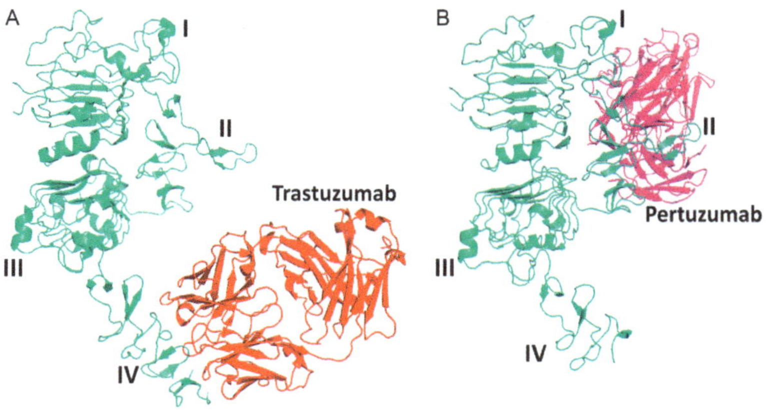

Fig. 13 Binding of (A) trastuzumab to domain IV of ECD of HER2 protein (PDB ID 1N8Z). (B) Binding of pertuzumab (PDB ID 1S78) to domain II of ECD of HER2 protein. PyMol was used to generate the figures.

6.3.2 Design Concept

The idea here is to inhibit domain IV of the ECDs of EGFR:HER2 and HER2:EHR3. Inhibition of the ECD of these proteins inhibits the phosphorylation of the kinase domain and downstream signaling for cancer cell growth. Thus, the growth of cancer tumors can be reduced. Trastuzumab is known to bind to domain IV of HER2 protein (Fig. 13A). However, its exact mechanism of action is not clear. Examination of EGFR homodimer, the crystal structure of the complex of trastuzumab and HER2, indicates that domain IV has hydrophobic hot spots. We used the structure of a complex of HER2 protein with trastuzumab for the design of a template structure. Although the antibody structure is large, the binding region to HER2 protein is relatively small. The binding region has hydrophobic amino acid residues such as Tyr, Trp, and Phe (Fig. 14A). This hydrophobic region is surrounded by electrostatic and hydrogen-bonding interactions. By measuring the distance between Cα atoms of amino acid residues Glu558, Phe573, and Lys593 of HER2, a peptide template structure was built with amino acids Arg-Tyr-Trp-Tyr-Gly (Fig. 14B). These residues correspond to the antibody trastuzumab that interacts with HER2. Evaluation of these peptides for antiproliferative activity showed that they were not potent in inhibiting cell growth of HER2-positive cells BT-474 and SKBR-3 (Satyanarayanajois, Villalba, Jianchao, & Lin, 2009). The designed peptide can also interact with PPI site of domain IV of HER2 (Fig. 15A). The peptide design was modified using a peptidomimetic approach with the incorporation of a beta-naphthyl group to fit into the hydrophobic core of the HER2 protein PPI site with trastuzumab (Fig. 15B). Since Arg participates in salt bridge, Arg on the left side, a hydrophobic core at the center, and Phe at the C-terminal were used to build a peptidomimetic (Fig. 15B). This peptidomimetic was evaluated for its antiproliferative activity using a cellular assay such as MTT or CellTiter-glo assay in breast cancer cells that overexpress HER2 protein. Furthermore, to find the specificity of the designed molecules for HER2-overexpressing cell lines, different cell lines that do not overexpress HER2 protein such as MCF-7 and HCT-116 were used. The molecule that has a beta-naphthyl group exhibited antiproliferative activity with an IC_{50} value of 0.4 μM in HER2-overexpressing cancer cell lines. However, in MCF-7 and HCT-116, the activity was 40 μM, suggesting the specificity of this compound for HER2-positive cancer cell lines. The binding of this compound was verified by fluorescence assay and SPR. It is known that other homologous proteins such as EGFR and HER3 are also important in different types of cancer. EGFR, HER3, and HER4 have a

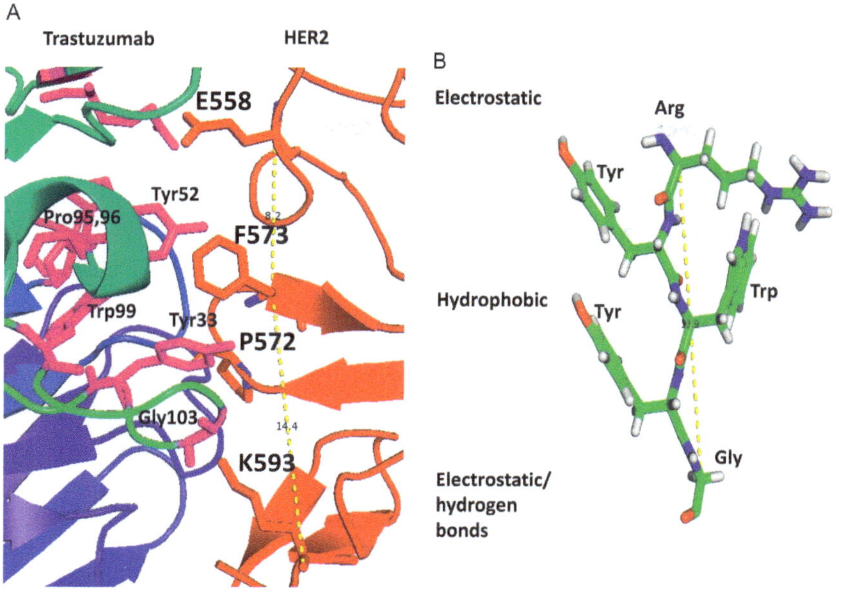

Fig. 14 (A) PPI region of trastuzumab-HER2 domain IV complex (PDB ID 1N8Z). Hydrophobic region of trastuzumab formed by amino acid residues Tyr52, Pro95,96, Trp99, and Tyr33 interacts with Phe573 and P572 of HER2 protein. On either side of the hydrophobic region, there is electrostatic interaction formed by Glu558 of HER2 with Arg59 of trastuzumab and from K593 of HER2 to Gly103. (B) This interaction can be used to design a template structure for a PPI inhibitor. Figure was generated using PyMol (Schrodinger LLC, OR).

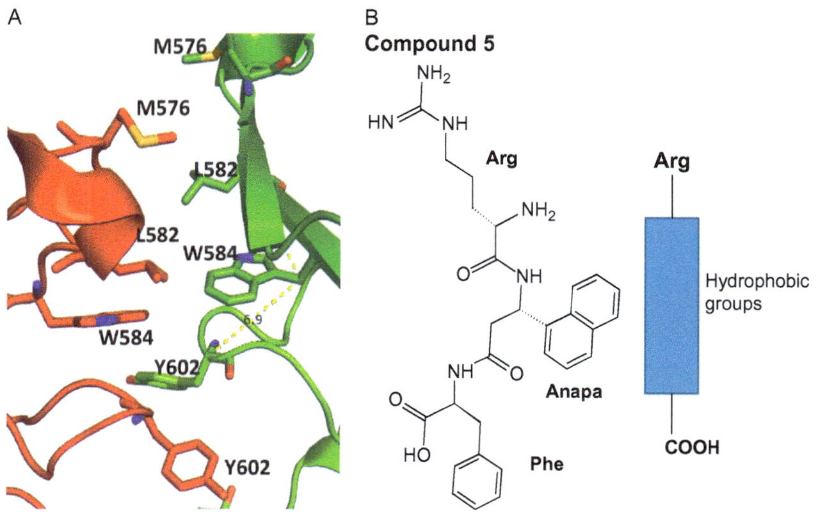

Fig. 15 The structure of the designed template compound **5** and a peptidomimetic (B) along with the PPI site (A) of the EGFR homodimer of domain IV (PDB ID 1N8Z and 3NJP). Note that the PPI site is dominated by hydrophobic residues. Peptidomimetic can bind to this PPI site. PyMol (Schrodinger LLC, OR) was used to generate the structure of the protein complexes.

sequence homology of nearly 50%, and all of them have similar 3D structures. To show that the designed compound binds specifically to HER2, SPR studies were carried out; it was shown that the designed compound **5** binds only to HER2 protein ECD (Banappagari et al., 2012).

To improve the activity of the compound, several modifications were performed. An Asp residue was introduced at the C-terminal, making the compound more specific for HER2-overexpressed cells and giving it better activity (Banappagari, Ronald, & Satyanarayanajois, 2011; Kanthala et al., 2015; Kanthala, Gauthier, & Satyanarayanajois, 2014). In addition, using the PPI of domain IV of EGFR and HER2 proteins (compound **18**, patent application: WO/2015/175299), a conformationally constrained cyclic peptidomimetic compound was designed. The resulting compound exhibited antiproliferative activity around 200 nM in breast cancer cell lines and 18 nM in HER2-positive lung cancer Calu-3 cell lines. Once the structural aspects of the compound were optimized for pharmacological action, it was investigated for PPI inhibition activity. Using SPR analysis, enzyme fragment complementation assay, and PLA assay, it was shown that the compound designed inhibited not only EGFR:HER2 dimerization but also HER2:HER3 dimerization (Banappagari, Ronald, & Satyanarayanajois, 2010; Kanthala et al., 2015, 2014). All of these studies are related to cell-based studies. To determine whether the compound inhibits the heterodimerization of HER2:HER3 and has any pharmacological action in vivo, a xenograft model of breast cancer mice was used. In vivo studies indicated that the compound was able to suppress the growth of a tumor in a xenograft model of breast cancer, inhibit the phosphorylation of HER2 kinase, and inhibit HER2:HER3 dimerization in breast cancer tumors in mice (Kanthala et al., 2015). The compound was shown to be stable for more than 24 h in mouse serum. Overall, the example provided here illustrates the way that a PPI inhibitor can be designed and modified to improve selectivity and activity based on the structure of protein complexes.

6.4 PD-1–PD-L1 Pathway

In the two-signal model, T cells interact with antigen-presenting cells (APCs) with two important signals—the first from the T-cell receptor (TCR) and major histocompatibility (MHC) complex and the second from costimulatory molecules (Bretscher, 1999). If these two signals are generated, immune response takes place. The strength and duration of the TCR–MHC signal depend on the costimulatory molecules and cytokines generated based on these signaling proteins. However, this signal, although necessary, is alone not sufficient to induce T-cell proliferation. There are several costimulatory molecules that generate signaling in an orchestrated process to assist the immunological synapse (IS) and control the immune function (Chen & Flies, 2013). In the case of Signal 2, important costimulatory molecules are of two receptors on T lymphocytes CD28 and cytotoxic T-lymphocyte-associated antigen-4 (CTLA-4) that interact with B7 family (CD80 and CD86) on APCs. If Signals 1 and 2 are provided, the result is T-cell clonal expansion and the induction of effector functions such as lymphokine production, T-cell proliferation, and cytokine secretion. Costimulation is necessary for a productive immune response to take place. The lack of costimulation after engagement of T-cell receptor by antigen results in a state of antigen-specific unresponsiveness termed anergy (Schwartz, 2003). Thus, a costimulatory pathway plays a potential role in the treatment of autoimmune diseases, organ rejection, and tumor immunity. CD28 is generally expressed on the surface of T cells and is one of the major costimulatory molecules for initial T-cell activation. Interaction of TCR CD28 ligation results in the augmentation of many aspects of T-cell-mediated immunity. The immune stimulatory features of the CD28 pathway are triggered by engagement of two well-described ligands found on APC. CD28 along with CTLA-4 and B7-1, B7-2 balance the costimulation and coinhibition process. When CD28 binds to its ligands B7-1 and B7-2, it delivers the T-cell stimulatory signal resulting in T-cell proliferation, expansion, and differentiation (Fig. 16). On the other hand, when CTLA-4 on activated T cells binds to B7 ligands, it delivers negative signals (Beyersdorf, Kerkau, & Hunig, 2015; Dilek et al., 2013; Gardner, Jeffery, & Sansom, 2014). Thus, these molecules maintain the immunological tolerance and response (Fig. 16). Programmed death-1 (PD-1) plays an important role in negatively regulating immune responses (Buchbinder & Desai, 2016; Greenwald, Freeman, & Sharpe, 2005). PD-1 is a CD28 family member expressed on activated T cells, B cells, and myeloid cells. PD-1 delivers a coinhibitory signal upon binding to ligands PD-L1 or PD-L2. PD-1 binding to its ligand produces tyrosine phosphorylation of the PD-1 cytoplasmic domain and recruitment of phosphatases, particularly Src homology phosphatase 2 (SHP2) (Fig. 16). This results in dephosphorylation of TCR-proximal signaling molecules, including ZAP70, PKC-θ, and CD3-ζ, leading to attenuation of the TCR/CD28 signal.

The PD-1/PD-L1 pathway delivers inhibitory signals that regulate both peripheral and central tolerance. Disruption of the *Pdcd1* gene can accelerate

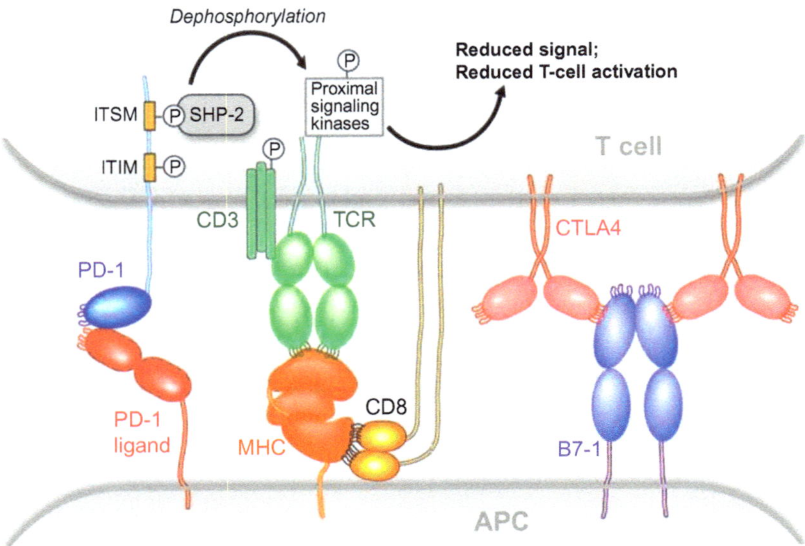

Fig. 16 Structures of the B7/CD28 family. Structures are modeled on the crystal determinations. Loops have been added to one end of the IgV domains to emphasize the orientation of the CDR-like loops and their interaction with ligand or lack thereof. *Reproduced with permission from Freeman, G. J. (2008). Structures of PD-1 with its ligands: Sideways and dancing cheek to cheek.* Proceedings of the National Academy of Sciences of the United States of America, 105*(30), 10275–10276. https://doi.org/10.1073/pnas.0805459105. Copyright (2008) National Academy of Sciences, USA.*

autoimmune diseases in mice, including a lupus-like disorder in *lpr* mice or diabetes in nonobese diabetic mice (Freeman, 2008). Cancer cells are known to escape the immune system in the body. PD-L1 is expressed on a wide variety of tumors and participates in the immunosuppressive activity of cancer cells (Sharpe, Wherry, Ahmed, & Freeman, 2007; Zang & Allison, 2007). PD-L1 on tumors inhibits T-cell activation and lysis of tumor cells; it is also known that PD-L1 interaction with its receptor leads to the death of tumor-specific T cells (Brown et al., 2003). Monoclonal antibodies have been developed against PD-L1 or its ligands for therapeutic purposes (Dolan & Gupta, 2014). The structures of PD-1–PDL-1 and PD-1–PD-L2 have been elucidated using X-ray crystallography (Lazar-Molnar et al., 2008; Lin et al., 2008). PD-1, a type I transmembrane protein of the Ig superfamily, consists of an extracellular N-terminal IgV-like domain, a TMD, and a cytoplasmic tail (Ishida, Agata, Shibahara, & Honjo, 1992; Zhang et al., 2004) that is involved in inhibitory signal transmission. The PD-1 ectodomain contains a single IgV domain typical of the CD28 family, whereas PD-L1 and PD-L2 belong to the B7 family and are composed of

IgV and IgC domains. In the 1:1 complex of PD-1 and PD-L2, a binding interface is formed by the front β-sheets of both the PD-1 and PD-L2 IgV domains. Interaction of PD-1 on activated T cells with PD-L1 or L2 diminishes the effector T-cell activity in peripheral organs and tissues during inflammation. This is an important step to protect against tissue damage when the immune system is activated in response to infection. However, in cancer, this pathway is used for cancer cell survival from immune surveillance to mask the cancer cells from the immune system. Thus, blocking PD-1–PD-L1 pathway can lead to T-cell activation against cancer cells (Tang & Heng, 2013). It is reported that in cancers such as melanoma, hepatocellular carcinoma, glioblastoma, lung, kidney, breast, ovarian, pancreatic, and esophageal, as well as hematological malignancies, positive expression of PD-L1 was seen clearly, indicating that cancer cells use the PD-1–PD-L1 pathway for their survival from T-cell immune response against them (Zitvogel & Kroemer, 2012). Thus, modulation of the PD-1–PD-L1 pathway by modulation of PPI of these proteins has therapeutic value.

Monoclonal antibodies targeted to PD-1 or PD-L1 can prevent PD-1-mediated T-cell inhibition, leading to antitumor immune responses. However, selecting a particular antibody for one of these molecules is important. If antibodies are directed against PD-1, they block PD-1 binding to both of its ligands, PD-L1 and PD-L2, whereas anti-PD-L1 antibodies should be selective in preventing PD-1 binding to PD-L1, maintaining the interactive and binding ability of PD-1 to PD-L2. Medina and Adams (2016) reported a study where both CTLA-4 and PD-1 blockade were used in a combined manner. Blocking of both the pathways, CTLA-2 and PD-1 resulted in stronger antitumor effect than blockade of either pathway alone.

Several antibody molecules, including MDX-1106/BMS-936558/ONO-4538, CT-011, MK-3475, MPDL3280A/RG7446, BMS-936559, and AMP-224, have been developed to modulate PPI of the PD-1–PD-L1 pathway. Antibodies nivolumab and pembrolizumab that are targeted to PD1 are approved by FDA for the treatment of patients with unresectable or metastatic melanoma as well as metastatic squamous and nonsquamous NSCLC, with progression on or after platinum-based chemotherapy. However, administration of these antibodies to cancer patients needs careful monitoring and analysis of changes in the immune response in patients. Thus, clinicians will need to have an understanding of the science of how inhibition of PD-1 can lead to tumor reduction with associated immune-mediated adverse events.

6.5 CD-28/CD-80

CD28 is a stimulatory cell surface receptor of the Ig superfamily. Other members in this family include cytotoxic T-lymphocyte-associated antigen-4 (CTLA-4), CD152-inducible costimulator (ICOS) (Boomer & Green, 2010; Tezuka et al., 2000), programmed death receptor 1 (PD-1) (Ishida et al., 1992), and B- and T-lymphocyte attenuator (Watanabe et al., 2003). The ligands for CD28 are CD80 (B7/BB1 or B7-1) and CD86 (B7-2). In addition to binding to CD28, both CD80 and CD86 also bind to the inhibitory protein CTLA-4, which is a CD28 homolog expressed on activated T cells and, especially, regulatory T cells (Tregs) (Walker, 2013). CD28 is also known to bind to ICOS. Studies have indicated that CD28 is involved in several functions, and that CD28 and CTL-4 have integrated functions (Fig. 17A). CD28 costimulation provides the T cells to control unwanted (antiself) signaling and triggering wanted (antimicrobial) immunity (Gardner et al., 2014). There are at least three mechanisms to ensure that CD28 is capable of generating only a costimulatory signal when the TCR is engaged, but does not activate the T cell by itself. Without TCR stimulation, CD28 binds to its ligand CD80 monovalently (low affinity). In the presence of TCR–MHC ligation, the CD28 homodimer form binds to its ligand CD80 bivalently. Overall, CD28 induces a costimulatory signal in the T-cell upon coligation together with the TCR and amplifies the TCR signal (Beyersdorf et al., 2015). Upon TCR and CD28 stimulation, the T cell forms an IS with the APC (Brzostek, Gascoigne, & Rybakin, 2016). CD28 signaling is important for the production of IL2, and it is known that CD28 costimulation increases the IL-2 production compared to without CD28 stimulation (Boomer & Green, 2010).

CD2 binds to its ligands and initiates signaling in the cytoplasmic domain. Details of molecular mechanism and binding epitopes involved in CD28 signaling indicated that CD28 initiates two signaling cascades using two motifs in the cytoplasmic domain of CD28. Each of these motifs binds to distinct set of proteins. One of the signaling motifs consists of YMNM sequence that involves phosphorylation of a tyrosine residue (Fig. 17B). The other pathway is initiated by the more distal proline-rich regions (Boomer & Green, 2010; Rudd, Taylor, & Schneider, 2009; Sharpe, 2009). However, the involvement of these motifs in in vivo studies is not clear. In vivo studies using knockout mice studies suggested that mutation of the proximal YMNM motif did not have any effect on signaling

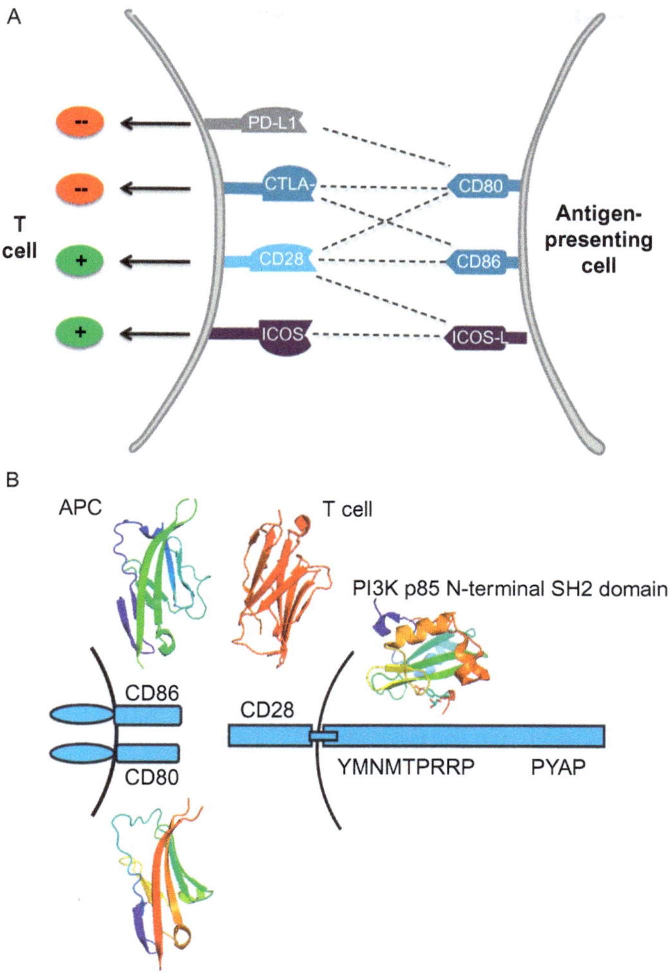

Fig. 17 (A) Complexities of CD28 costimulatory pathways. The CD28 costimulatory receptor can be ligated by CD80, CD86, and ICOS-L (B7-H2). The CTLA-4 coinhibitor competes with CD28 for binding to CD80 and CD86. However, CD80 can also bind to PD-L1 (B7-H1) and deliver a coinhibitory signal. ICOS competes with CD28 for binding to B7-H2. (B) Schematic representation of PPI interactions of CD28 with CD80/CD86. CD28 cytoplasmic domain motifs YMNM and PYAP that are important in signaling are indicated. Crystal structures of domains of CD28 (1YJD), CD80 (4RWH), CD86 (1NCN), and p53 (5GJI) domain important in binding to CD28 signaling are shown. *Panel (A): Reproduced with permission from Ford, M. L., Adams, A. B., & Pearson, T. C. (2014). Targeting co-stimulatory pathways: Transplantation and autoimmunity.* Nature Reviews Nephrology, 10*(1), 14–24. https://doi.org/10.1038/nrneph.2013.183. Copyright (2013) Nature Publishing Group.*

(Dodson et al., 2009). However, mutation of the distal proline motif (PYAP) (Fig. 17B) resulted in impaired CD28-dependent functions (Friend et al., 2006). Thus, the distal proline motif is involved in a critical, nonredundant signaling pathway required for CD28 function, whereas the proximal tyrosine-based motif signaling is not clearly understood.

CD28 PPI inhibition represents an example of indirect inhibition of PPI for drug design. The mechanism of PPI of CD28 is quite complicated, and inhibition of CD28 can lead to unwanted effects. CTLA-4 is also known to bind to B7 ligands (CD80, CD86) and send inhibitory signaling to APCs by counteracting signaling generated by CD28. Thus, CTLA-4 binding to CD80 or CD86 (B7 molecules) inhibits T-cell activation and proliferation. This strategy can be used to target CD28 for designing immunosuppressive drugs. A fusion protein consisting of the ECD of human CTLA-4 linked to the Fc domain of human IgG1 (CTLA-4-Ig, abatacept) was designed to block CD28–CD80/CD86 interactions. Abatacept has a high affinity for CD80 and CD86, blocks CD28-dependent costimulation, and inhibits T-cell proliferation in vitro (Linsley & Nadler, 2009). Based on these encouraging results, abatacept was evaluated in preclinical animal models of autoimmune diseases. Evaluation of abatacept in collagen-induced arthritis (CIA) of the rat showed that it prevented the onset of CIA, indicating that the fusion protein mediated blockade of T-cell costimulation in vivo. The drug was approved for RA in 2006. Modification of abatacept with two amino acids resulted in 10-fold higher in vitro potency (Larsen et al., 2005). This modification resulted in a second-generation CTLA-Ig protein belatacept (Larsen et al., 2005). Belatacept selectively inhibits T-cell activation by preventing CD28 activation and by binding its ligands B7-1 and B7-2. This prevents the stimulation of CD28 antagonizing CD80 and CD86 on APCs and thus blocks the signals of the transduction pathway.

There were attempts to directly inhibit CD28 signaling with its ligands (Hunig, 2007; Poirier, Blancho, & Vanhove, 2011); however, such attempts were not successful. Ford, Adams, and Pearson (2014) describe the way a selective blockade of CD28 was attempted in a phase I trial of an agonistic anti-CD28 monoclonal antibody, TGN1412. However, a fatal immunological reaction called "cytokine storm" caused by significant T-cell activation was observed in patients, suggesting that direct manipulation of CD28 signaling is harmful because CD28 is involved in myriad signaling pathways. To avoid this overwhelming T-cell activation, novel domain antibodies, in which the Fc portion is completely removed, were produced. These antibodies to CD28 that lack a Fc region have permitted the development of novel blocking, nonactivating reagents that can safely and specifically block CD28 costimulatory signals but leave the coinhibitory signaling due to CTLA-4 intact. A monovalent CD28-specific fusion antibody sc28AT, a novel nonactivating single-chain Fv-based reagent, was developed (Zhang et al., 2011). Evaluation of this fusion protein in a nonhuman primate model indicated that sc28AT modestly prolongs cardiac and renal allograft survival (Poirier et al., 2010). To improve the pharmacokinetic profile (Poirier et al., 2012), a pegylated anti-CD28 SCFV named FR104, which was derived from the anti-CD28 mAb clone, was developed. The pegylated reagent has been shown to prevent graft rejection in monkeys. Furthermore, there are attempts to search for small molecules that can target CD28 and inhibit PPI of CD28 with its ligands (Uvebrant et al., 2007).

6.6 CD2–CD58 Interactions and the Design of Multicyclic Peptides

T-cell adhesion to APCs and the subsequent immune response is important in its pathogenesis. The two-signal hypothesis proposes that T-cell activation requires recognition of an antigen by the TCR (Signal 1) and a concomitant signal provided by adhesion/costimulatory molecules (Signal 2) to achieve full activation (Leitner et al., 2010). Among the adhesion/costimulatory molecules, the most abundant is CD2, a transmembrane protein in T cells, which binds to its ligand CD58 on APC. CD58, also known as leukocyte function-associated antigen 3 (LFA-3), is a cell adhesion molecule with only one known ligand, CD2 (Chen & Flies, 2013; Davis et al., 2003; van der Merwe & Davis, 2003). Ligation of CD2 on T cells to CD58 on APC facilitates T-cell–APC adhesion and is important in the early stages of immune response. This interaction results in the induction of IFN-γ and subsequent regulation of human leukocyte antigen–antigen D related (HLA-DR), intracellular adhesion molecule-1 (ICAM-1), and B-7 molecules on APC, causing an amplification of the signal for immune response. In vitro studies have indicated that inhibition of CD2 and CD58 interactions using an LFA-3 fusion protein (alefacept) inhibits T-cell activation (da Silva et al., 2002). Alefacept is a recombinant human CD58-Ig fusion protein that effectively binds to CD2 and prevents CD2 interaction with CD58. It has been successfully used clinically to treat plaque psoriasis (Chamian et al., 2005). Modulation of adhesion interaction between cell adhesion molecules has been shown to be effective in treating autoimmune diseases such as RA (Chen & Flies, 2013; Ford et al., 2014; Papoutsaki & Costanzo, 2013). Current treatments for autoimmune diseases with biologics include antibodies and fusion proteins (Papp et al., 2012; Webber, Hirose, & Vincenti, 2011). However, they have limitations in terms of stability (shelf-life), administration, and immunogenicity (Hansel, Kropshofer, Singer, Mitchell, & George, 2010). We have designed peptides that block cell adhesion between T cells expressing CD2 and Caco-2 cells expressing CD58 (Gokhale, Weldeghiorghis, Taneja, & Satyanarayanajois, 2011). The CD58-binding domain of CD2 consists of β-strands with charged residues (Fig. 18A and B). From our studies, it is very clear that peptides designed from β-strands exhibit cell adhesion inhibition activity between T cells and epithelial cells. The peptides could block the anti-CD58 binding to CD58 expressed on Caco-2 cells, indicating that peptides bind to the CD58 protein. In rodents, the homolog of CD58 is CD48. It is postulated that CD48 and CD58 have the same evolutionary origin. In mice, CD2 binds to its ligand CD48 to generate an immune response (Ianelli, Edson, & Thorley-Lawson, 1997). CD48 has a high degree of homology to CD58 and is similar in the 3D structure (Velikovsky et al., 2007; Wang et al., 1999). The adhesion domains of the two proteins overlap with backbone rmsd of 1.35 Å. We have shown that peptides designed from human CD2 protein bind to mouse CD48. Furthermore, one of the designed peptide (peptide 6) that is known to bind to

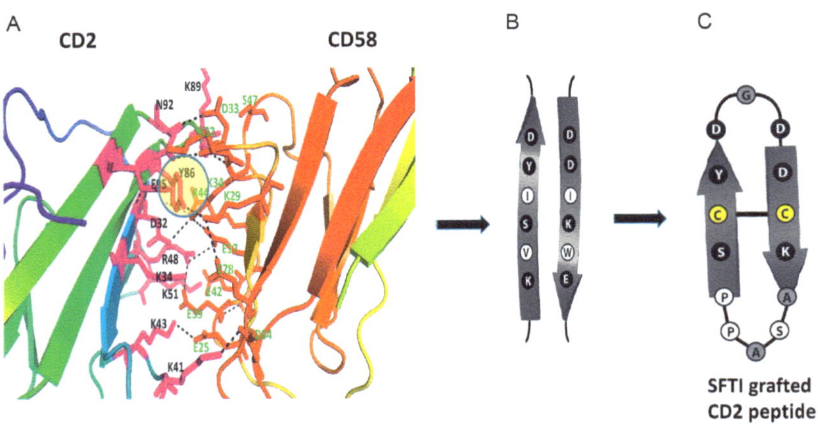

Fig. 18 (A) PPI interface of CD2–CD58 showing electrostatic and hydrophobic interactions. There are 10 salt bridges, 5 hydrogen bonds, and 1 hydrophobic interaction (PDB ID 1QA9). The design of peptides and multicyclic grafted peptide from CD2 adhesion domain based on the interface residues of CD2 of CD2–CD58 interactions. (B) F and C strands from the CD2 protein that interact with CD58. (C) Grafting of CD2 F and C strand residues onto sunflower trypsin inhibitor peptide to generate stable peptides that inhibit PPI of CD2–CD58 for immunomodulation (Sable et al., 2016 ACS Chemical Biology).

CD58 was able to suppress the progression of RA in the CIA model. We have also shown that peptide 6 is not immunogenic in mice (Gokhale, Kanthala, Latendresse, Taneja, & Satyanarayanajois, 2013; Gokhale et al., 2011).

6.6.1 Multicyclic Peptide Approach

Although the designed peptides have shown in vitro and in vivo activity, in general, many of these peptides have limitations in terms of in vivo stability (Vlieghe, Lisowski, Martinez, & Khrestchatisky, 2010). Multicyclic structures have unusually stable properties; they are generally nonimmunogenic and can be used as templates for grafting (Poth, Chan, & Craik, 2013). In order to improve the stability of the peptides, we grafted the CD2 peptide epitope onto multicyclic structures (Fig. 18C). Structural comparisons using molecular modeling suggested that sunflower trypsin inhibitor (SFTI) is suitable for grafting the CD2 peptides that we designed. In the multicyclic SFTI structure, the two β-strands are placed nearly 5 Å apart, which is suitable for grafting CD2 peptide 6 as we reported earlier (Gokhale et al., 2011). The β-strands are stabilized by disulfide bridges and cyclization that resembles peptide 6. We have grafted the CD2-derived peptides onto SFTI and have shown that the designed molecule inhibits CD2–CD58 interaction in cell adhesion assays at nanomolar concentrations. The peptide SFTI-1a was shown to bind to CD58 protein using SPR and inhibit PPI between CD2 and CD58. Furthermore, the peptide was shown to suppress T-cell immune response in an in vitro assay using the cells derived from a transgenic animal model of RA. Since these peptides are intended to exhibit thermal and enzymatic stability, SFTI-1a was evaluated for its thermal and serum stability. The peptide exhibited thermal stability up to 80°C and was stable in human serum up to 48 h (Sable et al., 2016). This study illustrates the design of peptides to inhibit PPI and translation of the pharmacophore template from the peptide to multicyclic stable structure that can be used for PPI inhibition and as a therapeutic agent for autoimmune diseases such as RA.

PPI drug discovery has been slow in the last decade, but has steadily increased in the past few years. More than 40 PPI have been targeted for drug discovery, and some of them have reached clinical trials (Petta et al., 2016). Only a few drugs that are on the market are PPI inhibitors. For example, tirofiban, an antiplatelet drug that is an inhibitor of glycoprotein IIb/IIIa (Arkin et al., 2014), is already on the market. Lifitegrast inhibits the interaction between leukocyte function associated antigen-1 (LFA-1) (CD11a/α2, CD18/β2) and ICAM-1. LFA-1 is a β2 integrin receptor

Table 2 Some of the Protein–Protein Interaction Inhibitors That Are in the Clinical Trial

Drug Name	Proprietary	Clinical Trial Stage	Indications	Website URL[a]
BRD4 inhibitors				
I-BET762 (GSK525762)	GSK	Phase II	Solid tumor	https://clinicaltrials.gov/ct2/show/NCT01587703
CPI-0610	Constellation Pharmaceutical, Cambridge, MA	Phase I	Lymphoma	https://clinicaltrials.gov/ct2/show/NCT01949883?term=CPI-0610&rank=2
Ten-010	Hoffmann-La Roche	Phase I	Acute myeloid leukemia	https://clinicaltrials.gov/ct2/show/NCT02308761?term=Ten-010&rank=1
OTX015 (MK-8628)	OncoEthix	Phase I	Cancer	https://clinicaltrials.gov/ct2/show/NCT01713582?term=OTX015&rank=2
BRD3 inhibitors				
RVX-208	Resverlogix Corp	Phase II	Diabetes, atherosclerosis	https://clinicaltrials.gov/ct2/show/NCT01728467?term=RVX-208&rank=1
IAP inhibitors				
LCL161	Novartis	Phase I	Advanced solid tumors	https://clinicaltrials.gov/ct2/show/NCT01098838?term=LCL161&rank=5
GDC-0917/CUDC-427	Genentech	Phase I	Solid cancer	https://clinicaltrials.gov/ct2/show/NCT01226277?term=GDC-0917&rank=1
GDC-0152 (RG7419)	Genentech	Phase I	Solid cancer	https://clinicaltrials.gov/ct2/show/NCT00977067?term=GDC-0152&rank=1
Bivalent inhibitors-SMAC mimetics				
Birinapant (TL-32711)	TetraLogic Pharmaceuticals	Phase II	Cancer	https://clinicaltrials.gov/ct2/show/NCT01188499?term=Birinapant&rank=5

[a]Date accessed May 17, 2017.

found on leukocytes and is involved in T-cell activation through binding to its ligand ICAM-1 (Long, 2011). Lifitegrast successfully passed clinical trials for the treatment of keratoconjunctivitis sicca inflammatory dry eye syndrome (Sheppard et al., 2014) and was approved by the FDA in 2016. Although there are challenges in PPI drug design, it is a rapidly growing area with many pharmaceutical companies participating in this new area of drug discovery (Mullard, 2012). Table 2 provides a list of PPI inhibitors that are in clinical trials.

REFERENCES

Acuner Ozbabacan, S. E., Engin, H. B., Gursoy, A., & Keskin, O. (2011). Transient protein-protein interactions. *Protein Engineering, Design & Selection*, *24*(9), 635–648. https://doi.org/10.1093/protein/gzr025.

Adams, P. D., Seeholzer, S., & Ohh, M. (2002). Identification of associated proteins by coimmunoprecipitation. In *Protein–protein interactions: A molecular cloning manual* (pp. 59–74). New York: Cold Spring Harbor Laboratory Press, Cold Spring Harbor.

Aglan, M., Amr, K., Ismail, S., Ashour, A., Otaify, G. A., Mehrez, M. A., et al. (2015). Clinical and molecular characterization of seven Egyptian families with autosomal recessive robinow syndrome: Identification of four novel ROR2 gene mutations. *American Journal of Medical Genetics. Part A*, *167A*(12), 3054–3061. https://doi.org/10.1002/ajmg.a.37287.

Arkin, M. R., Tang, Y., & Wells, J. A. (2014). Small-molecule inhibitors of protein-protein interactions: Progressing toward the reality. *Chemistry & Biology*, *21*(9), 1102–1114. https://doi.org/10.1016/j.chembiol.2014.09.001.

Bakail, M., & Ochsenbein, F. (2016). Targeting protein–protein interactions, a wide open field for drug design. *Comptes Rendus Chimie*, *19*(1), 19–27.

Banappagari, S., Corti, M., Pincus, S., & Satyanarayanajois, S. (2012). Inhibition of protein-protein interaction of HER2-EGFR and HER2-HER3 by a rationally designed peptidomimetic. *Journal of Biomolecular Structure & Dynamics*, *30*(5), 594–606. https://doi.org/10.1080/07391102.2012.687525.

Banappagari, S., Ronald, S., & Satyanarayanajois, S. D. (2010). A conformationally constrained peptidomimetic binds to the extracellular region of HER2 protein. *Journal of Biomolecular Structure & Dynamics*, *28*(3), 289–308.

Banappagari, S., Ronald, S., & Satyanarayanajois, S. D. (2011). Structure-activity relationship of conformationally constrained peptidomimetics for antiproliferative activity in HER2-overexpressing breast cancer cell lines. *MedChemComm*, *2*(8), 752–759. https://doi.org/10.1039/C1MD00126D.

Baselga, J., & Swain, S. M. (2009). Novel anticancer targets: Revisiting ERBB2 and discovering ERBB3. *Nature Reviews. Cancer*, *9*(7), 463–475. https://doi.org/10.1038/nrc2656.

Berg, T. (2005). Methods in molecular biology. In H. Fu (Ed.), *Protein–protein interactions: Methods and applications: Vol. 261* (pp. 446–447). Weinheim: WILEY-VCH Verlag GmbH & Co. KGaA. https://doi.org/10.1002/cbic.200400235. Wiley Online Library.

Beyersdorf, N., Kerkau, T., & Hunig, T. (2015). CD28 co-stimulation in T-cell homeostasis: A recent perspective. *ImmunoTargets and Therapy*, *4*, 111–122. https://doi.org/10.2147/ITT.S61647.

Biskobing, D. M. (2007). Update on bazedoxifene: A novel selective estrogen receptor modulator. *Clinical Interventions in Aging*, *2*(3), 299–303.

Bledsoe, R. K., Montana, V. G., Stanley, T. B., Delves, C. J., Apolito, C. J., McKee, D. D., et al. (2002). Crystal structure of the glucocorticoid receptor ligand binding domain reveals a novel mode of receptor dimerization and coactivator recognition. *Cell*, *110*(1), 93–105.

Bogan, A. A., & Thorn, K. S. (1998). Anatomy of hot spots in protein interfaces. *Journal of Molecular Biology*, *280*(1), 1–9. https://doi.org/10.1006/jmbi.1998.1843.

Bollag, G., Hirth, P., Tsai, J., Zhang, J., Ibrahim, P. N., Cho, H., et al. (2010). Clinical efficacy of a RAF inhibitor needs broad target blockade in BRAF-mutant melanoma. *Nature*, *467*(7315), 596–599. https://doi.org/10.1038/nature09454.

Boomer, J. S., & Green, J. M. (2010). An enigmatic tail of CD28 signaling. *Cold Spring Harbor Perspectives in Biology*, *2*(8), a002436. https://doi.org/10.1101/cshperspect.a002436.

Bouvier, M. (2001). Oligomerization of G-protein-coupled transmitter receptors. *Nature Reviews. Neuroscience*, *2*(4), 274–286. https://doi.org/10.1038/35067575.

Brabender, J., Danenberg, K. D., Metzger, R., Schneider, P. M., Park, J., Salonga, D., et al. (2001). Epidermal growth factor receptor and HER2-neu mRNA expression in non-small cell lung cancer is correlated with survival. *Clinical Cancer Research: An Official Journal of the American Association for Cancer Research*, 7(7), 1850–1855.

Braisted, A. C., Oslob, J. D., Delano, W. L., Hyde, J., McDowell, R. S., Waal, N., et al. (2003). Discovery of a potent small molecule IL-2 inhibitor through fragment assembly. *Journal of the American Chemical Society*, *125*(13), 3714–3715.

Bretscher, P. A. (1999). A two-step, two-signal model for the primary activation of precursor helper T cells. *Proceedings of the National Academy of Sciences of the United States of America*, *96*(1), 185–190.

Brogi, S., Tafi, A., Desaubry, L., & Nebigil, C. G. (2014). Discovery of GPCR ligands for probing signal transduction pathways. *Frontiers in Pharmacology*, *5*, 255. https://doi.org/10.3389/fphar.2014.00255.

Brown, J. A., Dorfman, D. M., Ma, F. R., Sullivan, E. L., Munoz, O., Wood, C. R., et al. (2003). Blockade of programmed death-1 ligands on dendritic cells enhances T cell activation and cytokine production. *Journal of Immunology*, *170*(3), 1257–1266.

Brzostek, J., Gascoigne, N. R., & Rybakin, V. (2016). Cell type-specific regulation of immunological synapse dynamics by B7 ligand recognition. *Frontiers in Immunology*, 7, 24. https://doi.org/10.3389/fimmu.2016.00024.

Buchbinder, E. I., & Desai, A. (2016). CTLA-4 and PD-1 pathways: Similarities, differences, and implications of their inhibition. *American Journal of Clinical Oncology*, *39*(1), 98–106. https://doi.org/10.1097/COC.0000000000000239.

Burgess, A. W., Cho, H. S., Eigenbrot, C., Ferguson, K. M., Garrett, T. P., Leahy, D. J., et al. (2003). An open-and-shut case? Recent insights into the activation of EGF/ErbB receptors. *Molecular Cell*, *12*(3), 541–552.

Chabre, M., Cone, R., & Saibil, H. (2003). Biophysics: Is rhodopsin dimeric in native retinal rods? *Nature*, *426*(6962), 30–31. discussion 31 https://doi.org/10.1038/426030b.

Chakrabarti, P., & Janin, J. (2002). Dissecting protein-protein recognition sites. *Proteins*, *47*(3), 334–343.

Chamian, F., Lowes, M. A., Lin, S. L., Lee, E., Kikuchi, T., Gilleaudeau, P., et al. (2005). Alefacept reduces infiltrating T cells, activated dendritic cells, and inflammatory genes in psoriasis vulgaris. *Proceedings of the National Academy of Sciences of the United States of America*, *102*(6), 2075–2080. https://doi.org/10.1073/pnas.0409569102.

Chen, L., & Flies, D. B. (2013). Molecular mechanisms of T cell co-stimulation and co-inhibition. *Nature Reviews. Immunology*, *13*(4), 227–242. https://doi.org/10.1038/nri3405.

Chene, P. (2006). Drugs targeting protein-protein interactions. *ChemMedChem*, *1*(4), 400–411. https://doi.org/10.1002/cmdc.200600004.

Cho, H. S., Mason, K., Ramyar, K. X., Stanley, A. M., Gabelli, S. B., Denney, D. W., Jr., et al. (2003). Structure of the extracellular region of HER2 alone and in complex with the Herceptin Fab. *Nature*, *421*(6924), 756–760. https://doi.org/10.1038/nature01392.

Clackson, T., & Wells, J. A. (1995). A hot spot of binding energy in a hormone-receptor interface. *Science, 267*(5196), 383–386.

da Silva, A. J., Brickelmaier, M., Majeau, G. R., Li, Z., Su, L., Hsu, Y. M., et al. (2002). Alefacept, an immunomodulatory recombinant LFA-3/IgG1 fusion protein, induces CD16 signaling and CD2/CD16-dependent apoptosis of CD2(+) cells. *Journal of Immunology, 168*(9), 4462–4471.

Davis, S. J., Ikemizu, S., Evans, E. J., Fugger, L., Bakker, T. R., & van der Merwe, P. A. (2003). The nature of molecular recognition by T cells. *Nature Immunology, 4*(3), 217–224. https://doi.org/10.1038/ni0303-217.

Dayhoff, J. E., Shoemaker, B. A., Bryant, S. H., & Panchenko, A. R. (2010). Evolution of protein binding modes in homooligomers. *Journal of Molecular Biology, 395*(4), 860–870. https://doi.org/10.1016/j.jmb.2009.10.052.

Dechant, G., & Barde, Y. A. (2002). The neurotrophin receptor p75(NTR): Novel functions and implications for diseases of the nervous system. *Nature Neuroscience, 5*(11), 1131–1136. https://doi.org/10.1038/nn1102-1131.

DeLano, W. L. (2002). Unraveling hot spots in binding interfaces: Progress and challenges. *Current Opinion in Structural Biology, 12*(1), 14–20.

de Mol, N. J., & Fischer, M. J. (2010). Surface plasmon resonance: A general introduction. *Methods in Molecular Biology, 627*, 1–14. https://doi.org/10.1007/978-1-60761-670-2_1.

De Strooper, B., Vassar, R., & Golde, T. (2010). The secretases: Enzymes with therapeutic potential in Alzheimer disease. *Nature Reviews. Neurology, 6*(2), 99–107. https://doi.org/10.1038/nrneurol.2009.218.

Dilek, N., Poirier, N., Hulin, P., Coulon, F., Mary, C., Ville, S., et al. (2013). Targeting CD28, CTLA-4 and PD-L1 costimulation differentially controls immune synapses and function of human regulatory and conventional T-cells. *PLoS One, 8*(12), e83139. https://doi.org/10.1371/journal.pone.0083139.

Dodson, L. F., Boomer, J. S., Deppong, C. M., Shah, D. D., Sim, J., Bricker, T. L., et al. (2009). Targeted knock-in mice expressing mutations of CD28 reveal an essential pathway for costimulation. *Molecular and Cellular Biology, 29*(13), 3710–3721. https://doi.org/10.1128/MCB.01869-08.

Dolan, D. E., & Gupta, S. (2014). PD-1 pathway inhibitors: Changing the landscape of cancer immunotherapy. *Cancer Control, 21*(3), 231–238.

Dunbar, A. Y., Kamada, Y., Jenkins, G. J., Lowe, E. R., Billecke, S. S., & Osawa, Y. (2004). Ubiquitination and degradation of neuronal nitric-oxide synthase in vitro: Dimer stabilization protects the enzyme from proteolysis. *Molecular Pharmacology, 66*(4), 964–969. https://doi.org/10.1124/mol.104.000125.

Eggert, S., Midthune, B., Cottrell, B., & Koo, E. H. (2009). Induced dimerization of the amyloid precursor protein leads to decreased amyloid-beta protein production. *The Journal of Biological Chemistry, 284*(42), 28943–28952. https://doi.org/10.1074/jbc.M109.038646.

Endres, N. F., Das, R., Smith, A. W., Arkhipov, A., Kovacs, E., Huang, Y., et al. (2013). Conformational coupling across the plasma membrane in activation of the EGF receptor. *Cell, 152*(3), 543–556. https://doi.org/10.1016/j.cell.2012.12.032.

Erlanson, D. A. (2006). Fragment-based lead discovery: A chemical update. *Current Opinion in Biotechnology, 17*(6), 643–652. https://doi.org/10.1016/j.copbio.2006.10.007.

Ferguson, K. M. (2004). Active and inactive conformations of the epidermal growth factor receptor. *Biochemical Society Transactions, 32*(Pt. 5), 742–745. https://doi.org/10.1042/BST0320742.

Ferguson, K. M. (2008). Structure-based view of epidermal growth factor receptor regulation. *Annual Review of Biophysics, 37*, 353–373. https://doi.org/10.1146/annurev.biophys.37.032807.125829.

Ferguson, K. M., Berger, M. B., Mendrola, J. M., Cho, H. S., Leahy, D. J., & Lemmon, M. A. (2003). EGF activates its receptor by removing interactions that autoinhibit ectodomain dimerization. *Molecular Cell*, *11*(2), 507–517.
Fichter, C. D., Timme, S., Braun, J. A., Gudernatsch, V., Schopflin, A., Bogatyreva, L., et al. (2014). EGFR, HER2 and HER3 dimerization patterns guide targeted inhibition in two histotypes of esophageal cancer. *International Journal of Cancer*, *135*(7), 1517–1530. https://doi.org/10.1002/ijc.28771.
Ford, M. L., Adams, A. B., & Pearson, T. C. (2014). Targeting co-stimulatory pathways: Transplantation and autoimmunity. *Nature Reviews. Nephrology*, *10*(1), 14–24. https://doi.org/10.1038/nrneph.2013.183.
Fox, S., Farr-Jones, S., Sopchak, L., Boggs, A., Nicely, H. W., Khoury, R., et al. (2006). High-throughput screening: Update on practices and success. *Journal of Biomolecular Screening*, *11*(7), 864–869. https://doi.org/10.1177/1087057106292473.
Franco, R., Casado, V., Cortes, A., Ferrada, C., Mallol, J., Woods, A., et al. (2007). Basic concepts in G-protein-coupled receptor homo- and heterodimerization. *ScientificWorldJournal*, 7, 48–57. https://doi.org/10.1100/tsw.2007.197.
Franco, R., Martinez-Pinilla, E., Lanciego, J. L., & Navarro, G. (2016). Basic pharmacological and structural evidence for class A G-protein-coupled receptor heteromerization. *Frontiers in Pharmacology*, 7, 76. https://doi.org/10.3389/fphar.2016.00076.
Fredriksson, S., Gullberg, M., Jarvius, J., Olsson, C., Pietras, K., Gustafsdottir, S. M., et al. (2002). Protein detection using proximity-dependent DNA ligation assays. *Nature Biotechnology*, *20*(5), 473–477. https://doi.org/10.1038/nbt0502-473.
Fredriksson, R., Lagerstrom, M. C., Lundin, L. G., & Schioth, H. B. (2003). The G-protein-coupled receptors in the human genome form five main families. Phylogenetic analysis, paralogon groups, and fingerprints. *Molecular Pharmacology*, *63*(6), 1256–1272. https://doi.org/10.1124/mol.63.6.1256.
Freeman, G. J. (2008). Structures of PD-1 with its ligands: Sideways and dancing cheek to cheek. *Proceedings of the National Academy of Sciences of the United States of America*, *105*(30), 10275–10276. https://doi.org/10.1073/pnas.0805459105.
Friend, L. D., Shah, D. D., Deppong, C., Lin, J., Bricker, T. L., Juehne, T. I., et al. (2006). A dose-dependent requirement for the proline motif of CD28 in cellular and humoral immunity revealed by a targeted knockin mutant. *The Journal of Experimental Medicine*, *203*(9), 2121–2133. https://doi.org/10.1084/jem.20052230.
Fry, D., Huang, K. S., Di Lello, P., Mohr, P., Muller, K., So, S. S., et al. (2013). Design of libraries targeting protein-protein interfaces. *ChemMedChem*, *8*(5), 726–732. https://doi.org/10.1002/cmdc.201200540.
Gardner, D., Jeffery, L. E., & Sansom, D. M. (2014). Understanding the CD28/CTLA-4 (CD152) pathway and its implications for costimulatory blockade. *American Journal of Transplantation*, *14*(9), 1985–1991. https://doi.org/10.1111/ajt.12834.
Garrett, T. P., McKern, N. M., Lou, M., Elleman, T. C., Adams, T. E., Lovrecz, G. O., et al. (2002). Crystal structure of a truncated epidermal growth factor receptor extracellular domain bound to transforming growth factor alpha. *Cell*, *110*(6), 763–773.
Gentry, J. J., Rutkoski, N. J., Burke, T. L., & Carter, B. D. (2004). A functional interaction between the p75 neurotrophin receptor interacting factors, TRAF6 and NRIF. *The Journal of Biological Chemistry*, *279*(16), 16646–16656. https://doi.org/10.1074/jbc.M309209200.
Gokhale, A., Kanthala, S., Latendresse, J., Taneja, V., & Satyanarayanajois, S. (2013). Immunosuppression by co-stimulatory molecules: Inhibition of CD2-CD48/CD58 interaction by peptides from CD2 to suppress progression of collagen-induced arthritis in mice. *Chemical Biology & Drug Design*, *82*(1), 106–118. https://doi.org/10.1111/cbdd.12138.
Gokhale, A. S., & Satyanarayanajois, S. (2014). Peptides and peptidomimetics as immunomodulators. *Immunotherapy*, *6*(6), 755–774. https://doi.org/10.2217/imt.14.37.

Gokhale, A., Weldeghiorghis, T. K., Taneja, V., & Satyanarayanajois, S. D. (2011). Conformationally constrained peptides from CD2 to modulate protein-protein interactions between CD2 and CD58. *Journal of Medicinal Chemistry*, *54*(15), 5307–5319. https://doi.org/10.1021/jm200004e.

Goodsell, D. S., & Olson, A. J. (2000). Structural symmetry and protein function. *Annual Review of Biophysics and Biomolecular Structure*, *29*, 105–153. https://doi.org/10.1146/annurev.biophys.29.1.105.

Graham, J., Muhsin, M., & Kirkpatrick, P. (2004). Cetuximab. *Nature Reviews. Drug Discovery*, *3*(7), 549–550. https://doi.org/10.1038/nrd1445.

Greenwald, R. J., Freeman, G. J., & Sharpe, A. H. (2005). The B7 family revisited. *Annual Review of Immunology*, *23*, 515–548. https://doi.org/10.1146/annurev.immunol.23.021704.115611.

Guo, W., Wisniewski, J. A., & Ji, H. (2014). Hot spot-based design of small-molecule inhibitors for protein-protein interactions. *Bioorganic & Medicinal Chemistry Letters*, *24*(11), 2546–2554. https://doi.org/10.1016/j.bmcl.2014.03.095.

Hall, R. A. (2005). Co-immunoprecipitation as a strategy to evaluate receptor-receptor or receptor-protein interactions. *G Protein-Coupled Receptor-Protein Interactions*, *9*. 165À178.

Hanold, L. E., Oruganty, K., Ton, N. T., Beedle, A. M., Kannan, N., & Kennedy, E. J. (2015). Inhibiting EGFR dimerization using triazolyl-bridged dimerization arm mimics. *PLoS One*, *10*(3), e0118796. https://doi.org/10.1371/journal.pone.0118796.

Hansel, T. T., Kropshofer, H., Singer, T., Mitchell, J. A., & George, A. J. (2010). The safety and side effects of monoclonal antibodies. *Nature Reviews. Drug Discovery*, *9*(4), 325–338. https://doi.org/10.1038/nrd3003.

Hikita, H., Takehara, T., Shimizu, S., Kodama, T., Shigekawa, M., Iwase, K., et al. (2010). The Bcl-xL inhibitor, ABT-737, efficiently induces apoptosis and suppresses growth of hepatoma cells in combination with sorafenib. *Hepatology*, *52*(4), 1310–1321. https://doi.org/10.1002/hep.23836.

Hiller, C., Kuhhorn, J., & Gmeiner, P. (2013). Class A G-protein-coupled receptor (GPCR) dimers and bivalent ligands. *Journal of Medicinal Chemistry*, *56*(17), 6542–6559. https://doi.org/10.1021/jm4004335.

Hirsch, F. R., Jotte, R. M., Berry, C. A., Mencia, W. A., Stowell, S. A., & Gardner, A. J. (2014). Quality of care of patients with non-small-cell lung cancer: A report of a performance improvement initiative. *Cancer Control: Journal of the Moffitt Cancer Center*, *21*(1), 90–97.

Hirsch, F. R., Varella-Garcia, M., & Cappuzzo, F. (2009). Predictive value of EGFR and HER2 overexpression in advanced non-small-cell lung cancer. *Oncogene*, *28*(Suppl. 1), S32–37. https://doi.org/10.1038/onc.2009.199.

Hunig, T. (2007). Manipulation of regulatory T-cell number and function with CD28-specific monoclonal antibodies. *Advances in Immunology*, *95*, 111–148. https://doi.org/10.1016/S0065-2776(07)95004-X.

Ianelli, C. J., Edson, C. M., & Thorley-Lawson, D. A. (1997). A ligand for human CD48 on epithelial cells. *Journal of Immunology*, *159*(8), 3910–3920.

Ishida, Y., Agata, Y., Shibahara, K., & Honjo, T. (1992). Induced expression of PD-1, a novel member of the immunoglobulin gene superfamily, upon programmed cell death. *The EMBO Journal*, *11*(11), 3887–3895.

Ispolatov, I., Yuryev, A., Mazo, I., & Maslov, S. (2005). Binding properties and evolution of homodimers in protein-protein interaction networks. *Nucleic Acids Research*, *33*(11), 3629–3635. https://doi.org/10.1093/nar/gki678.

Iyer, V. V. (2016). A review of stapled peptides and small molecules to inhibit protein-protein interactions in cancer. *Current Medicinal Chemistry*, *23*(27), 3025–3043.

Jacoby, E., Bouhelal, R., Gerspacher, M., & Seuwen, K. (2006). The 7 TM G-protein-coupled receptor target family. *ChemMedChem*, *1*(8), 761–782. https://doi.org/10.1002/cmdc.200600134.

James, J. R., Oliveira, M. I., Carmo, A. M., Iaboni, A., & Davis, S. J. (2006). A rigorous experimental framework for detecting protein oligomerization using bioluminescence resonance energy transfer. *Nature Methods*, *3*(12), 1001–1006. https://doi.org/10.1038/nmeth978.

Jencks, W. P. (1981). On the attribution and additivity of binding energies. *Proceedings of the National Academy of Sciences of the United States of America*, *78*(7), 4046–4050.

Jin, L., Wang, W., & Fang, G. (2014). Targeting protein-protein interaction by small molecules. *Annual Review of Pharmacology and Toxicology*, *54*, 435–456. https://doi.org/10.1146/annurev-pharmtox-011613-140028.

Jones, C. D., Jevnikar, M. G., Pike, A. J., Peters, M. K., Black, L. J., Thompson, A. R., et al. (1984). Antiestrogens. 2. Structure-activity studies in a series of 3-aroyl-2-arylbenzo[b]thiophene derivatives leading to [6-hydroxy-2-(4-hydroxyphenyl)benzo[b]thien-3-yl][4-[2-(1-piperidinyl)ethoxy]-phenyl]methanone hydrochloride (LY156758), a remarkably effective estrogen antagonist with only minimal intrinsic estrogenicity. *Journal of Medicinal Chemistry*, *27*(8), 1057–1066.

Jubb, H., Higueruelo, A. P., Winter, A., & Blundell, T. L. (2012). Structural biology and drug discovery for protein-protein interactions. *Trends in Pharmacological Sciences*, *33*(5), 241–248. https://doi.org/10.1016/j.tips.2012.03.006.

Jung, J. I., Ran, Y., Cruz, P. E., Rosario, A. M., Ladd, T. B., Kukar, T. L., et al. (2014). Complex relationships between substrate sequence and sensitivity to alterations in gamma-secretase processivity induced by gamma-secretase modulators. *Biochemistry*, *53*(12), 1947–1957. https://doi.org/10.1021/bi401521t.

Kaake, R. M., Wang, X., Burke, A., Yu, C., Kandur, W., Yang, Y., et al. (2014). A new in vivo cross-linking mass spectrometry platform to define protein-protein interactions in living cells. *Molecular & Cellular Proteomics*, *13*(12), 3533–3543. https://doi.org/10.1074/mcp.M114.042630.

Kadmiel, M., & Cidlowski, J. A. (2013). Glucocorticoid receptor signaling in health and disease. *Trends in Pharmacological Sciences*, *34*(9), 518–530. https://doi.org/10.1016/j.tips.2013.07.003.

Kanthala, S., Banappagari, S., Gokhale, A., Liu, Y. Y., Xin, G., Zhao, Y., et al. (2015). Novel peptidomimetics for inhibition of HER2:HER3 heterodimerization in HER2-positive breast cancer. *Chemical Biology & Drug Design*, *85*, 702–714. https://doi.org/10.1111/cbdd.12453.

Kanthala, S., Gauthier, T., & Satyanarayanajois, S. (2014). Structure-activity relationships of peptidomimetics that inhibit PPI of HER2-HER3. *Biopolymers*, *101*(6), 693–702. https://doi.org/10.1002/bip.22441.

Karpova, T., & McNally, J. G. (2006). Detecting protein-protein interactions with CFP-YFP FRET by acceptor photobleaching. *Current Protocols in Cytometry*. Chapter 12, Unit12.7. https://doi.org/10.1002/0471142956.cy1207s35.

Khalifa, N. B., Van Hees, J., Tasiaux, B., Huysseune, S., Smith, S. O., Constantinescu, S. N., et al. (2010). What is the role of amyloid precursor protein dimerization? *Cell Adhesion & Migration*, *4*(2), 268–272.

Kim, M., Sun, Z. Y., Byron, O., Campbell, G., Wagner, G., Wang, J., et al. (2001). Molecular dissection of the CD2-CD58 counter-receptor interface identifies CD2 Tyr86 and CD58 Lys34 residues as the functional "hot spot". *Journal of Molecular Biology*, *312*(4), 711–720. https://doi.org/10.1006/jmbi.2001.4980.

Kocan, M., & Pfleger, K. D. (2011). Study of GPCR-protein interactions by BRET. *Methods in Molecular Biology*, *746*, 357–371. https://doi.org/10.1007/978-1-61779-126-0_20.

Kozakov, D., Grove, L. E., Hall, D. R., Bohnuud, T., Mottarella, S. E., Luo, L., et al. (2015). The FTMap family of web servers for determining and characterizing ligand-binding hot spots of proteins. *Nature Protocols*, *10*(5), 733–755. https://doi.org/10.1038/nprot.2015.043.

Krishna, S. S., & Aravind, L. (2010). The bridge-region of the Ku superfamily is an atypical zinc ribbon domain. *Journal of Structural Biology, 172*(3), 294–299. https://doi.org/10.1016/j.jsb.2010.05.011.

Larsen, C. P., Pearson, T. C., Adams, A. B., Tso, P., Shirasugi, N., Strobert, E., et al. (2005). Rational development of LEA29Y (belatacept), a high-affinity variant of CTLA4-Ig with potent immunosuppressive properties. *American Journal of Transplantation, 5*(3), 443–453. https://doi.org/10.1111/j.1600-6143.2005.00749.x.

Lazar-Molnar, E., Yan, Q., Cao, E., Ramagopal, U., Nathenson, S. G., & Almo, S. C. (2008). Crystal structure of the complex between programmed death-1 (PD-1) and its ligand PD-L2. *Proceedings of the National Academy of Sciences of the United States of America, 105*(30), 10483–10488. https://doi.org/10.1073/pnas.0804453105.

Lee-Hoeflich, S. T., Crocker, L., Yao, E., Pham, T., Munroe, X., Hoeflich, K. P., et al. (2008). A central role for HER3 in HER2-amplified breast cancer: Implications for targeted therapy. *Cancer Research, 68*(14), 5878–5887. https://doi.org/10.1158/0008-5472.CAN-08-0380.

Leitner, J., Kuschei, W., Grabmeier-Pfistershammer, K., Woitek, R., Kriehuber, E., Majdic, O., et al. (2010). T cell stimulator cells, an efficient and versatile cellular system to assess the role of costimulatory ligands in the activation of human T cells. *Journal of Immunological Methods, 362*(1–2), 131–141. https://doi.org/10.1016/j.jim.2010.09.020.

Lensink, M. F., Velankar, S., Kryshtafovych, A., Huang, S. Y., Schneidman-Duhovny, D., Sali, A., et al. (2016). Prediction of homoprotein and heteroprotein complexes by protein docking and template-based modeling: A CASP-CAPRI experiment. *Proteins, 84*(Suppl. 1), 323–348. https://doi.org/10.1002/prot.25007.

Levy, E. D., Pereira-Leal, J. B., Chothia, C., & Teichmann, S. A. (2006). 3D complex: A structural classification of protein complexes. *PLoS Computational Biology, 2*(11), e155. https://doi.org/10.1371/journal.pcbi.0020155.

Li, H., Xiao, H., Lin, L., Jou, D., Kumari, V., Lin, J., et al. (2014). Drug design targeting protein-protein interactions (PPIs) using multiple ligand simultaneous docking (MLSD) and drug repositioning: Discovery of raloxifene and bazedoxifene as novel inhibitors of IL-6/GP130 interface. *Journal of Medicinal Chemistry, 57*(3), 632–641. https://doi.org/10.1021/jm401144z.

Lin, D. Y., Tanaka, Y., Iwasaki, M., Gittis, A. G., Su, H. P., Mikami, B., et al. (2008). The PD-1/PD-L1 complex resembles the antigen-binding Fv domains of antibodies and T cell receptors. *Proceedings of the National Academy of Sciences of the United States of America, 105*(8), 3011–3016. https://doi.org/10.1073/pnas.0712278105.

Lin, Z., Tann, J. Y., Goh, E. T., Kelly, C., Lim, K. B., Gao, J. F., et al. (2015). Structural basis of death domain signaling in the p75 neurotrophin receptor. *eLife, 4*, e11692. https://doi.org/10.7554/eLife.11692.

Linsley, P. S., & Nadler, S. G. (2009). The clinical utility of inhibiting CD28-mediated costimulation. *Immunological Reviews, 229*(1), 307–321. https://doi.org/10.1111/j.1600-065X.2009.00780.x.

Liu, Y., Ross, J. F., Bodine, P. V., & Billiard, J. (2007). Homodimerization of Ror2 tyrosine kinase receptor induces 14-3-3(beta) phosphorylation and promotes osteoblast differentiation and bone formation. *Molecular Endocrinology, 21*(12), 3050–3061. https://doi.org/10.1210/me.2007-0323.

Lo Conte, L., Chothia, C., & Janin, J. (1999). The atomic structure of protein-protein recognition sites. *Journal of Molecular Biology, 285*(5), 2177–2198.

London, N., Raveh, B., & Schueler-Furman, O. (2013). Druggable protein-protein interactions—From hot spots to hot segments. *Current Opinion in Chemical Biology, 17*(6), 952–959. https://doi.org/10.1016/j.cbpa.2013.10.011.

Long, E. O. (2011). ICAM-1: Getting a grip on leukocyte adhesion. *Journal of Immunology, 186*(9), 5021–5023. https://doi.org/10.4049/jimmunol.1100646.
Lowe, J., Li, H., Downing, K. H., & Nogales, E. (2001). Refined structure of alpha beta-tubulin at 3.5 A resolution. *Journal of Molecular Biology, 313*(5), 1045–1057. https://doi.org/10.1006/jmbi.2001.5077.
Lu, C., Mi, L. Z., Grey, M. J., Zhu, J., Graef, E., Yokoyama, S., et al. (2010). Structural evidence for loose linkage between ligand binding and kinase activation in the epidermal growth factor receptor. *Molecular and Cellular Biology, 30*(22), 5432–5443. https://doi.org/10.1128/MCB.00742-10.
Luttrell, L. M. (2006). Transmembrane signaling by G protein-coupled receptors. *Methods in Molecular Biology, 332*, 3–49. https://doi.org/10.1385/1-59745-048-0:1.
Madauss, K. P., Bledsoe, R. K., McLay, I., Stewart, E. L., Uings, I. J., Weingarten, G., et al. (2008). The first X-ray crystal structure of the glucocorticoid receptor bound to a non-steroidal agonist. *Bioorganic & Medicinal Chemistry Letters, 18*(23), 6097–6099. https://doi.org/10.1016/j.bmcl.2008.10.021.
Marianayagam, N. J., Sunde, M., & Matthews, J. M. (2004). The power of two: Protein dimerization in biology. *Trends in Biochemical Sciences, 29*(11), 618–625. https://doi.org/10.1016/j.tibs.2004.09.006.
Masters, C. L., Simms, G., Weinman, N. A., Multhaup, G., McDonald, B. L., & Beyreuther, K. (1985). Amyloid plaque core protein in Alzheimer disease and Down syndrome. *Proceedings of the National Academy of Sciences of the United States of America, 82*(12), 4245–4249.
Mayr, L. M., & Fuerst, P. (2008). The future of high-throughput screening. *Journal of Biomolecular Screening, 13*(6), 443–448. https://doi.org/10.1177/1087057108319644.
Medina, P. J., & Adams, V. R. (2016). PD-1 pathway inhibitors: Immuno-oncology agents for restoring antitumor immune responses. *Pharmacotherapy, 36*(3), 317–334. https://doi.org/10.1002/phar.1714.
Mi, L. Z., Grey, M. J., Nishida, N., Walz, T., Lu, C., & Springer, T. A. (2008). Functional and structural stability of the epidermal growth factor receptor in detergent micelles and phospholipid nanodiscs. *Biochemistry, 47*(39), 10314–10323. https://doi.org/10.1021/bi801006s.
Mi, L. Z., Lu, C., Li, Z., Nishida, N., Walz, T., & Springer, T. A. (2011). Simultaneous visualization of the extracellular and cytoplasmic domains of the epidermal growth factor receptor. *Nature Structural & Molecular Biology, 18*(9), 984–989. https://doi.org/10.1038/nsmb.2092.
Mineev, K. S., Bocharov, E. V., Pustovalova, Y. E., Bocharova, O. V., Chupin, V. V., & Arseniev, A. S. (2010). Spatial structure of the transmembrane domain heterodimer of ErbB1 and ErbB2 receptor tyrosine kinases. *Journal of Molecular Biology, 400*(2), 231–243. https://doi.org/10.1016/j.jmb.2010.05.016.
Moreira, I. S., Fernandes, P. A., & Ramos, M. J. (2007). Hot spots—A review of the protein-protein interface determinant amino-acid residues. *Proteins, 68*(4), 803–812. https://doi.org/10.1002/prot.21396.
Morelli, X., Bourgeas, R., & Roche, P. (2011). Chemical and structural lessons from recent successes in protein-protein interaction inhibition (2P2I). *Current Opinion in Chemical Biology, 15*(4), 475–481. https://doi.org/10.1016/j.cbpa.2011.05.024.
Mullard, A. (2012). Protein-protein interaction inhibitors get into the groove. *Nature Reviews. Drug Discovery, 11*(3), 173–175. https://doi.org/10.1038/nrd3680.
Munter, L. M., Voigt, P., Harmeier, A., Kaden, D., Gottschalk, K. E., Weise, C., et al. (2007). GxxxG motifs within the amyloid precursor protein transmembrane sequence are critical for the etiology of Abeta42. *The EMBO Journal, 26*(6), 1702–1712. https://doi.org/10.1038/sj.emboj.7601616.

Nadezhdin, K. D., Garcia-Carpio, I., Goncharuk, S. A., Mineev, K. S., Arseniev, A. S., & Vilar, M. (2016). Structural basis of p75 transmembrane domain dimerization. *The Journal of Biological Chemistry*, *291*(23), 12346–12357. https://doi.org/10.1074/jbc.M116.723585.
Nguyen, T. N., & Goodrich, J. A. (2006). Protein-protein interaction assays: Eliminating false positive interactions. *Nature Methods*, *3*(2), 135–139. https://doi.org/10.1038/nmeth0206-135.
Nykjaer, A., Willnow, T. E., & Petersen, C. M. (2005). p75NTR—Live or let die. *Current Opinion in Neurobiology*, *15*(1), 49–57. https://doi.org/10.1016/j.conb.2005.01.004.
Oakley, R. H., & Cidlowski, J. A. (2011). Cellular processing of the glucocorticoid receptor gene and protein: New mechanisms for generating tissue-specific actions of glucocorticoids. *The Journal of Biological Chemistry*, *286*(5), 3177–3184. https://doi.org/10.1074/jbc.R110.179325.
Oasa, S., Sasaki, A., Yamamoto, J., Mikuni, S., & Kinjo, M. (2015). Homodimerization of glucocorticoid receptor from single cells investigated using fluorescence correlation spectroscopy and microwells. *FEBS Letters*, *589*(17), 2171–2178. https://doi.org/10.1016/j.febslet.2015.07.003.
Ogiso, H., Ishitani, R., Nureki, O., Fukai, S., Yamanaka, M., Kim, J. H., et al. (2002). Crystal structure of the complex of human epidermal growth factor and receptor extracellular domains. *Cell*, *110*(6), 775–787.
Oltersdorf, T., Elmore, S. W., Shoemaker, A. R., Armstrong, R. C., Augeri, D. J., Belli, B. A., et al. (2005). An inhibitor of Bcl-2 family proteins induces regression of solid tumours. *Nature*, *435*(7042), 677–681. https://doi.org/10.1038/nature03579.
Oxnard, G. R., Arcila, M. E., Chmielecki, J., Ladanyi, M., Miller, V. A., & Pao, W. (2011). New strategies in overcoming acquired resistance to epidermal growth factor receptor tyrosine kinase inhibitors in lung cancer. *Clinical Cancer Research: An Official Journal of the American Association for Cancer Research*, *17*(17), 5530–5537. https://doi.org/10.1158/1078-0432.CCR-10-2571.
Papoutsaki, M., & Costanzo, A. (2013). Treatment of psoriasis and psoriatic arthritis. *BioDrugs*, *27*(Suppl. 1), 3–12. https://doi.org/10.1007/BF03325637.
Papp, K. A., Dekoven, J., Parsons, L., Pirzada, S., Robern, M., Robertson, L., et al. (2012). Biologic therapy in psoriasis: Perspectives on associated risks and patient management. *Journal of Cutaneous Medicine and Surgery*, *16*(3), 153–168. https://doi.org/10.1177/120347541201600305.
Petta, I., Lievens, S., Libert, C., Tavernier, J., & De Bosscher, K. (2016). Modulation of protein-protein interactions for the development of novel therapeutics. *Molecular Therapy*, *24*(4), 707–718. https://doi.org/10.1038/mt.2015.214.
Piccart-Gebhart, M. J., Procter, M., Leyland-Jones, B., Goldhirsch, A., Untch, M., Smith, I., et al. (2005). Trastuzumab after adjuvant chemotherapy in HER2-positive breast cancer. *The New England Journal of Medicine*, *353*(16), 1659–1672. https://doi.org/10.1056/NEJMoa052306.
Poger, D., & Mark, A. E. (2014). Activation of the epidermal growth factor receptor: A series of twists and turns. *Biochemistry*, *53*(16), 2710–2721. https://doi.org/10.1021/bi401632z.
Poirier, N., Azimzadeh, A. M., Zhang, T., Dilek, N., Mary, C., Nguyen, B., et al. (2010). Inducing CTLA-4-dependent immune regulation by selective CD28 blockade promotes regulatory T cells in organ transplantation. *Science Translational Medicine*, *2*(17), 17ra10. https://doi.org/10.1126/scitranslmed.3000116.
Poirier, N., Blancho, G., & Vanhove, B. (2011). A more selective costimulatory blockade of the CD28-B7 pathway. *Transplant International*, *24*(1), 2–11. https://doi.org/10.1111/j.1432-2277.2010.01176.x.

Poirier, N., Mary, C., Dilek, N., Hervouet, J., Minault, D., Blancho, G., et al. (2012). Preclinical efficacy and immunological safety of FR104, an antagonist anti-CD28 monovalent Fab' antibody. *American Journal of Transplantation*, *12*(10), 2630–2640. https://doi.org/10.1111/j.1600-6143.2012.04164.x.

Poth, A. G., Chan, L. Y., & Craik, D. J. (2013). Cyclotides as grafting frameworks for protein engineering and drug design applications. *Biopolymers*, *100*(5), 480–491. https://doi.org/10.1002/bip.22284.

Reichmann, D., Rahat, O., Cohen, M., Neuvirth, H., & Schreiber, G. (2007). The molecular architecture of protein-protein binding sites. *Current Opinion in Structural Biology*, *17*(1), 67–76. https://doi.org/10.1016/j.sbi.2007.01.004.

Richter, L., Munter, L. M., Ness, J., Hildebrand, P. W., Dasari, M., Unterreitmeier, S., et al. (2010). Amyloid beta 42 peptide (Abeta42)-lowering compounds directly bind to Abeta and interfere with amyloid precursor protein (APP) transmembrane dimerization. *Proceedings of the National Academy of Sciences of the United States of America*, *107*(33), 14597–14602. https://doi.org/10.1073/pnas.1003026107.

Roarty, K., Shore, A. N., Creighton, C. J., & Rosen, J. M. (2015). Ror2 regulates branching, differentiation, and actin-cytoskeletal dynamics within the mammary epithelium. *The Journal of Cell Biology*, *208*(3), 351–366. https://doi.org/10.1083/jcb.201408058.

Roskoski, R., Jr. (2014). The ErbB/HER family of protein-tyrosine kinases and cancer. *Pharmacological Research*, *79*, 34–74. https://doi.org/10.1016/j.phrs.2013.11.002.

Rudd, C. E., Taylor, A., & Schneider, H. (2009). CD28 and CTLA-4 coreceptor expression and signal transduction. *Immunological Reviews*, *229*(1), 12–26. https://doi.org/10.1111/j.1600-065X.2009.00770.x.

Sable, R., Durek, T., Taneja, V., Craik, D. J., Pallerla, S., Gauthier, T., et al. (2016). Constrained cyclic peptides as immunomodulatory inhibitors of the CD2:CD58 protein-protein interaction. *ACS Chemical Biology*, *11*(8), 2366–2374. https://doi.org/10.1021/acschembio.6b00486.

Sable, R., & Jois, S. (2015). Surfing the protein-protein interaction surface using docking methods: Application to the design of PPI inhibitors. *Molecules*, *20*(6), 11569–11603. https://doi.org/10.3390/molecules200611569.

Saleh, A. Z., Greenman, K. L., Billings, S., Van Vranken, D. L., & Krolewski, J. J. (2005). Binding of madindoline A to the extracellular domain of gp130. *Biochemistry*, *44*(32), 10822–10827. https://doi.org/10.1021/bi050439+.

Sato, T., Kienlen-Campard, P., Ahmed, M., Liu, W., Li, H., Elliott, J. I., et al. (2006). Inhibitors of amyloid toxicity based on beta-sheet packing of Abeta40 and Abeta42. *Biochemistry*, *45*(17), 5503–5516. https://doi.org/10.1021/bi052485f.

Sato, T., Tang, T. C., Reubins, G., Fei, J. Z., Fujimoto, T., Kienlen-Campard, P., et al. (2009). A helix-to-coil transition at the epsilon-cut site in the transmembrane dimer of the amyloid precursor protein is required for proteolysis. *Proceedings of the National Academy of Sciences of the United States of America*, *106*(5), 1421–1426. https://doi.org/10.1073/pnas.0812261106.

Satyanarayanajois, S., Villalba, S., Jianchao, L., & Lin, G. M. (2009). Design, synthesis, and docking studies of peptidomimetics based on HER2-herceptin binding site with potential antiproliferative activity against breast cancer cell lines. *Chemical Biology & Drug Design*, *74*(3), 246–257. https://doi.org/10.1111/j.1747-0285.2009.00855.x.

Scheuermann, S., Hambsch, B., Hesse, L., Stumm, J., Schmidt, C., Beher, D., et al. (2001). Homodimerization of amyloid precursor protein and its implication in the amyloidogenic pathway of Alzheimer's disease. *The Journal of Biological Chemistry*, *276*(36), 33923–33929. https://doi.org/10.1074/jbc.M105410200.

Schwartz, R. H. (2003). T cell anergy. *Annual Review of Immunology*, *21*, 305–334. https://doi.org/10.1146/annurev.immunol.21.120601.141110.

Scott, D. E., Coyne, A. G., Hudson, S. A., & Abell, C. (2012). Fragment-based approaches in drug discovery and chemical biology. *Biochemistry, 51*(25), 4990–5003. https://doi.org/10.1021/bi3005126.

Shankaran, H., Wiley, H. S., & Resat, H. (2006). Modeling the effects of HER/ErbB1-3 coexpression on receptor dimerization and biological response. *Biophysical Journal, 90*(11), 3993–4009. https://doi.org/10.1529/biophysj.105.080580.

Sharpe, A. H. (2009). Mechanisms of costimulation. *Immunological Reviews, 229*(1), 5–11. https://doi.org/10.1111/j.1600-065X.2009.00784.x.

Sharpe, A. H., Wherry, E. J., Ahmed, R., & Freeman, G. J. (2007). The function of programmed cell death 1 and its ligands in regulating autoimmunity and infection. *Nature Immunology, 8*(3), 239–245. https://doi.org/10.1038/ni1443.

Sheppard, J. D., Torkildsen, G. L., Lonsdale, J. D., D'Ambrosio, F. A., Jr., McLaurin, E. B., Eiferman, R. A., et al. (2014). Lifitegrast ophthalmic solution 5.0% for treatment of dry eye disease: Results of the OPUS-1 phase 3 study. *Ophthalmology, 121*(2), 475–483. https://doi.org/10.1016/j.ophtha.2013.09.015.

Shuker, S. B., Hajduk, P. J., Meadows, R. P., & Fesik, S. W. (1996). Discovering high-affinity ligands for proteins: SAR by NMR. *Science, 274*(5292), 1531–1534.

Skwarczynska, M., & Ottmann, C. (2015). Protein-protein interactions as drug targets. *Future Medicinal Chemistry*, 7(16), 2195–2219. https://doi.org/10.4155/fmc.15.138.

Soba, P., Eggert, S., Wagner, K., Zentgraf, H., Siehl, K., Kreger, S., et al. (2005). Homo- and heterodimerization of APP family members promotes intercellular adhesion. *The EMBO Journal, 24*(20), 3624–3634. https://doi.org/10.1038/sj.emboj.7600824.

Sowmya, G., Breen, E. J., & Ranganathan, S. (2015). Linking structural features of protein complexes and biological function. *Protein Science, 24*(9), 1486–1494. https://doi.org/10.1002/pro.2736.

Sperl, B., Seifert, M. H., & Berg, T. (2009). Natural product inhibitors of protein-protein interactions mediated by Src-family SH2 domains. *Bioorganic & Medicinal Chemistry Letters, 19*(12), 3305–3309. https://doi.org/10.1016/j.bmcl.2009.04.083.

Stamos, J., Sliwkowski, M. X., & Eigenbrot, C. (2002). Structure of the epidermal growth factor receptor kinase domain alone and in complex with a 4-anilinoquinazoline inhibitor. *The Journal of Biological Chemistry*, 277(48), 46265–46272. https://doi.org/10.1074/jbc.M207135200.

Stumpf, M. P., Thorne, T., de Silva, E., Stewart, R., An, H. J., Lappe, M., et al. (2008). Estimating the size of the human interactome. *Proceedings of the National Academy of Sciences of the United States of America, 105*(19), 6959–6964. https://doi.org/10.1073/pnas.0708078105.

Swapna, L. S., Srikeerthana, K., & Srinivasan, N. (2012). Extent of structural asymmetry in homodimeric proteins: Prevalence and relevance. *PLoS One*, 7(5), e36688. https://doi.org/10.1371/journal.pone.0036688.

Szidonya, L., Cserzo, M., & Hunyady, L. (2008). Dimerization and oligomerization of G-protein-coupled receptors: Debated structures with established and emerging functions. *The Journal of Endocrinology, 196*(3), 435–453. https://doi.org/10.1677/JOE-07-0573.

Tampellini, D., Magrane, J., Takahashi, R. H., Li, F., Lin, M. T., Almeida, C. G., et al. (2007). Internalized antibodies to the Abeta domain of APP reduce neuronal Abeta and protect against synaptic alterations. *The Journal of Biological Chemistry, 282*(26), 18895–18906. https://doi.org/10.1074/jbc.M700373200.

Tang, P. A., & Heng, D. Y. (2013). Programmed death 1 pathway inhibition in metastatic renal cell cancer and prostate cancer. *Current Oncology Reports, 15*(2), 98–104. https://doi.org/10.1007/s11912-012-0284-2.

Tao, R. H., & Maruyama, I. N. (2008). All EGF(ErbB) receptors have preformed homo- and heterodimeric structures in living cells. *Journal of Cell Science*, *121*(Pt. 19), 3207–3217. https://doi.org/10.1242/jcs.033399.
Tezuka, K., Tsuji, T., Hirano, D., Tamatani, T., Sakamaki, K., Kobayashi, Y., et al. (2000). Identification and characterization of rat AILIM/ICOS, a novel T-cell costimulatory molecule, related to the CD28/CTLA4 family. *Biochemical and Biophysical Research Communications*, *276*(1), 335–345. https://doi.org/10.1006/bbrc.2000.3466.
Tilley, J. W., Chen, L., Fry, D. C., Emerson, S. D., Powers, G. D., Biondi, D., et al. (1997). Identification of a small molecule inhibitor of the IL-2/IL-2Rα receptor interaction which binds to IL-2. *Journal of the American Chemical Society*, *119*(32), 7589–7590.
Trifilieff, P., Rives, M. L., Urizar, E., Piskorowski, R. A., Vishwasrao, H. D., Castrillon, J., et al. (2011). Detection of antigen interactions ex vivo by proximity ligation assay: Endogenous dopamine D2-adenosine A2A receptor complexes in the striatum. *BioTechniques*, *51*(2), 111–118. https://doi.org/10.2144/000113719.
Tyndall, J. D., & Sandilya, R. (2005). GPCR agonists and antagonists in the clinic. *Medicinal Chemistry*, *1*(4), 405–421.
Uvebrant, K., da Graca Thrige, D., Rosen, A., Akesson, M., Berg, H., Walse, B., et al. (2007). Discovery of selective small-molecule CD80 inhibitors. *Journal of Biomolecular Screening*, *12*(4), 464–472. https://doi.org/10.1177/1087057107300464.
van der Merwe, P. A., & Davis, S. J. (2003). Molecular interactions mediating T cell antigen recognition. *Annual Review of Immunology*, *21*, 659–684. https://doi.org/10.1146/annurev.immunol.21.120601.141036.
Vassar, R., Bennett, B. D., Babu-Khan, S., Kahn, S., Mendiaz, E. A., Denis, P., et al. (1999). Beta-secretase cleavage of Alzheimer's amyloid precursor protein by the transmembrane aspartic protease BACE. *Science*, *286*(5440), 735–741.
Vassilev, L. T., Vu, B. T., Graves, B., Carvajal, D., Podlaski, F., Filipovic, Z., et al. (2004). In vivo activation of the p53 pathway by small-molecule antagonists of MDM2. *Science*, *303*(5659), 844–848. https://doi.org/10.1126/science.1092472.
Velazquez-Campoy, A., Leavitt, S. A., & Freire, E. (2004). Characterization of protein-protein interactions by isothermal titration calorimetry. *Methods in Molecular Biology*, *261*, 35–54. https://doi.org/10.1385/1-59259-762-9:035.
Velikovsky, C. A., Deng, L., Chlewicki, L. K., Fernandez, M. M., Kumar, V., & Mariuzza, R. A. (2007). Structure of natural killer receptor 2B4 bound to CD48 reveals basis for heterophilic recognition in signaling lymphocyte activation molecule family. *Immunity*, *27*(4), 572–584. https://doi.org/10.1016/j.immuni.2007.08.019.
Verlinde, C. L., Rudenko, G., & Hol, W. G. (1992). In search of new lead compounds for trypanosomiasis drug design: A protein structure-based linked-fragment approach. *Journal of Computer-Aided Molecular Design*, *6*(2), 131–147.
Vilar, M., Sung, T. C., Chen, Z., Garcia-Carpio, I., Fernandez, E. M., Xu, J., et al. (2014). Heterodimerization of p45-p75 modulates p75 signaling: Structural basis and mechanism of action. *PLoS Biology*, *12*(8), e1001918. https://doi.org/10.1371/journal.pbio.1001918.
Vilasco, M., Communal, L., Mourra, N., Courtin, A., Forgez, P., & Gompel, A. (2011). Glucocorticoid receptor and breast cancer. *Breast Cancer Research and Treatment*, *130*(1), 1–10. https://doi.org/10.1007/s10549-011-1689-6.
Vlieghe, P., Lisowski, V., Martinez, J., & Khrestchatisky, M. (2010). Synthetic therapeutic peptides: Science and market. *Drug Discovery Today*, *15*(1–2), 40–56. https://doi.org/10.1016/j.drudis.2009.10.009.
Walker, L. S. (2013). Treg and CTLA-4: Two intertwining pathways to immune tolerance. *Journal of Autoimmunity*, *45*, 49–57. https://doi.org/10.1016/j.jaut.2013.06.006.
Wang, J. H., Smolyar, A., Tan, K., Liu, J. H., Kim, M., Sun, Z. Y., et al. (1999). Structure of a heterophilic adhesion complex between the human CD2 and CD58 (LFA-3) counterreceptors. *Cell*, *97*(6), 791–803.

Watanabe, N., Gavrieli, M., Sedy, J. R., Yang, J., Fallarino, F., Loftin, S. K., et al. (2003). BTLA is a lymphocyte inhibitory receptor with similarities to CTLA-4 and PD-1. *Nature Immunology*, *4*(7), 670–679. https://doi.org/10.1038/ni944.

Webber, A., Hirose, R., & Vincenti, F. (2011). Novel strategies in immunosuppression: Issues in perspective. *Transplantation*, *91*(10), 1057–1064. https://doi.org/10.1097/TP.0b013e3182145306.

Wells, J. A., & McClendon, C. L. (2007). Reaching for high-hanging fruit in drug discovery at protein-protein interfaces. *Nature*, *450*(7172), 1001–1009. https://doi.org/10.1038/nature06526.

Wilson, W. D. (2002). Tech.Sight. Analyzing biomolecular interactions. *Science*, *295*(5562), 2103–2105. https://doi.org/10.1126/science.295.5562.2103.

Wilson, A. J. (2009). Inhibition of protein-protein interactions using designed molecules. *Chemical Society Reviews*, *38*(12), 3289–3300. https://doi.org/10.1039/b807197g.

Xu, L., Josan, J. S., Vagner, J., Caplan, M. R., Hruby, V. J., Mash, E. A., et al. (2012). Heterobivalent ligands target cell-surface receptor combinations in vivo. *Proceedings of the National Academy of Sciences of the United States of America*, *109*(52), 21295–21300. https://doi.org/10.1073/pnas.1211762109.

Yang, C. M., Ji, S., Li, Y., Fu, L. Y., Jiang, T., & Meng, F. D. (2017). Ror2, a developmentally regulated kinase, is associated with tumor growth, apoptosis, migration, and invasion in renal cell carcinoma. *Oncology Research*, *25*(2), 195–205. https://doi.org/10.3727/096504016X14732772150424.

Yang, R. Y., Yang, K. S., Pike, L. J., & Marshall, G. R. (2010). Targeting the dimerization of epidermal growth factor receptors with small-molecule inhibitors. *Chemical Biology & Drug Design*, *76*(1), 1–9. https://doi.org/10.1111/j.1747-0285.2010.00986.x.

Yin, H., Chu, A., Li, W., Wang, B., Shelton, F., Otero, F., et al. (2009). Lipid G protein-coupled receptor ligand identification using beta-arrestin PathHunter assay. *The Journal of Biological Chemistry*, *284*(18), 12328–12338. https://doi.org/10.1074/jbc.M806516200.

Zang, X., & Allison, J. P. (2007). The B7 family and cancer therapy: Costimulation and coinhibition. *Clinical Cancer Research*, *13*(18 Pt 1), 5271–5279. https://doi.org/10.1158/1078-0432.CCR-07-1030.

Zhang, W., Ben-David, M., & Sidhu, S. S. (2016). Engineering cell signaling modulators from native protein-protein interactions. *Current Opinion in Structural Biology*, *45*, 25–35. https://doi.org/10.1016/j.sbi.2016.11.002.

Zhang, T., Fresnay, S., Welty, E., Sangrampurkar, N., Rybak, E., Zhou, H., et al. (2011). Selective CD28 blockade attenuates acute and chronic rejection of murine cardiac allografts in a CTLA-4-dependent manner. *American Journal of Transplantation*, *11*(8), 1599–1609. https://doi.org/10.1111/j.1600-6143.2011.03624.x.

Zhang, X., Schwartz, J. C., Guo, X., Bhatia, S., Cao, E., Lorenz, M., et al. (2004). Structural and functional analysis of the costimulatory receptor programmed death-1. *Immunity*, *20*(3), 337–347.

Zhanhua, C., Gan, J. G., Lei, L., Sakharkar, M. K., & Kangueane, P. (2005). Protein subunit interfaces: Heterodimers versus homodimers. *Bioinformation*, *1*(2), 28–39.

Zitvogel, L., & Kroemer, G. (2012). Targeting PD-1/PD-L1 interactions for cancer immunotherapy. *Oncoimmunology*, *1*(8), 1223–1225. https://doi.org/10.4161/onci.21335.

CHAPTER TWO

Targeting Intramembrane Protein–Protein Interactions: Novel Therapeutic Strategy of Millions Years Old

Alexander B. Sigalov[1]
SignaBlok, Inc., Shrewsbury, MA, USA
[1]Corresponding author: e-mail address: sigalov@signablok.com

Contents

Abstract

Intramembrane protein–protein interactions (PPIs) are involved in transmembrane signal transduction mediated by cell surface receptors and play an important role in health and disease. Recently, receptor-specific modulatory peptides rationally designed using a general platform of transmembrane signaling, the signaling chain homooligomerization (SCHOOL) model, have been proposed to therapeutically target these interactions in a variety of serious diseases with unmet needs including cancer, sepsis, arthritis, retinopathy, and thrombosis. These peptide drug candidates use ligand-independent mechanisms of action (SCHOOL mechanisms) and demonstrate potent efficacy in vitro and in vivo. Recent studies surprisingly revealed that in order to modify and/or escape the host immune response, human viruses use similar mechanisms and

Advances in Protein Chemistry and Structural Biology, Volume 111
ISSN 1876-1623
https://doi.org/10.1016/bs.apcsb.2017.06.004
© 2018 Elsevier Inc.
All rights reserved.

modulate cell surface receptors by targeting intramembrane PPIs in a ligand-independent manner. Here, I review these intriguing mechanistic similarities and discuss how the viral strategies optimized over a billion years of the coevolution of viruses and their hosts can help to revolutionize drug discovery science and develop new, disruptive therapies. Examples are given.

1. INTRODUCTION

Protein–protein interactions (PPIs) are involved in the vast majority of biological processes (Fletcher & Hamilton, 2006; Fry, 2006; Ryan & Matthews, 2005; Toogood, 2002; Zinzalla & Thurston, 2009). Nearly every important pathway in health and disease includes and is critically influenced by PPIs (Fry, 2006). In this context, the specific and controlled modulation (inhibition) of PPIs provides a promising avenue for rational drug design, as revealed by recent progress in the design of inhibitory antibodies, peptides, and small molecules (Berg, 2008; Che, Brooks, & Marshall, 2006; Fry, 2006; Pagliaro et al., 2004; Robinson, 2009; Sillerud & Larson, 2005; Veselovsky & Archakov, 2007).

Cell surface receptors are integral membrane proteins that translate extracellular information into intracellular signaling sequences and further into physiological cell response in the complex fundamental process called transmembrane signal transduction. This process plays a crucial role in health and disease (Kadmiel & Cidlowski, 2013; Rudd, 2006; Sabroe et al., 2003; Sigalov, 2008a, 2010g), which makes therapeutic control of PPIs involved in this process of both fundamental and clinical importance. However, until recently, the lack of a mechanistic understanding of what PPIs are actually taking place during transmembrane signal transduction has significantly impeded progress not only in fundamental studies in biology and life sciences but also in molecular target-driven drug discovery. The result of this lack of understanding is that the most widely used therapeutic strategies of receptor modulation attempt to prevent binding of receptors to their cognate ligands using clinically relevant antibodies (Hansel, Kropshofer, Singer, Mitchell, & George, 2010) or soluble receptor domains (Bouchon, Facchetti, Weigand, & Colonna, 2001; Molloy, Sewell, & Jakobsen, 2005). However, these large protein entities possess severe disadvantages including a long and costly development process, low activity per mass, manufacturing difficulties, unintended immune response, side effects, low organ and tissue penetration, and a lack of oral bioavailability (Hansel et al., 2010).

As integral membrane proteins, cell surface receptors consist of three domains: (1) extracellular ligand-binding domains that function to detect and bind external stimuli, (2) transmembrane domains that function as anchors to the cell membrane and mediators of signal transduction, and (3) cytoplasmic signaling domains that initiate an intracellular signal transduction cascades. Depending on whether extracellular and intracellular domains are located on the same or separate protein chains, unrelated and functionally diverse cell receptors can be divided into two main structural families: single-chain and multichain receptors (SRs and MRs), respectively (Fig. 1) (Keegan & Paul, 1992; Sigalov, 2004, 2010c). MRs most commonly represent immune receptors and are often referred to as multichain immune recognition receptors (MIRRs) (Keegan & Paul, 1992; Sigalov, 2004). The signature feature of MIRRs is the presence of one or more copies of the immunoreceptor tyrosine-based activation motif (ITAM) regions (Reth, 1989) or the YxxM motif (Wu, Cherwinski, Spies, Phillips, & Lanier, 2000) in their cytoplasmic signaling domains. Upon receptor triggering, tyrosine residues of the ITAM/YxxM regions are phosphorylated in an early and obligatory event in the signaling cascade.

A general platform for receptor-mediated transmembrane signal transduction, the signaling chain homooligomerization (SCHOOL) platform (Sigalov, 2004, 2006, 2008a, 2008b, 2008c, 2010b, 2010c) uncovers for the first time the major mechanisms coupling recognition and signaling functions. Based on the unusual biophysical phenomenon—the ability of the intrinsically disordered cytoplasmic domains of MIRR signaling subunits to form specific homodimers and higher homooligomers (Sigalov, Aivazian, & Stern, 2004; Sigalov, Zhuravleva, & Orekhov, 2007), this platform was originally suggested for MIRRs (Sigalov, 2004) and then—as a general platform for transmembrane signaling mediated by MIRRs and SRs (Sigalov, 2008a, 2010c). Conceptually, the SCHOOL platform suggests that the similar architecture of the receptors dictates similar mechanisms of receptor triggering. Mechanistically, according to the platform, multivalent ligand binding outside the cell induces (or tunes) receptor oligomerization (clustering) which is then translated across the membrane into formation of competent signaling homooligomers in the cytoplasmic milieu—a force that drives transmembrane signaling and is necessary and sufficient to trigger receptor activation (Fig. 1). This provides the similarity of the targets revealed at the level of specific PPIs—biochemical processes that can be influenced and controlled for therapeutic purposes (Sigalov, 2004, 2008a, 2010b, 2010c).

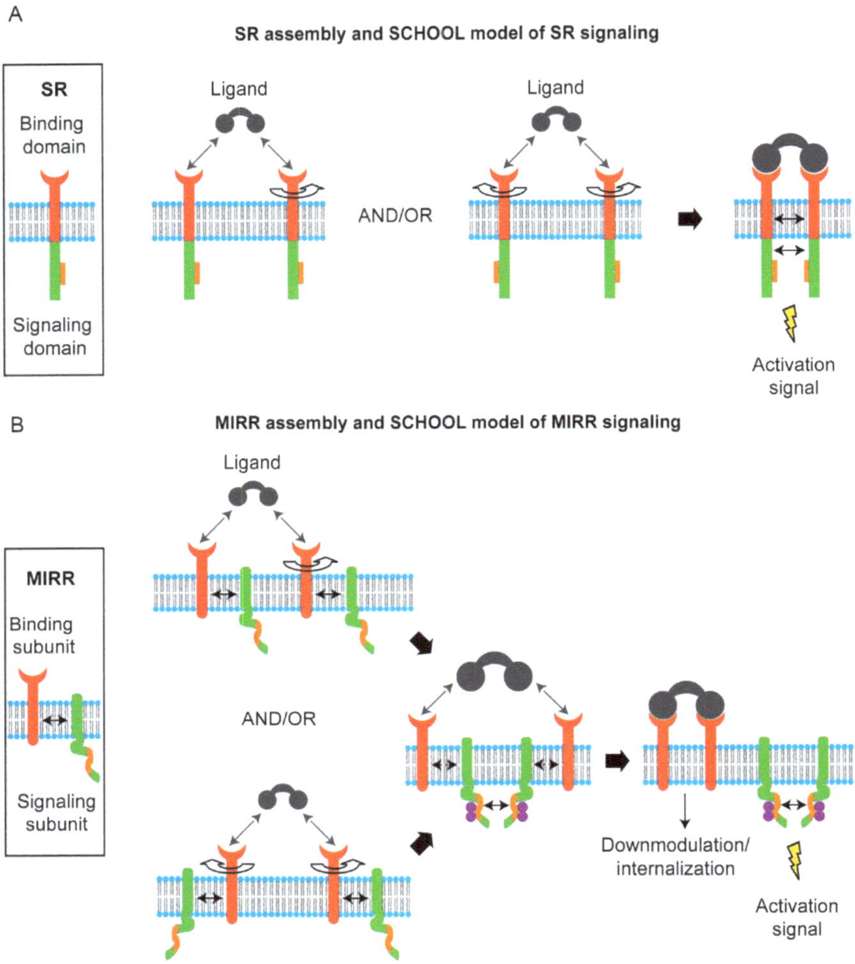

Fig. 1 Single and multichain activating receptor assembly and signaling. (A) In single-chain receptors (SRs), extracellular recognition domain (*red*), and intracellular signaling domain (*green*) with a signaling sequence (*orange rectangles*) are located on the same protein chain. The signaling chain homooligomerization (SCHOOL) model of SR signaling model proposes that receptor homooligomerization in the cytoplasmic milieu plays a central role in triggering SRs. Ligand-induced SR clustering and reorientation (and/or receptor reorientation in preexisting SR clusters) results in SR oligomerization mediated by transmembrane interactions between charged residues. In these oligomers, receptors are in sufficient proximity and adopt a correct interreceptor relative orientation and geometry to promote homointeractions between cytoplasmic domains. Formation of competent signaling oligomers results in generation of the activation signal (for receptor tyrosine kinases, this means transautophosphorylation of Tyr residues in cytoplasmic signaling sequences) and thus triggers downstream signaling pathways. Protein–protein interactions are shown by *solid black arrows*. (B) In multichain immune

2. PPIs IN TRANSMEMBRANE SIGNALING

The SCHOOL platform considers homooligomerization of cytoplasmic signaling domains of cell surface receptors (both SRs and MIRRs) as a driving force of transmembrane signaling and suggests that bringing these domains in the correct orientation and close enough proximity to one other to promote homotypic PPIs between them is an obligatory step to trigger the receptor (Sigalov, 2004, 2010c). This defines the process of ligand-induced receptor triggering and activation as an outcome of the interplay between three major driving forces: extracellular PPIs between receptors and their cognate ligands, inter- (SR) and intrareceptor (MIRR) PPIs in the cell membrane, and interreceptor homotypic PPIs in the cytoplasmic milieu. Interestingly, in transmembrane signaling mediated by receptor tyrosine kinases (RTKs), members of the SR family, a weak dimerization propensity for all RTK transmembrane domains allows for a tight control of the ratio between receptor monomers and dimers (Finger, Escher, & Schneider, 2009; Li & Hristova, 2006). In MIRR-mediated signaling, three major PPIs indicated earlier all are characterized by micromolar affinity and relatively rapid kinetics (Bormann & Engelman, 1992; Davis et al., 1998; Finger et al., 2006; Sigalov et al., 2004). This conjugated and well-balanced system of PPIs involved in receptor triggering explains the molecular mechanisms underlying the ability of MIRRs to transduce the diverse recognition

recognition receptors (MIRRs), the extracellular recognition domains and intracellular ITAM-containing signaling domains are located on separate subunits bound together by noncovalent intramembrane interactions (*solid arrow*). ITAMs are colored *orange*. The signaling chain homooligomerization (SCHOOL) model of MIRR signaling. The model proposes that the homooligomerization of signaling subunits in the cytoplasmic milieu plays a key role in triggering MIRRs. Ligand-induced MIRR clustering and reorientation (and/or receptor reorientation in preexisting MIRR clusters) lead to formation of a dimeric/oligomeric intermediate. In this intermediate, receptors are in sufficient proximity and adopt the correct relative orientation and geometry to promote trans-homointeractions between cytoplasmic domains of signaling subunits resulting in formation of competent signaling oligomers. In these oligomers, protein tyrosine kinases phosphorylate the ITAM tyrosine residues, leading to the generation of activation signal(s), dissociation of signaling oligomers, and internalization of the engaged MIRR ligand-binding subunits. *Circular arrows* indicate ligand-induced receptor reorientation. All interchain interactions in a dimeric intermediate are shown by *dotted black arrows* reflecting their transition state. *Curved lines* depict disorder of the cytoplasmic domains of MIRR signaling subunits. Phosphate groups are shown as *purple circles*. Abbreviations: *ITAM*, immunoreceptor tyrosine-based activation motif.

(discrimination) information across the cell membrane and translate it into different signaling cascades, thus triggering different intracellular pathways and providing different cell responses. In the immune defense, this system of PPIs explains mechanistically how immune cells can have high specificity, selectivity, and sensitivity in the recognition and discrimination of different antigens (ligands) and how this recognition (discrimination) results in different functional outcomes.

In contrast to well-established strategies that target interaction of receptors with their ligands in order to modulate receptor signaling, the SCHOOL strategy is to target other PPIs involved in receptor-mediated transmembrane signal transduction—intramembrane and cytoplasmic homotypic PPIs. This "Freedom to Bind not to Signal" strategy allows for effective and selective therapeutic targeting using small molecule inhibitors and modulatory peptides and peptidomimetics.

2.1 Intramembrane and Cytoplasmic Interactions

2.1.1 Single-Chain Receptors

According to the SCHOOL platform (Sigalov, 2004, 2010c), upon binding to multivalent ligand, intramembrane PPIs between SRs mediate dimerization (oligomerization) of SRs to bring the receptors into close proximity and the correct interreceptor geometry to promote cytoplasmic homotypic PPIs and thus triggers the receptor (Fig. 1A). Interestingly, RTKs and some other SRs such as members of the tumor necrosis factor (TNF) receptor superfamily (Chan, 2007; Chan, Siegel, & Lenardo, 2000) can exist as preassembled dimers (oligomers) on the surface of resting cells. In this scenario, the platform suggests that binding to multivalent ligand leads to reorientation of receptors in these dimers (oligomers) to adopt the geometry competent to promote dimerization (oligomerization) of cytoplasmic signaling domains that trigger the receptor (Fig. 1A). Indeed, in RTK-mediated signaling, multivalent ligand binding results in a conformational change in the receptor extracellular domain, leading to the rotation of the whole receptor (Jiang & Hunter, 1999; Moriki, Maruyama, & Maruyama, 2001). Triggering of the Neu RTK occurs only for a specific transmembrane dimer interface, the rotation of which leads to periodic oscillations in kinase activity (Bell et al., 2000), further confirming the necessity of the correct interreceptor orientation in the receptor dimers (oligomers) formed upon binding to multivalent ligand. In addition, the rotation of the RTK domain with respect to its transmembrane domain by inserting residues into the C-terminal transmembrane flanking region restores the RTK activity. Furthermore, upon binding of the epidermal growth factor receptor (EGFR) to its cognate ligand, the EGFR extracellular domain rotates and this rotation is transmitted to the receptor transmembrane domain (Moriki et al., 2001). Interreceptor orientation in receptor dimers plays a critical role in erythropoietin receptor-mediated transmembrane signaling (Livnah et al., 1998; Syed et al., 1998), suggesting a tight coupling of the extracellular domain orientation to the cytoplasmic signaling events. Furthermore, signal transduction via interleukin (IL)-6 requires not only gp130 homodimerization but also the correct relative orientation of the gp130 cytoplasmic domains in a ligand-specific receptor dimer, suggesting that subtle changes in the orientation of the receptor chains relative to each other might result in very different responses (Greiser, Stross, Heinrich, Behrmann, & Hermanns, 2002). Dimerization of the cytokine receptors by monoclonal antibodies is in most cases not enough to induce signal transduction (Autissier et al., 1998). Formation of homooligomers of Fas cytoplasmic tails plays an important role in receptor triggering (Siegel et al., 2004). Similarly, cytoplasmic domain-mediated dimerization of toll-like receptor-4 (TLR-4) is necessary for TLR-4 triggering and to activate signaling cascades (Lee, Dunzendorfer, & Tobias, 2004).

Thus, these and other studies of SRs strongly support the general SCHOOL platform concept and demonstrate that ligand-induced receptor dimerization is translated into protein dimerization in the cell membrane milieu and that the receptor dimer intramembrane interface contains the critical structural information that positions the receptor cytoplasmic signaling domains in a way competent for oligomerization and receptor triggering (Fig. 1A).

2.1.2 Multichain Receptors

Transmembrane signaling mediated by MIRRs can be roughly divided into four stages: (1) recognition and binding of the receptor to its multivalent cognate ligand outside the cell, (2) transmembrane transduction of this information to the cytoplasmic milieu, (3) phosphorylation of the ITAM or YxxM tyrosine residues by protein tyrosine kinases followed by activation of specific intracellular signaling cascades, and (4) activation of genes in the nucleus. While the molecular mechanisms underlying the stages 1, 3, and 4 are understood in significant detail, the mechanisms by which the MIRRs transduce recognition (discrimination) information via receptor transmembrane and juxtamembrane regions into intracellular biochemical events (stage 2) have been a long-standing mystery. It was also not clear how this putative mechanism could explain the intriguing ability of immune cells to discern and differentially respond to slightly different ligands in the immune defense. Ultimately, the specific PPIs involved in this process have remained largely unknown preventing the development of novel therapeutic approaches.

By uncovering a crucial physiological role of homotypic interactions between intrinsically disordered cytoplasmic domains of MIRR signaling subunits (Sigalov, 2011c; Sigalov et al., 2004, 2007), the biophysically unique phenomenon that still remains a matter of debate whether it exists or not (Nourse & Mittag, 2014; Sigalov, 2016; Uversky & Dunker, 2013), the SCHOOL model suggests that formation of competent MIRR signaling subunit oligomers is necessary and sufficient to trigger the receptor and induce downstream signaling cascades (Fig. 1B) (Sigalov, 2004, 2005, 2006, 2008c, 2010b). This is consistent with the structural hypothesis of cross-phosphorylation (Ortega, 1995; Pribluda, Pribluda, & Metzger, 1994) that assumes that the kinase(s) responsible for catalyzing ITAM Tyr residue phosphorylation exist associated with the receptors, and that for steric reasons, these kinases cannot phosphorylate tyrosine residues on ITAM-containing signaling chains of the same receptor complex. In contrast, in MIRR dimers (oligomers) formed upon binding to their multivalent cognate ligands, the kinases phosphorylate the tyrosines of a distinct receptor complex (cross-phosphorylation or transphosphorylation), thus triggering the receptor (Ortega, 1995).

In unstimulated (resting) cells, intramembrane PPIs between transmembrane helices of recognition and signaling MIRR subunits maintain receptor integrity and determine the relative positions of these subunits in the receptor complex (angles, distances, etc.), thus dictating the overall geometry and topology of MIRRs (Biassoni et al., 2000; Call, Pyrdol, Wiedmann, & Wucherpfennig, 2002; Daeron, 1997; DeFranco et al., 1994; Kim et al., 2003; Klesney-Tait, Turnbull, & Colonna, 2006; Moroi & Jung, 2004; Sigalov, 2004, 2005, 2008c). The SCHOOL platform suggests that MIRR engagement by multivalent ligand or anti-MIRR antibodies (e.g., antibodies against CD3ε and T cell receptor β (TCRβ) chains of TCR or anti-Igβ antibodies for B cell receptor) leads to receptor dimerization (oligomerization) coupled with a multistep structural reorganization aiming to promote homotypic PPIs between MIRR signaling subunits in the cytoplasmic milieu that trigger the receptor and initiate downstream signaling cascades (Fig. 1B). According to the SCHOOL model, sufficiently close interreceptor proximity and correct relative orientation in MIRR dimers (oligomers) as well as long enough duration of the MIRR–ligand interaction and sufficient lifetime of an individual receptor in these MIRR clusters all play important roles for productive MIRR triggering as strongly supported by a growing body of evidence (Carreno, Gonzalez, & Kalergis, 2006; Holler, Lim, Cho, Rund, & Kranz, 2001; Holowka, Sil, Torigoe, & Baird, 2007; Kiessling, Gestwicki, & Strong, 2006; Klemm, Schreiber, & Crabtree, 1998; Metzger, 2004; Miura, Takahashi, Jung, & Moroi, 2002; Ortega, Schweitzer-Stenner, & Pecht, 1988; Posner et al., 2002; Puffer, Pontrello, Hollenbeck, Kink, & Kiessling, 2007; Radaev & Sun, 2003; Reth & Wienands, 1997; Schweitzer-Stenner, Tamir, & Pecht, 1997; Tamir & Cambier, 1998; Yamasaki, Ishikawa, Kohno, & Saito, 2004).

Some MIRRs such as TCR and major platelet collagen receptor, glycoprotein VI (GPVI), can exist as preassembled oligomers on the surface of resting cells (Berlanga et al., 2007; Jung, Tsuji, & Moroi, 2009; Schamel et al., 2005). In this scenario, similarly to that of SRs, the SCHOOL model suggests reorientation of MIRRs in these dimers (oligomers) to adopt the geometry competent to promote homotypic PPIs between cytoplasmic signaling domains that trigger the receptor.

In the immune defense, the SCHOOL mechanisms suggest that the diversity of the immune cell response is largely provided by the combinatorial nature of MIRR-mediated signaling. More specifically, signal diversification may be achieved through different patterns of MIRR signaling subunit oligomerization (Sigalov, 2004, 2005, 2006) in combination with distinct activation signals provided by different MIRR signaling modules (Jensen et al., 1997; Pike, Baig, & Ratcliffe, 2004; Pitcher & van Oers, 2003; Sanchez-Mejorada & Rosales, 1998) and/or different ITAMs located on the same signaling module (e.g., TCRζ chain) (Chae et al., 2004). Thus, according to the model, the diversity of cell functional outcomes in response to stimulation by different ligands is increasing with the number of different signaling subunits that the MIRR complex has.

2.2 Transmembrane Signaling and Viral Evasion Strategies: Evolutionary and Molecular Aspects

Structural consideration of transmembrane signaling mediated by a variety of functionally unrelated receptors expressed on various cells leads us to the important observation that recognition and signaling functions can be combined on one protein chain (SRs) or separated between different protein chains (MIRRs) (Fig. 1). Another observation is that while extracellular (ligand recognition), transmembrane and cytoplasmic (signaling) domains of SRs are all well structured (Sigalov, 2010c, 2011a, 2012b), cytoplasmic domains of MIRR signaling subunits are all intrinsically disordered (i.e., these domains lack a well-defined ordered structure under physiological conditions in vitro) in contrast to the well-structured MIRR extracellular (ligand recognition) and transmembrane domains (Sigalov, 2011a, 2011c, 2012b; Sigalov et al., 2004, 2007; Sigalov, Aivazian, Uversky, & Stern, 2006). Ultimately, these observations raise two intriguing questions (Sigalov, 2011a, 2012a). First, why did nature separate recognition and signaling functions for MIRRs, thereby increasing the risk of malfunction and potential attack by pathogens, and second, why did nature select protein disorder for MIRRs to translate recognition of distinct ligands into appropriate activation signals that would induce distinct specific functional outcomes (Fig. 2)? Answering the first question, one can suggest that the modular assembly of MIRRs brings multiple benefits the most important of which is the capability to diversify and vary signal transduction (Sigalov, 2010c, 2010d, 2011b, 2012a). Ultimately, this provides the mechanistic basis for

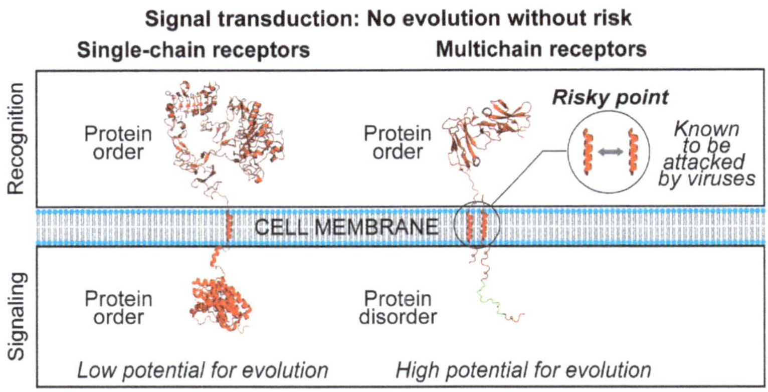

Fig. 2 Evolutionary and molecular aspects of transmembrane signal transduction mediated by single- and multichain cell surface receptors. Images were created using PyMol (www.pymol.org) from Protein Data Bank entries 1NQL and 3GOP for the EGFR extra- and intracellular (juxtamembrane and kinase) domains, respectively (shown as an example of structure of a single-chain receptor), and entry 1UCT for the FcαRI extracellular domain (shown as an example of structure of a multichain receptor recognition subunit). For illustrative purposes, the cytoplasmic domain of a multichain receptor-associated signaling subunit is shown as a monomer and using arbitrary idealized structural elements to represent the ensemble of unfolded conformations of an IDR. The immunoreceptor tyrosine-based activation motif (ITAM) of multichain receptors is depicted in *green*. Intramembrane interactions between recognition and signaling subunits of multichain receptor are shown by a *gray arrow*. Abbreviations: *EGFR*, epidermal growth factor receptor; *FcαRI*, Fc receptor I for IgA.

the diversity and variability of the immune response. According to Charles Darwin, diversity and variability are at the very core of evolutionary processes (Darwin, 1861). One can conclude that nature takes the risks associated with separation of functions in MIRRs in exchange for high potential for evolution of signal transduction (Figs. 2 and 3). In the context of immune signaling, this is of great importance in the development and evolution of the immune defense. Protein disorder contributes to the diversity and variability of signal transduction without compromising efficiency and specificity of signaling (Sigalov, 2010c, 2010d, 2011b, 2012a, 2012b). Thus, evolutionary and developmental benefits of immune signaling-related protein disorder outweigh attendant disadvantages which are largely compensated for by increasing the host defense.

There exist two types of immune receptors: the innate receptors, the germ line-encoded receptors that detect a limited set of conserved antigens; and the adaptive receptors, the somatically generated antigen receptors of the T and B cells (TCR and BCR, respectively) (Table 1). From the evolutionary standpoint, the adaptive immune response evolved long after the innate mechanisms of self-defense and provided significant added value in

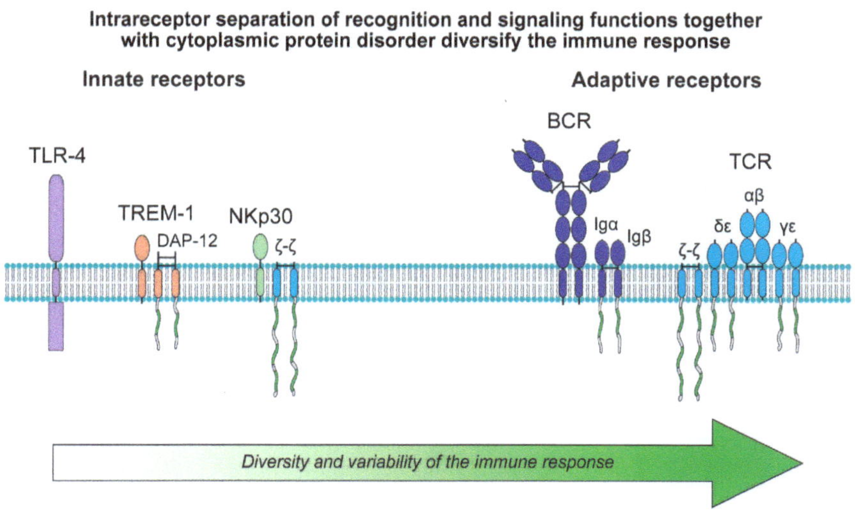

Fig. 3 Diversity and variability of the immune response. The immunoreceptor tyrosine-based activation motif (ITAM) is shown in *green*. *Curved lines* depict intrinsic disorder of the cytoplasmic domains of MIRR signaling subunits. Abbreviations: *BCR*, B cell receptor; *DAP-12*, DNAX adapter protein of 12 kDa; *Ig*, immunoglobulin; *MIRR*, multichain immune recognition receptor; *NK*, natural killer cell; *TCR*, T cell receptor; *TLR-4*, toll-like receptor 4; *TREM-1*, triggering receptor expressed on myeloid cells 1.

Table 1 The Characteristics of the Innate and Adaptive Receptors

Innate Receptors	**Adaptive Receptors**
Germline encoded	Encoded in multiple gene segments
Nonclonal distribution	Clonal distribution
Do not require gene rearrangement	Require gene rearrangement
Trigger immediate response	Trigger delayed response
Broad specificity: recognize pathogen-associated molecular patterns	Narrow specificity: recognize a particular epitope

promoting survival (Cooper & Alder, 2006; Kimbrell & Beutler, 2001; Litman & Cooper, 2007). Structurally, the adaptive receptors belong to the MIRR family with two (BCR) and four (TCR) signaling chains (Fig. 3). In contrast, the innate receptors belong to the SR family (e.g., TLRs) or represent MIRRs that contain only one signaling chain (e.g., triggering receptors expressed on myeloid cells, TREMs, or natural killer cell receptors, NK receptors) (Fig. 3). Importantly, functions of the ITAM modules, including those located on different (Jensen et al., 1997; Pike et al., 2004; Pitcher & van Oers, 2003; Sanchez-Mejorada & Rosales, 1998; Sigalov, 2004, 2006, 2010c, 2011a, 2012b) or the same (e.g., three different ITAMs on the ζ chain) (Chae et al., 2004) signaling chains and those located on the same chain are rather distinct instead of redundant. This provides a molecular explanation for significantly higher diversity and variability of the adaptive response as compared to the innate response. Interestingly, following this logic, one can conclude that the extent of diversity and variability of NKp30-mediated signaling should be an intermediate between the innate (e.g., TLRs) and adaptive receptors (Fig. 3). This is indeed consistent with the idea that NK cells represent an "evolutionary bridge" between innate and adaptive immunity (Sun & Lanier, 2009).

As biochemical processes that can be influenced and controlled (Archakov et al., 2003; Arkin & Wells, 2004; Berg, 2003; Fletcher & Hamilton, 2007; Pagliaro et al., 2004), intramembrane PPIs between MIRR recognition and signaling subunits as well as homotypic interactions between cytoplasmic domains of MIRR signaling subunits represent potential points of attack by pathogens in order to modulate the MIRR-mediated signaling and thus to modulate the immune response (Sigalov, 2009, 2010f, 2012a).

Examples are viruses that in the billion-years-long struggle for their existence, in order to establish a successful infection, replicate, and persist in the host, have evolved numerous strategies to counter and evade host antiviral immune responses as well as to exploit them for productive viral replication. Several different viruses such as human immunodeficiency virus (HIV), cytomegalovirus (CMV), severe acute respiratory syndrome coronavirus (SARS-CoV), and human herpesvirus 6, that are pathogenic for humans uniformly target members of the MIRR family, including innate and adaptive receptors. Intriguingly, these viruses use either a modular assembly of MIRRs to disrupt receptor-mediated signaling or cytoplasmic protein disorder of MIRRs to surprisingly augment cell activation as required for self-preservation (Cohen, Cohen, Antonovsky, Cohen, & Shai, 2010; Shen & Sigalov, 2016; Sigalov, 2007a, 2008a, 2009, 2010f, 2012a). The example of the coevolution of viruses and their hosts confirms that there is no evolution and development without risks. Although we cannot avoid these risks, we can turn this lemon into lemonade and learn from nature how to efficiently target the immune system for therapeutic purposes (Kim & Sigalov, 2008; Shen & Sigalov, 2016; Sigalov, 2010a, 2010f). This will be further illustrated in the next section on the example of intramembrane PPIs as a therapeutic target.

3. INTRAMEMBRANE PPIs AS A THERAPEUTIC TARGET

Although there is a growing line of experimental evidence indicating that targeting intramembrane PPIs to inhibit (modulate) SR-mediated cell activation might represent a promising therapeutic approach (Bennasroune et al., 2004, 2005; Hebert et al., 1996; Sigalov, 2010a, 2010e; Smith et al., 2002; Tarasova, Rice, & Michejda, 1999), in this work, I will focus on therapeutic intramembrane PPIs using short synthetic peptides (SCHOOL peptides) in order to inhibit MIRR-mediated transmembrane signaling for therapeutic purposes.

Importantly, the receptor-specific SCHOOL peptides employ the molecular mechanisms of receptor inhibition that are not dependent upon binding of the receptor to its cognate ligand. This is especially important in those situations when ligands are still unknown. Example is TREM-1 receptor. Despite some recent evidence that peptidoglycan (PGN) recognition protein 1 (PGLYRP1) may potentially act as a ligand for TREM-1 (Read et al., 2015), the actual nature of the TREM-1 ligand(s) is still not yet well understood, impeding the development of clinically relevant inhibitors of TREM-1. This receptor functions as a potent amplifier of inflammation and mediates production of proinflammatory cytokines such as TNFα, IL-1β, and IL-6 (Bouchon et al., 2001; Murakami et al., 2009; Schenk, Bouchon, Seibold, & Mueller, 2007). Importantly, TREM-1 activation preferentially induces expression of functional macrophage colony-stimulating factor (M-CSF, also known as CSF-1) (Dower, Ellis, Saraf, Jelinsky, & Lin, 2008) and in contrast to cytokine blockers, blockade of TREM-1 can blunt excessive inflammation while preserving the capacity for microbial control (Weber et al., 2014). Collectively, these findings implicate TREM-1 as a promising therapeutic target in a variety of inflammation-associated diseases (Bosco, Raggi, & Varesio, 2016; Ford & McVicar, 2009; Tammaro et al., 2017).

3.1 Cancer

3.1.1 Nonsmall Cell Lung Cancer

Lung cancer is the leading cause of cancer deaths worldwide. There are two types of this cancer—nonsmall cell lung cancer (NSCLC) and small cell lung cancer. Despite advances made in chemotherapy (Felip, Santarpia, & Rosell, 2007), NSCLC kills over 1.1 million people annually worldwide, and the 5-year survival rate for patients with NSCLC is only 15% (Jemal et al., 2011). Similar to most solid tumors, the infiltrate of NSCLC, contains tumor-associated macrophages (TAMs) (Solinas, Germano, Mantovani, & Allavena, 2009). By secreting a variety of growth factors, cytokines, chemokines, and enzymes, TAMs regulate tumor growth, angiogenesis, invasion, and metastasis (Shih, Yuan, Chen, & Yang, 2006). High TAM content correlates with the promotion of tumor growth and metastasis (Solinas et al., 2009) and is associated with poor prognosis in NSCLC (Welsh et al., 2005). TAM recruitment, activation, growth, and differentiation are regulated by M-CSF (Elgert, Alleva, & Mullins, 1998). Increased pretreatment serum M-CSF level is a significant independent predictor of poor survival in patients with NSCLC (Kaminska et al., 2006).

In patients with NSCLC, TREM-1 expression on TAMs is associated with cancer recurrence and poor survival: patients with low TREM-1 expression have a 4-year survival rate of over 60%, compared with less than 20% for patients with high TREM-1 expression (Ho et al., 2008), which suggests inhibition of the TREM-1 pathway may be a promising target for the development of a targeted anticancer therapy.

Considering the unknown nature of TREM-1 ligand(s), in our recent study (Sigalov, 2014), we used the SCHOOL platform to design a novel, ligand-independent TREM-1 inhibitory peptide sequence GF9, and demonstrated that the GF9 peptide specifically blocks TREM-1-mediated transmembrane signaling in vitro and in vivo. Utilizing two human NSCLC xenograft nude mouse models (H292 and A549), we demonstrated for the first time that blockade of TREM-1 function using GF9 substantially decreases cytokine production in vitro and significantly delays tumor growth in vivo. We also showed that the SCHOOL peptide inhibitor of TREM-1 GF9 can be formulated into self-assembling macrophage-specific lipopeptide and lipoprotein complexes that mimic human high density lipoproteins (HDLs) for peptide half-life extension and targeted delivery (Sigalov, 2014). This incorporation significantly (up to 10-fold) increases GF9 therapeutic efficacy (Sigalov, 2014).

3.1.2 Pancreatic Cancer

Pancreatic cancer (PC, 85% of which are pancreatic ductal adenocarcinomas) is the fourth leading cause of cancer-related mortality across the world with very poor clinical outcome (Hariharan, Saied, & Kocher, 2008). Current treatments of PC all only marginally prolong survival or relieve symptoms in patients with PC (Schneider, Siveke, Eckel, & Schmid, 2005). There has been no significant progress in the field of targeted therapy for PC (Walker & Ko, 2014) and despite tremendous efforts, the 5-year survival rate remains less than 5%.

As lung cancer and many other solid tumors, PC is characterized by a marked infiltration of macrophages into the stromal compartment (Shih et al., 2006; Solinas et al., 2009). In patients with PC, macrophage infiltration begins during the preinvasive stage of the disease and increases progressively (Clark et al., 2007). The number of TAMs is significantly higher in patients with metastases (Gardian, Janczewska, Olszewski, & Durlik, 2012). Presence of TAMs in the PC stroma correlates with increased angiogenesis (Esposito et al., 2004), a known predictor of poor prognosis (Kuwahara et al., 2003). Similar to NSCLC, high pretreatment serum M-CSF is a strong independent predictor of poor survival in PC patients (Groblewska et al., 2007). Interestingly, M-CSF blockade not only suppresses tumor angiogenesis and lymphangiogenesis (Kubota et al., 2009) but also upregulates T cell checkpoint molecules, including programmed cell death protein 1 (PD-1) and cytotoxic T lymphocyte antigen-4 (CTLA-4), thereby overcoming limitations of PD-1 and CTLA-4 antagonists to achieve regression in even well-established tumors (Zhu et al., 2014). Importantly, continuous M-CSF inhibition affects only pathological angiogenesis but not healthy vascular and lymphatic systems outside tumors (Kubota et al., 2009). In contrast to blockade of vascular endothelial growth factor (VEGF), interruption of M-CSF inhibition does not promote rapid vascular regrowth (Kubota et al., 2009). TREM-1 activation is known to enhance release of most cytokines that are increased in patients with PC (Tjomsland et al., 2011; Yako, Kruger, Smith, & Brand, 2016) and play a vital role in creating and sustaining inflammation in the tumor favorable microenvironment, thus affecting patient survival. These include monocyte chemoattractant protein-1 (MCP-1), TNFα, IL-1α, IL-1β, IL-6, and M-CSF (Lagler et al., 2009; Schenk et al., 2007; Sigalov, 2014). Collectively, these findings suggest that a therapy that targets TREM-1-mediated intratumoral inflammation, angiogenesis, and vascularization can represent a promising strategy for treating PC.

Recently, we tested the therapeutic potential of TREM-1 inhibitory GF9 peptide sequences in three human PC xenograft mouse models (AsPC-1, BxPC-3, and Capan-1) (Zu T. Shen and Alexander B. Sigalov, unpublished). Administration of these SCHOOL peptides resulted in a strong antitumor effect achieving an optimal-treated/control (T/C) value of 19% depending on the xenograft and formulation used and persisting even after treatment was halted. The effect correlated significantly with increased survival, reduced angiogenesis, and suppressed TAM infiltration. The peptides were well tolerated when deployed in either free form or formulated into HDL-mimicking macrophage-targeted lipopeptide complexes. Further, blockade of TREM-1 significantly reduced serum levels of IL-1α, IL-6, and M-CSF, but not VEGF, suggesting an M-CSF-dependent inhibitory effect on angiogenesis.

Taken together with our previous study (Sigalov, 2014), these data suggest that SCHOOL TREM-1-specific peptide inhibitors have a cancer type independent, therapeutically beneficial antitumor activity and can be potentially used as a stand-alone therapy or as a component of combinational therapy for PC, NSCLC, and other solid tumors.

3.2 Autoimmune Diseases

Autoimmune diseases are a fascinating but poorly understood group of diseases that are characterized by a specific adaptive immune response that is mounted against self-antigens (Davidson & Diamond, 2001). As a result, the effector pathways of immunity cause chronic inflammatory injury to tissues, which may often prove lethal. Examples are rheumatoid arthritis (RA), systemic lupus erythematosus, inflammatory bowel disease, multiple sclerosis, atopic dermatitis (AD), and other disorders. T cells and macrophages, two main subtypes of immune cells, are known to be implicated in mediating many aspects of autoimmune inflammation.

RA is a chronic, systemic inflammatory disorder that causes chronic inflammation of the joints and affects about 1% of the world's population, with women affected three times more often than men. The economic burden created by RA is enormous with annual direct and indirect costs per patient of $2500–14,500 and $1500–45,000, respectively, explaining the high demand for new therapies (Kukar, Petryna, & Efthimiou, 2009). Typically, new approaches to RA involve the use of monoclonal antibodies, peptides, antigen mimetics, T cell vaccinations, infusion/injection of antiinflammatory cytokines, or cytokine inhibitors (Kukar et al., 2009; Kurosaka, Ali, Byth, & Manolios, 2007). However, current RA treatments including TNF blockers such as Humira, all have multiple shortcomings including a high level of serious side effects and insufficient efficacy, thus highlighting the need for new and improved treatments. The pathogenesis of RA suggests a key role for aberrant pathways of T cell activation in the initiation and/or perpetuation of disease (Cope, Schulze-Koops, & Aringer, 2007; Sakaguchi, Benham, Cope, & Thomas, 2012). This makes inhibition of pathways involved in T cell activation including TCR-mediated cell activation a promising therapeutic strategy for novel RA therapies (Lorenz, 2003; Won & Lee, 2008).

AD is an inflammatory skin autoimmune disease characterized by impaired epidermal barrier function and cutaneous inflammation. The prevalence of AD has steadily increased during the past few decades. AD affects 1%–3% of adults and up to 20% of children, in 85% of them the disease onset is before the age of 5 years (Novak, 2009). Novel therapeutic approaches are required, as most of the treatments of AD are limited to symptomatic therapies. Activation and skin-selective homing of peripheral blood T cells, and effector functions in the skin, are involved in the pathogenesis of AD (Akdis, Akdis, Trautmann, & Blaser, 2000). Numerous studies have pointed to the role of activated CD4+ T cells in AD (Nomura & Kabashima, 2016). Topical immunomodulators that affect T cells and block TCR-mediated cell activation may therefore have a role in the treatment of AD (Nomura & Kabashima, 2016).

Furthermore, macrophages are central to the pathogenesis of autoimmune diseases including RA and AD (Kasraie & Werfel, 2013; Kinne, Brauer, Stuhlmuller, Palombo-Kinne, & Burmester, 2000). The abundance and activation of macrophages in the inflamed synovial membrane significantly correlates with the severity of RA (Kinne et al., 2000; Mulherin, Fitzgerald, & Bresnihan, 1996). In order to be clinically effective, RA therapies should reduce the number of synovial sublining macrophages (Bresnihan et al., 2007). Major macrophage-related targets in RA and AD include TNFα, IL-1, IL-6, and M-CSF (Li, Hsu, & Mountz, 2012). M-CSF-dependent cells are known to be essential for collagen-induced arthritis (CIA) development (Campbell, Rich, Bischof, & Hamilton, 2000). These and other findings (Colonna & Facchetti, 2003; Genua, Rutella, Correale, & Danese, 2014; Kasraie & Werfel, 2013; Kuai et al., 2009) suggest targeting macrophage activation including TREM-1-mediated cell activation as another promising strategy to treat RA and AD.

3.2.1 Targeting T Cell Receptor

Structurally, TCR is a member of the MIRR family and has α and β antigen-binding subunits (TCRα and TCRβ, respectively) that are bound by electrostatic intramembrane PPIs with three signaling homo- and heterodimers: ζζ, CD3εδ, and CD3εγ (Fig. 3). Short synthetic peptides capable of inhibiting TCR-mediated cell activation in a ligand-independent manner are known since 1997 (Manolios et al., 1997) when TCR-targeted inhibitory activity was first reported for a synthetic peptide corresponding to the sequence of the TCRα transmembrane domain (the so-called core peptide, CP). In the TCR complex, this domain is known to interact with the transmembrane domains of CD3εδ and ζ (Call et al., 2002; Manolios, Bonifacino, & Klausner, 1990). Later, the TCRα CP has been shown to reduce inflammation and inhibits the progression of experimental arthritis (Amon et al., 2006; Kurosaka, Ali, et al., 2007). In patients with AD, psoriasis, or lichen planus, a daily topical application of this peptide over three consecutive days has been demonstrated to inhibit inflammation and ameliorate the disease equally as well as betamethasone (Enk & Knop, 2000; Gollner, Muller, Alt, Knop, & Enk, 2000).

Interestingly, similar ligand-independent TCR inhibitory activity was later reported for HIV fusion peptide (FP) found in the N terminus of the HIV envelope glycoprotein 41 (gp41) (Bloch et al., 2007; Quintana, Gerber, Kent, Cohen, & Shai, 2005). The patterns of inhibition of TCR-mediated cell activation exhibited by TCR CP and HIV gp41 FP were intriguingly similar: both peptides inhibit antigen—but not anti-CD3-stimulated T cell activation (Quintana et al., 2005; Wang, Djordjevic, Kurosaka, Schibeci, Lee, Williamson, et al., 2002). Similar to

TCRα CP, HIV gp41 FP was shown to reduce inflammation and ameliorate T cell-mediated autoimmune arthritis in animal models (Quintana et al., 2005).

However, despite extensive studies (Ali et al., 2006; Amon et al., 2006; Bloch et al., 2007; Collier, Bolte, & Manolios, 2006; Kurosaka, Bolte, Ali, & Manolios, 2007; Manolios et al., 1997; Quintana et al., 2005; Wang, Djordjevic, Bender, & Manolios, 2002), the mode of action of these clinically relevant peptides was enigmatic until the SCHOOL model was first introduced and applied to this field (Sigalov, 2004, 2006, 2007a). According to the model (Sigalov, 2004, 2005, 2006, 2010c, 2010e), TCRα CP and HIV gp41 both compete with the TCRα chain for binding to CD3δε and ζζ and disrupt the corresponding intramembrane PPIs. This results in disconnection/predissociation of the affected signaling subunits from the remaining receptor complex and upon stimulation by multivalent antigen, leads to inhibition of antigen—but not antibody-mediated TCR triggering and cell activation. Importantly, TCR assembly and cell surface expression are not affected by treatment with TCRα CP (Kurosaka, Bolte, et al., 2007). This finding as well as colocalization of TCR CP with TCR in the membrane of resting cells (Wang, Djordjevic, Bender, et al., 2002) directly prove the hypothesis about "pre-" rather than full dissociation state of the unstimulated TCR complex in the presence of TCRα CP, whereas upon stimulation, the affected signaling subunits, ζζ and CD3εδ, become physically disconnected from the remaining receptor complex. It should be noted that the proposed SCHOOL mechanism is the only mechanism consistent with all experimental and clinical data reported up to date for transmembrane peptides of TCR and other MIRRs as well as for lipid and/or sugar conjugates of these peptides (Ali, Amon, Bender, & Manolios, 2005; Ali, De Planque, Huynh, Manolios, & Separovic, 2002; Amon et al., 2006, 2008; Bender, Ali, Amon, Diefenbach, & Manolios, 2004; Collier et al., 2006; Enk & Knop, 2000; Gerber, Quintana, Bloch, Cohen, & Shai, 2005; Gollner et al., 2000; Huynh, Ffrench, Boadle, & Manolios, 2003; Kurosaka, Bolte, et al., 2007; Manolios, Huynh, & Collier, 2002; Sigalov, 2007b, 2008a; Wang, Djordjevic, Bender, et al., 2002; Wang, Djordjevic, Kurosaka, et al., 2002).

Later, the use of the SCHOOL model and comparative primary sequence analysis of proven and predicted immunomodulatory sequences of viral fusion protein regions allowed not only to suggest the specific molecular mechanisms of inhibition of TCR-mediated cell activation by TCRα CP and HIV gp41 FP (Sigalov, 2007a, 2008b, 2009) but also to predict similar immunomodulatory activity for other viral FPs such as SARS-CoV FP (Sigalov, 2009). Our recent study (Shen & Sigalov, 2016) provided compelling experimental in vivo evidence in support of this hypothesis. As demonstrated, a synthetic 11 amino acid-long peptide-derived from SARS CoV FP predicted using the SCHOOL model to disrupt intramembrane PPIs in the TCR complex was able to reduce inflammation in DBA/1J mice with CIA and protect mice against bone and cartilage damage (Shen & Sigalov, 2016). Formulation of the peptide into self-assembling macrophage-targeted lipopeptide nanoparticles that mimic native human HDLs significantly increased peptide dosage efficacy (Shen & Sigalov, 2016). Collectively, these findings further confirm that viral immune evasion strategies evolved during the host–virus coevolution can be transferred to therapeutic strategies that require similar functionalities (e.g., in the treatment of autoimmune diseases).

3.2.2 Targeting TREM-1

TREM-1 expressed on macrophages is a promising target for the development of new rational therapies for autoimmune diseases including RA (Kuai et al., 2009). In addition, targeted delivery of antirheumatic drugs to macrophages is another highly desirable strategy for the treatment of RA because it would not only strike the cells that mediate or amplify most of the permanent tissue destruction but also spare other cells that do not affect joint damage (Garrood & Pitzalis, 2006; Kinne, Stuhlmuller, & Burmester, 2007).

Recently, we demonstrated that a TREM-1-specific inhibitory nonapeptide GF9 designed using the SCHOOL model and capable of inhibiting TREM-1-mediated macrophage activation in a ligand-independent manner, suppresses release of proinflammatory cytokines and M-CSF, decreases inflammation and protects against bone and cartilage destruction in experimental arthritis (Shen & Sigalov, 2017). Administration of GF9 reduced the plasma levels of M-CSF, TNFα, IL-1, and IL-6 (Shen & Sigalov, 2017). Interestingly, 31-mer peptides with sequences from GF9 and helices 4 (GE31) and 6 (GA31) of the major HDL protein, apolipoprotein A–I, were able to perform three functions: assist in the self-assembly of GA/E31-HDL, target these particles to macrophages, and block TREM-1 signaling (Shen & Sigalov, 2017). Similar to our cancer studies (Sigalov, 2014), formulation of GF9 alone or as a part of GE31 and GA31 peptides into HDL significantly increased its therapeutic efficacy. Colocalization of GF9 with TREM-1 in the macrophage membrane observed using confocal microscopy indicates that GF9 self-inserts into the cell membrane and disrupts intramembrane

PPIs between TREM-1 and its signaling partner, DAP-12, further confirming the SCHOOL mechanisms of action of this peptide (Shen & Sigalov, 2017). Collectively, these findings suggest that TREM-1 inhibitory SCHOOL sequences may be promising alternatives for the treatment of RA and possibly, other TREM-1-mediated autoimmune diseases.

3.3 Sepsis

Septic shock is a complex clinical syndrome that results from the systemic response to infection and is characterized by overwhelming production of proinflammatory cytokines that leads not only to tissue damage, but also to hemodynamic changes, multiple organ failure, and ultimately death (Cohen, 2002). Despite the use of potent antibiotics and advanced resuscitative equipment costing $17 billion annually, sepsis still kills about 375,000 Americans each year and mortality rates for patients with septic shock are nearly 50% (Martin, Mannino, Eaton, & Moss, 2003). The only approved sepsis drug, Xigris by Ely Lilly and Company, has been withdrawn from all markets in 2011 and currently, no approved drugs are available. In addition, over 30 drug candidates have failed late-stage clinical trials. Together, this highlights an urgent unmet medical need for new sepsis treatments.

Macrophages are known to secrete high levels of TNFα, IL-6, and IL-1 in septic mice (Ayala & Chaudry, 1996). Further, in patients with sepsis, M-CSF is overproduced (Oren, Duman, Abacioglu, Ozkan, & Irken, 2001). Furthermore, elevated levels of TNFα along with other proinflammatory cytokines are closely linked with poor patient outcome (Gogos, Drosou, Bassaris, & Skoutelis, 2000). Importantly, in patients with sepsis, TREM-1 expression on macrophages is markedly increased (Gibot, 2005; van Bremen et al., 2013). Initial findings established TREM-1 as an amplifier of the systemic inflammatory response syndrome associated with sepsis (Bouchon, Dietrich, & Colonna, 2000; Bouchon et al., 2001). Blockade of TREM-1 in septic mice lowers expression levels of TNF-α and IL-6 and increases survival from 5% to 10% in control animals to 70%–80% in treated animals (Bouchon et al., 2001; Gibot et al., 2007; Wang et al., 2012). Taken together, this suggests TREM-1-mediated macrophage activation as a promising therapeutic target for this syndrome.

However, as discussed earlier, the development of conventional inhibitors of TREM-1 that attempt to block binding of TREM-1 to its ligand (Bouchon et al., 2001; Gibot et al., 2007; Wang et al., 2012) is significantly complicated by the unknown nature of TREM-1 ligand(s) and hence the increased risk of failure of these approaches in clinical development. In this context, TREM-1-specific inhibitory SCHOOL peptide sequences that employ ligand-independent mechanism of receptor inhibition represent a promising alternative to conventional approaches. Ligand-independent blockade of the TREM-1 signaling pathway in mice with experimental lipopolysaccharide (LPS)-induced septic shock using TREM-1 inhibitory SCHOOL peptide GF9 was shown to substantially suppress proinflammatory cytokine production and prolong survival of septic mice (Sigalov, 2014). Targeted delivery of GF9 to macrophages using self-assembling lipopeptide complexes significantly increased peptide half-life and therapeutic efficacy (Sigalov, 2014). These data strongly support the hypothesis that ligand-independent modulation of TREM-1 function using small synthetic peptides might be a suitable treatment for sepsis.

3.4 Thrombosis

Damage to the integrity of the vessel triggers platelet adhesion and aggregation resulting in the formation of a thrombus, which prevents blood loss at sites of injury or leads to occlusion and irreversible tissue damage or infarction in diseased vessels (Leslie, 2010; Nieswandt et al., 2001). Despite intensive research and recent advances in antithrombotic drug discovery and development (Angiolillo, Bhatt, Gurbel, & Jennings, 2009), uncontrolled hemorrhage still remains the most common side effect associated with antithrombotic drugs that are currently in use on the about $11 billion market.

Engagement and clustering of GPVI, the major collagen receptor on platelets, play a crucial role in platelet adhesion, aggregation, and activation induced by collagen (Jung & Moroi, 2008). The selective inhibition of GPVI is believed to inhibit thrombosis without affecting hemostatic plug formation, thus providing new therapeutic strategies to fight platelet-mediated diseases (Li et al., 2007; Lockyer et al., 2006). Thus, in contrast to current drugs, GPVI-specific inhibitors may represent an ideal class of clinically suitable antithrombotics.

Despite intensive studies (Farndale, 2006; Gawaz, 2004; Gibbins, Okuma, Farndale, Barnes, & Watson, 1997; Moroi & Jung, 2004), the molecular mechanisms underlying GPVI triggering and platelet activation were not known until recently when the SCHOOL model was introduced and applied to GPVI triggering and transmembrane signal transduction (Sigalov, 2004, 2005, 2006, 2007b, 2007c, 2008a). This resulted in the development of a novel concept of platelet inhibition and the invention of new platelet inhibitors that employ the ligand-independent SCHOOL mechanisms of GPVI inhibition (Sigalov, 2007a, 2007b, 2008a).

Structurally, GPVI is a member of the MIRR family and signals through its signaling partner, the FcRγ chain (Moroi & Jung, 2004). Uncovering the molecular mechanisms of collagen-stimulated triggering of the GPVI-mediated signal cascade, the SCHOOL model reveals GPVI-FcRγ intramembrane PPIs as a novel therapeutic target for the prevention and treatment of platelet-mediated thrombotic events (Sigalov, 2006, 2007b, 2007c, 2008a, 2010a). Specific blockade or disruption of these PPIs causes a physical and functional disconnection of the GPVI and FcRγ subunits. Experimental data obtained using the GPVI-specific inhibitory SCHOOL peptide GV11 (Sigalov, 2007a, 2007b, 2008a) provided support for this novel concept of platelet inhibition and demonstrated that depending on the donor, incubation of whole blood samples with GV11 prior to addition of collagen (10 and 20 μg/mL) or convulxin (10 ng/mL) leads to a 30%–60% reduction in both the percentage of P-selectin-positive platelets and the expression of the platelet activation markers, P-selectin and PAC-1. This effect is specific: platelet activation via ADP (20 μM) is not affected by the peptide. Collectively, these findings suggest that targeting intramembrane PPIs using the SCHOOL technology opens exciting new avenues in innovative antithrombotic drug discovery and development.

3.5 Retinopathy

Pathologic retinal neovascularization (RNV) causes an angiogenesis-related vision impairment in retinopathy of prematurity (ROP), diabetic retinopathy (DR), and retinal vein occlusion (RVO), which are the most common causes of vision loss and blindness in each age group (Jo & Kim, 2010; Laouri, Chen, Looman, & Gallagher, 2011; Mutlu & Sarici, 2013). In premature infants, normal retinal vascular development is interrupted resulting in retinal ischemia and invasion of the vitreous by abnormal neovessels. In addition, vitreoretinal neovascularization can promote traction retinal detachment, leading to blindness (Al-Shabrawey et al., 2013). In the United States, 14,000–16,000 premature infants are affected by ROP annually and about 4.1 million adults over 40 years have DR (Bashinsky, 2017; Hartnett, 2017). Complications of conventional therapeutic options including laser ablation (corneal edema, anterior chamber reaction, intraocular hemorrhage, cataract formation, and intraocular pressure changes) (Mutlu & Sarici, 2013) and anti-VEGF therapy (damage of healthy vessels, potential side effects on neurons, rapid vascular regrowth upon interrupting the VEGF blockade, and limited effectiveness in some patients) (Maharaj et al., 2008; Mancuso et al., 2006; Pieramici & Rabena, 2008; Verheul & Pinedo, 2007) highlight an unmet need for new targeted therapies that can better address the pathogenesis of neovascular retinal diseases and improve their treatment. In addition, drug delivery to the retina via systemic administration is ideal but it is still a challenge due to the blood–retinal barrier (BRB) (Gaudana, Ananthula, Parenky, & Mitra, 2010).

Retinal microglia and blood-derived macrophages (BDM) that regulate angiogenesis are critically involved in the pathogenesis of RNV diseases (Checchin, Sennlaub, Levavasseur, Leduc, & Chemtob, 2006; Espinosa-Heidmann et al., 2003; Kubota et al., 2009). In ROP, the retina is infiltrated by activated leukocytes and macrophages (Davies, Eubanks, & Powers, 2006). While retinal microglia can become activated in response to retinal pathologies (Penfold, Madigan, Gillies, & Provis, 2001), infiltration and activation of BDM may provide a major contribution in retinal degeneration (Caicedo, Espinosa-Heidmann, Pina, Hernandez, & Cousins, 2005). In mice with oxygen-reduced retinopathy (OIR), macrophages promote the development of pathological RNV (Gao et al., 2016). Macrophages induce angiogenesis by secreting multiple proangiogenic factors, including VEGF (Liu et al., 2013), MCP-1 (Niu, Azfer, Zhelyabovska, Fatma, & Kolattukudy, 2008) and proinflammatory cytokines, such as TNFα and IL-1 (Nagai et al., 2001). Inhibition of MCP-1 suppresses RNV (Kim, Byeon, Hong, Han, & Jeong, 2005; Yoshida, Yoshida, Ishibashi, Elner, & Elner, 2003). M-CSF that acts as an "angiogenic switch" (Lin et al., 2006) is involved in the differentiation of tissue macrophages and microglia during postnatal development (Cecchini et al., 1994). In contrast to VEGF blockade, interruption of M-CSF inhibition does not promote rapid vascular regrowth (Kubota et al., 2009). In the OIR mice, M-CSF is required for pathological RNV but not for the recovery of normal vasculature (Kubota et al., 2009). In the retinas of diabetic rats and in the vitreous of patients with proliferative DR, M-CSF levels are significantly higher compared to control subjects (Liu, Xu, Jiang, & Da, 2009; Yoshida et al., 2015). This suggests targeting M-CSF either directly or via the specific inflammatory signaling pathway as a highly promising strategy for treating ocular neovascular diseases including ROP.

TREM-1 is upregulated under a variety of inflammatory conditions (Pelham & Agrawal, 2014). Upon activation, TREM-1 enhances the production of multiple cytokines and growth factors including MCP-1, TNFα,

IL-1α, IL-1β, IL-6, and M-CSF (Dower et al., 2008; Lagler et al., 2009; Schenk et al., 2007; Shen & Sigalov, 2017; Sigalov, 2014). However, despite the recent identification of TREM-1 as a novel hypoxic marker in vivo and in vitro (Bosco et al., 2011), the role of TREM-1 in RNV diseases including ROP remained unclear. Recently, we found that TREM-1 is highly overexpressed in the retina of the OIR mice compared to controls (Modesto A. Rojas, Zu T. Shen, Ruth B. Caldwell, and Alexander B. Sigalov; unpublished data). Blockade of TREM-1 using the TREM-1 inhibitory SCHOOL GF9 peptide sequences substantially suppressed retinal protein levels of TREM-1 and M-CSF, suggesting an M-CSF-dependent inhibitory effect on angiogenesis, and significantly (up to 90%) reduced the area of vitreoretinal neovascularization (Modesto A. Rojas, Zu T. Shen, Ruth B. Caldwell, and Alexander B. Sigalov, unpublished data). Collectively, these promising data suggest that targeting intramembrane PPIs in the TREM-1/DAP-12 receptor complex using TREM-1-specific SCHOOL peptide inhibitors represent a novel strategy to treat RNV diseases including ROP.

4. CONCLUSIONS

There is a growing interest in therapeutic targeting specific PPIs involved in receptor-mediated transmembrane signal transduction and our improved understanding of the molecular mechanisms underlying this process can significantly contribute into the development of novel pharmacological approaches to a variety of diseases.

Example is the SCHOOL platform that by uncovering the molecular mechanisms of transmembrane signaling mediated by unrelated and functionally diversed surface receptors expressed on various cells, revealed the specific PPIs involved in receptor triggering as key points of therapeutic control. Importantly, the SCHOOL platform suggests that within the SR and MIRR families, the similar structural architecture of the receptors dictates similar mechanisms of receptor triggering. This, in turn, suggests similarity of therapeutic targets in seemingly unrelated diseases, which makes possible the development of global pharmacological approaches as well as the transfer of our clinical knowledge, experience, and therapeutic strategies between these diseases. Successful application of the SCHOOL strategy to target in vivo intramembrane PPIs involved in triggering of such unrelated receptors as TCR, TREM-1, and GPVI expressed on T cells, macrophages, and platelets, respectively, provided compelling evidence to support this hypothesis and resulted in the discovery of new drug candidates for a variety of diseases including cancer, sepsis, RA, AD, and retinopathy.

Further, the SCHOOL platform significantly improved our understanding of the immunomodulatory activity of many human viruses. It appears that different viruses use their fusogenic peptides not only to fuse their membranes to their target host cells but also to disarm the immune system using the SCHOOL mechanisms of receptor inhibition and to escape the host immune response. This gives an interesting example of how our advances in mechanistic understanding of the fundamental natural processes and their interactions can elucidate the billions years old strategies nature uses for combinatorial control and optimization at different scales, from evolution to organismal function.

ACKNOWLEDGMENTS

This work was partially supported by the Defense Advanced Research Projects Agency (DARPA)/US Army Research Office (contract no. W911NF-12-C-0003), the National Cancer Institute of the National Institutes of Health (grant numbers R43CA167865 and R43CA195684), and the National Institute of Arthritis and Musculoskeletal and Skin Diseases of the National Institutes of Health (grant number R43AR066376). The additional funding has come from SignaBlok, Inc. The author thanks Dr. Zu Shen for his critical reading of the manuscript and excellent help with figures.

REFERENCES

Akdis, C. A., Akdis, M., Trautmann, A., & Blaser, K. (2000). Immune regulation in atopic dermatitis. *Current Opinion in Immunology*, *12*(6), 641–646.

Ali, M., Amon, M., Bender, V., & Manolios, N. (2005). Hydrophobic transmembrane-peptide lipid conjugations enhance membrane binding and functional activity in T-cells. *Bioconjugate Chemistry*, *16*(6), 1556–1563.

Ali, M., De Planque, M. R. R., Huynh, N. T., Manolios, N., & Separovic, F. (2002). Biophysical studies of a transmembrane peptide derived from the T cell antigen receptor. *Letters in Peptide Science*, *8*(3–5), 227–233.

Ali, M., Salam, N. K., Amon, M., Bender, V., Hibbs, D. E., & Manolios, N. (2006). T-cell antigen receptor-alpha chain transmembrane peptides: Correlation between structure and function. *International Journal of Peptide Research and Therapeutics*, *12*(3), 261–267.

Al-Shabrawey, M., Elsherbiny, M., Nussbaum, J., Othman, A., Megyerdi, S., & Tawfik, A. (2013). Targeting neovascularization in ischemic retinopathy: Recent advances. *Expert Review of Ophthalmology*, *8*(3), 267–286. https://doi.org/10.1586/eop.13.17.

Amon, M. A., Ali, M., Bender, V., Chan, Y. N., Toth, I., & Manolios, N. (2006). Lipidation and glycosylation of a T cell antigen receptor (TCR) transmembrane hydrophobic peptide dramatically enhances in vitro and in vivo function. *Biochimica et Biophysica Acta*, *1763*(8), 879–888.

Amon, M. A., Ali, M., Bender, V., Hall, K., Aguilar, M. I., Aldrich-Wright, J., et al. (2008). Kinetic and conformational properties of a novel T-cell antigen receptor transmembrane peptide in model membranes. *Journal of Peptide Science*, *14*(6), 714–724.

Angiolillo, D. J., Bhatt, D. L., Gurbel, P. A., & Jennings, L. K. (2009). Advances in antiplatelet therapy: Agents in clinical development. *The American Journal of Cardiology, 103*(3 Suppl), 40A–51A.
Archakov, A. I., Govorun, V. M., Dubanov, A. V., Ivanov, Y. D., Veselovsky, A. V., Lewi, P., et al. (2003). Protein–protein interactions as a target for drugs in proteomics. *Proteomics, 3*(4), 380–391.
Arkin, M. R., & Wells, J. A. (2004). Small-molecule inhibitors of protein–protein interactions: Progressing towards the dream. *Nature Reviews. Drug Discovery, 3*(4), 301–317.
Autissier, P., De Vos, J., Liautard, J., Tupitsyn, N., Jacquet, C., Chavdia, N., et al. (1998). Dimerization and activation of the common transducing chain (gp130) of the cytokines of the IL-6 family by mAb. *International Immunology, 10*(12), 1881–1889.
Ayala, A., & Chaudry, I. H. (1996). Immune dysfunction in murine polymicrobial sepsis: Mediators, macrophages, lymphocytes and apoptosis. *Shock, 6*(Suppl. 1), S27–S38.
Bashinsky, A. L. (2017). Retinopathy of prematurity. *North Carolina Medical Journal, 78*(2), 124–128. https://doi.org/10.18043/ncm.78.2.124.
Bell, C. A., Tynan, J. A., Hart, K. C., Meyer, A. N., Robertson, S. C., & Donoghue, D. J. (2000). Rotational coupling of the transmembrane and kinase domains of the Neu receptor tyrosine kinase. *Molecular Biology of the Cell, 11*(10), 3589–3599.
Bender, V., Ali, M., Amon, M., Diefenbach, E., & Manolios, N. (2004). T cell antigen receptor peptide–lipid membrane interactions using surface plasmon resonance. *The Journal of Biological Chemistry, 279*(52), 54002–54007.
Bennasroune, A., Fickova, M., Gardin, A., Dirrig-Grosch, S., Aunis, D., Cremel, G., et al. (2004). Transmembrane peptides as inhibitors of ErbB receptor signaling. *Molecular Biology of the Cell, 15*(7), 3464–3474.
Bennasroune, A., Gardin, A., Auzan, C., Clauser, E., Dirrig-Grosch, S., Meira, M., et al. (2005). Inhibition by transmembrane peptides of chimeric insulin receptors. *Cellular and Molecular Life Sciences, 62*(18), 2124–2131.
Berg, T. (2003). Modulation of protein–protein interactions with small organic molecules. *Angewandte Chemie (International Ed. in English), 42*(22), 2462–2481.
Berg, T. (2008). Small-molecule inhibitors of protein–protein interactions. *Current Opinion in Drug Discovery & Development, 11*(5), 666–674.
Berlanga, O., Bori-Sanz, T., James, J. R., Frampton, J., Davis, S. J., Tomlinson, M. G., et al. (2007). Glycoprotein VI oligomerization in cell lines and platelets. *Journal of Thrombosis and Haemostasis, 5*(5), 1026–1033.
Biassoni, R., Cantoni, C., Falco, M., Pende, D., Millo, R., Moretta, L., et al. (2000). Human natural killer cell activating receptors. *Molecular Immunology, 37*(17), 1015–1024.
Bloch, I., Quintana, F. J., Gerber, D., Cohen, T., Cohen, I. R., & Shai, Y. (2007). T-cell inactivation and immunosuppressive activity induced by HIV gp41 via novel interacting motif. *The FASEB Journal, 21*(2), 393–401.
Bormann, B. J., & Engelman, D. M. (1992). Intramembrane helix–helix association in oligomerization and transmembrane signaling. *Annual Review of Biophysics and Biomolecular Structure, 21*, 223–242.
Bosco, M. C., Pierobon, D., Blengio, F., Raggi, F., Vanni, C., Gattorno, M., et al. (2011). Hypoxia modulates the gene expression profile of immunoregulatory receptors in human mature dendritic cells: Identification of TREM-1 as a novel hypoxic marker in vitro and in vivo. *Blood, 117*(9), 2625–2639. https://doi.org/10.1182/blood-2010-06-292136.
Bosco, M. C., Raggi, F., & Varesio, L. (2016). Therapeutic potential of targeting TREM-1 in inflammatory diseases and cancer. *Current Pharmaceutical Design, 22*(41), 6209–6233. https://doi.org/10.2174/1381612822666160826110539.
Bouchon, A., Dietrich, J., & Colonna, M. (2000). Cutting edge: Inflammatory responses can be triggered by TREM-1, a novel receptor expressed on neutrophils and monocytes. *Journal of Immunology, 164*(10), 4991–4995.

Bouchon, A., Facchetti, F., Weigand, M. A., & Colonna, M. (2001). TREM-1 amplifies inflammation and is a crucial mediator of septic shock. *Nature, 410*(6832), 1103–1107.

Bresnihan, B., Gerlag, D. M., Rooney, T., Smeets, T. J., Wijbrandts, C. A., Boyle, D., et al. (2007). Synovial macrophages as a biomarker of response to therapeutic intervention in rheumatoid arthritis: Standardization and consistency across centers. *The Journal of Rheumatology, 34*(3), 620–622.

Caicedo, A., Espinosa-Heidmann, D. G., Pina, Y., Hernandez, E. P., & Cousins, S. W. (2005). Blood-derived macrophages infiltrate the retina and activate Muller glial cells under experimental choroidal neovascularization. *Experimental Eye Research, 81*(1), 38–47. https://doi.org/10.1016/j.exer.2005.01.013.

Call, M. E., Pyrdol, J., Wiedmann, M., & Wucherpfennig, K. W. (2002). The organizing principle in the formation of the T cell receptor-CD3 complex. *Cell, 111*(7), 967–979.

Campbell, I. K., Rich, M. J., Bischof, R. J., & Hamilton, J. A. (2000). The colony-stimulating factors and collagen-induced arthritis: Exacerbation of disease by M-CSF and G-CSF and requirement for endogenous M-CSF. *Journal of Leukocyte Biology, 68*(1), 144–150.

Carreno, L. J., Gonzalez, P. A., & Kalergis, A. M. (2006). Modulation of T cell function by TCR/pMHC binding kinetics. *Immunobiology, 211*(1–2), 47–64.

Cecchini, M. G., Dominguez, M. G., Mocci, S., Wetterwald, A., Felix, R., Fleisch, H., et al. (1994). Role of colony stimulating factor-1 in the establishment and regulation of tissue macrophages during postnatal development of the mouse. *Development, 120*(6), 1357–1372.

Chae, W. J., Lee, H. K., Han, J. H., Kim, S. W., Bothwell, A. L., Morio, T., et al. (2004). Qualitatively differential regulation of T cell activation and apoptosis by T cell receptor zeta chain ITAMs and their tyrosine residues. *International Immunology, 16*(9), 1225–1236.

Chan, F. K. (2007). Three is better than one: Pre-ligand receptor assembly in the regulation of TNF receptor signaling. *Cytokine, 37*(2), 101–107.

Chan, K. F., Siegel, M. R., & Lenardo, J. M. (2000). Signaling by the TNF receptor superfamily and T cell homeostasis. *Immunity, 13*(4), 419–422.

Che, Y., Brooks, B. R., & Marshall, G. R. (2006). Development of small molecules designed to modulate protein–protein interactions. *Journal of Computer-Aided Molecular Design, 20*(2), 109–130.

Checchin, D., Sennlaub, F., Levavasseur, E., Leduc, M., & Chemtob, S. (2006). Potential role of microglia in retinal blood vessel formation. *Investigative Ophthalmology and Visual Science, 47*(8), 3595–3602. https://doi.org/10.1167/iovs.05-1522.

Clark, C. E., Hingorani, S. R., Mick, R., Combs, C., Tuveson, D. A., & Vonderheide, R. H. (2007). Dynamics of the immune reaction to pancreatic cancer from inception to invasion. *Cancer Research, 67*(19), 9518–9527. https://doi.org/10.1158/0008-5472.CAN-07-0175.

Cohen, J. (2002). The immunopathogenesis of sepsis. *Nature, 420*(6917), 885–891. https://doi.org/10.1038/nature01326.

Cohen, T., Cohen, S. J., Antonovsky, N., Cohen, I. R., & Shai, Y. (2010). HIV-1 gp41 and TCRalpha trans-membrane domains share a motif exploited by the HIV virus to modulate T-cell proliferation. *PLoS Pathogens, 6*(9), e1001085. https://doi.org/10.1371/journal.ppat.1001085.

Collier, S., Bolte, A., & Manolios, N. (2006). Discrepancy in CD3-transmembrane peptide activity between in vitro and in vivo T-cell inhibition. *Scandinavian Journal of Immunology, 64*(4), 388–391.

Colonna, M., & Facchetti, F. (2003). TREM-1 (triggering receptor expressed on myeloid cells): A new player in acute inflammatory responses. *The Journal of Infectious Diseases, 187*(Suppl. 2), S397–S401.

Cooper, M. D., & Alder, M. N. (2006). The evolution of adaptive immune systems. *Cell*, *124*(4), 815–822. https://doi.org/10.1016/j.cell.2006.02.001.
Cope, A. P., Schulze-Koops, H., & Aringer, M. (2007). The central role of T cells in rheumatoid arthritis. Clinical and Experimental Rheumatology, 25(5 Suppl. 46), S4–11. https://doi.org/2142 (pii).
Daeron, M. (1997). Fc receptor biology. *Annual Review of Immunology*, *15*, 203–234.
Darwin, C. R. (1861). *On the origin of species by means of natural selection or the preservation of favoured races in the struggle for life*. London: John Murray.
Davidson, A., & Diamond, B. (2001). Autoimmune diseases. *The New England Journal of Medicine*, *345*(5), 340–350. https://doi.org/10.1056/NEJM200108023450506.
Davies, M. H., Eubanks, J. P., & Powers, M. R. (2006). Microglia and macrophages are increased in response to ischemia-induced retinopathy in the mouse retina. *Molecular Vision*, *12*, 467–477.
Davis, M. M., Boniface, J. J., Reich, Z., Lyons, D., Hampl, J., Arden, B., et al. (1998). Ligand recognition by alpha beta T cell receptors. *Annual Review of Immunology*, *16*, 523–544.
DeFranco, A. L., Mittelstadt, P. R., Blum, J. H., Stevens, T. L., Law, D. A., Chan, V. W., et al. (1994). Mechanism of B cell antigen receptor function: Transmembrane signaling and triggering of apoptosis. *Advances in Experimental Medicine and Biology*, *365*, 9–22.
Dower, K., Ellis, D. K., Saraf, K., Jelinsky, S. A., & Lin, L. L. (2008). Innate immune responses to TREM-1 activation: Overlap, divergence, and positive and negative cross-talk with bacterial lipopolysaccharide. *Journal of Immunology*, *180*(5), 3520–3534.
Elgert, K. D., Alleva, D. G., & Mullins, D. W. (1998). Tumor-induced immune dysfunction: The macrophage connection. *Journal of Leukocyte Biology*, *64*(3), 275–290.
Enk, A. H., & Knop, J. (2000). T cell receptor mimic peptides and their potential application in T-cell-mediated disease. *International Archives of Allergy and Immunology*, *123*(4), 275–281.
Espinosa-Heidmann, D. G., Suner, I. J., Hernandez, E. P., Monroy, D., Csaky, K. G., & Cousins, S. W. (2003). Macrophage depletion diminishes lesion size and severity in experimental choroidal neovascularization. *Investigative Ophthalmology & Visual Science*, *44*(8), 3586–3592.
Esposito, I., Menicagli, M., Funel, N., Bergmann, F., Boggi, U., Mosca, F., et al. (2004). Inflammatory cells contribute to the generation of an angiogenic phenotype in pancreatic ductal adenocarcinoma. *Journal of Clinical Pathology*, *57*(6), 630–636.
Farndale, R. W. (2006). Collagen-induced platelet activation. *Blood Cells, Molecules & Diseases*, *36*(2), 162–165.
Felip, E., Santarpia, M., & Rosell, R. (2007). Emerging drugs for non-small-cell lung cancer. *Expert Opinion on Emerging Drugs*, *12*(3), 449–460.
Finger, C., Escher, C., & Schneider, D. (2009). The single transmembrane domains of human receptor tyrosine kinases encode self-interactions. *Science Signaling*, *2*(89), ra56.
Finger, C., Volkmer, T., Prodohl, A., Otzen, D. E., Engelman, D. M., & Schneider, D. (2006). The stability of transmembrane helix interactions measured in a biological membrane. *Journal of Molecular Biology*, *358*(5), 1221–1228.
Fletcher, S., & Hamilton, A. D. (2006). Targeting protein–protein interactions by rational design: Mimicry of protein surfaces. *Journal of the Royal Society Interface*, *3*(7), 215–233.
Fletcher, S., & Hamilton, A. D. (2007). Protein–protein interaction inhibitors: Small molecules from screening techniques. *Current Topics in Medicinal Chemistry*, 7(10), 922–927.
Ford, J. W., & McVicar, D. W. (2009). TREM and TREM-like receptors in inflammation and disease. *Current Opinion in Immunology*, *21*(1), 38–46.
Fry, D. C. (2006). Protein–protein interactions as targets for small molecule drug discovery. *Biopolymers*, *84*(6), 535–552.
Gao, X., Wang, Y. S., Li, X. Q., Hou, H. Y., Su, J. B., Yao, L. B., et al. (2016). Macrophages promote vasculogenesis of retinal neovascularization in an oxygen-induced retinopathy

model in mice. *Cell and Tissue Research*, *364*(3), 599–610. https://doi.org/10.1007/s00441-015-2353-y.
Gardian, K., Janczewska, S., Olszewski, W. L., & Durlik, M. (2012). Analysis of pancreatic cancer microenvironment: Role of macrophage infiltrates and growth factors expression. *Journal of Cancer*, *3*, 285–291. https://doi.org/10.7150/jca.4537.
Garrood, T., & Pitzalis, C. (2006). Targeting the inflamed synovium: The quest for specificity. *Arthritis & Rhematology*, *54*(4), 1055–1060. https://doi.org/10.1002/art.21720.
Gaudana, R., Ananthula, H. K., Parenky, A., & Mitra, A. K. (2010). Ocular drug delivery. *The AAPS Journal*, *12*(3), 348–360. https://doi.org/10.1208/s12248-010-9183-3.
Gawaz, M. (2004). Role of platelets in coronary thrombosis and reperfusion of ischemic myocardium. *Cardiovascular Research*, *61*(3), 498–511.
Genua, M., Rutella, S., Correale, C., & Danese, S. (2014). The triggering receptor expressed on myeloid cells (TREM) in inflammatory bowel disease pathogenesis. *Journal of Translational Medicine*, *12*, 293. https://doi.org/10.1186/s12967-014-0293-z.
Gerber, D., Quintana, F. J., Bloch, I., Cohen, I. R., & Shai, Y. (2005). D-enantiomer peptide of the TCRalpha transmembrane domain inhibits T-cell activation in vitro and in vivo. *The FASEB Journal*, *19*(9), 1190–1192.
Gibbins, J. M., Okuma, M., Farndale, R., Barnes, M., & Watson, S. P. (1997). Glycoprotein VI is the collagen receptor in platelets which underlies tyrosine phosphorylation of the Fc receptor gamma-chain. *FEBS Letters*, *413*(2), 255–259.
Gibot, S. (2005). Clinical review: Role of triggering receptor expressed on myeloid cells-1 during sepsis. *Critical Care*, *9*(5), 485–489.
Gibot, S., Massin, F., Marcou, M., Taylor, V., Stidwill, R., Wilson, P., et al. (2007). TREM-1 promotes survival during septic shock in mice. *European Journal of Immunology*, *37*(2), 456–466. https://doi.org/10.1002/eji.200636387.
Gogos, C. A., Drosou, E., Bassaris, H. P., & Skoutelis, A. (2000). Pro- versus anti-inflammatory cytokine profile in patients with severe sepsis: A marker for prognosis and future therapeutic options. *The Journal of Infectious Diseases*, *181*(1), 176–180. https://doi.org/10.1086/315214.
Gollner, G. P., Muller, G., Alt, R., Knop, J., & Enk, A. H. (2000). Therapeutic application of T cell receptor mimic peptides or cDNA in the treatment of T cell-mediated skin diseases. *Gene Therapy*, 7(12), 1000–1004.
Greiser, J. S., Stross, C., Heinrich, P. C., Behrmann, I., & Hermanns, H. M. (2002). Orientational constraints of the gp130 intracellular juxtamembrane domain for signaling. *The Journal of Biological Chemistry*, *277*(30), 26959–26965.
Groblewska, M., Mroczko, B., Wereszczynska-Siemiatkowska, U., Mysliwiec, P., Kedra, B., & Szmitkowski, M. (2007). Serum levels of granulocyte colony-stimulating factor (G-CSF) and macrophage colony-stimulating factor (M-CSF) in pancreatic cancer patients. *Clinical Chemistry and Laboratory Medicine*, *45*(1), 30–34. https://doi.org/10.1515/CCLM.2007.025.
Hansel, T. T., Kropshofer, H., Singer, T., Mitchell, J. A., & George, A. J. (2010). The safety and side effects of monoclonal antibodies. *Nature Reviews. Drug Discovery*, *9*(4), 325–338.
Hariharan, D., Saied, A., & Kocher, H. M. (2008). Analysis of mortality rates for pancreatic cancer across the world. *HPB (Oxford)*, *10*(1), 58–62. https://doi.org/10.1080/13651820701883148.
Hartnett, M. E. (2017). Advances in understanding and management of retinopathy of prematurity. *Survey of Ophthalmology*, *62*(3), 257–276. https://doi.org/10.1016/j.survophthal.2016.12.004.
Hebert, T. E., Moffett, S., Morello, J. P., Loisel, T. P., Bichet, D. G., Barret, C., et al. (1996). A peptide derived from a beta2-adrenergic receptor transmembrane domain inhibits both receptor dimerization and activation. *The Journal of Biological Chemistry*, *271*(27), 16384–16392.

Ho, C. C., Liao, W. Y., Wang, C. Y., Lu, Y. H., Huang, H. Y., Chen, H. Y., et al. (2008). TREM-1 expression in tumor-associated macrophages and clinical outcome in lung cancer. *American Journal of Respiratory and Critical Care Medicine*, *177*(7), 763–770.

Holler, P. D., Lim, A. R., Cho, B. K., Rund, L. A., & Kranz, D. M. (2001). CD8(−) T cell transfectants that express a high affinity T cell receptor exhibit enhanced peptide-dependent activation. *The Journal of Experimental Medicine*, *194*(8), 1043–1052.

Holowka, D., Sil, D., Torigoe, C., & Baird, B. (2007). Insights into immunoglobulin E receptor signaling from structurally defined ligands. *Immunological Reviews*, *217*, 269–279.

Huynh, N. T., Ffrench, R. A., Boadle, R. A., & Manolios, N. (2003). Transmembrane T-cell receptor peptides inhibit B- and natural killer-cell function. *Immunology*, *108*(4), 458–464.

Jemal, A., Bray, F., Center, M. M., Ferlay, J., Ward, E., & Forman, D. (2011). Global cancer statistics. *CA: A Cancer Journal for Clinicians*, *61*(2), 69–90. https://doi.org/10.3322/caac.20107.

Jensen, W. A., Pleiman, C. M., Beaufils, P., Wegener, A. M., Malissen, B., & Cambier, J. C. (1997). Qualitatively distinct signaling through T cell antigen receptor subunits. *European Journal of Immunology*, *27*(3), 707–716.

Jiang, G., & Hunter, T. (1999). Receptor signaling: When dimerization is not enough. *Current Biology*, *9*(15), R568–R571.

Jo, D. H., & Kim, J. H. (2010). How to overcome retinal neuropathy: The fight against angiogenesis-related blindness. *Archives of Pharmacal Research*, *33*(10), 1557–1565. https://doi.org/10.1007/s12272-010-1007-6.

Jung, S. M., & Moroi, M. (2008). Platelet glycoprotein VI. *Advances in Experimental Medicine and Biology*, *640*, 53–63.

Jung, S. M., Tsuji, K., & Moroi, M. (2009). Glycoprotein (GP) VI dimer as a major collagen-binding site of native platelets: Direct evidence obtained with dimeric GPVI-specific Fabs. *Journal of Thrombosis and Haemostasis*, 7(8), 1347–1355.

Kadmiel, M., & Cidlowski, J. A. (2013). Glucocorticoid receptor signaling in health and disease. *Trends in Pharmacological Sciences*, *34*(9), 518–530. https://doi.org/10.1016/j.tips.2013.07.003.

Kaminska, J., Kowalska, M., Kotowicz, B., Fuksiewicz, M., Glogowski, M., Wojcik, E., et al. (2006). Pretreatment serum levels of cytokines and cytokine receptors in patients with non-small cell lung cancer, and correlations with clinicopathological features and prognosis. M-CSF—An independent prognostic factor. *Oncology*, *70*(2), 115–125.

Kasraie, S., & Werfel, T. (2013). Role of macrophages in the pathogenesis of atopic dermatitis. *Mediators of Inflammation*, *2013*, 942375. https://doi.org/10.1155/2013/942375.

Keegan, A. D., & Paul, W. E. (1992). Multichain immune recognition receptors: Similarities in structure and signaling pathways. *Immunology Today*, *13*(2), 63–68.

Kiessling, L. L., Gestwicki, J. E., & Strong, L. E. (2006). Synthetic multivalent ligands as probes of signal transduction. *Angewandte Chemie (International Ed. in English)*, *45*(15), 2348–2368.

Kim, M. Y., Byeon, C. W., Hong, K. H., Han, K. H., & Jeong, S. (2005). Inhibition of the angiogenesis by the MCP-1 (monocyte chemoattractant protein-1) binding peptide. *FEBS Letters*, *579*(7), 1597–1601. https://doi.org/10.1016/j.febslet.2005.01.070.

Kim, M. K., Huang, Z. Y., Hwang, P. H., Jones, B. A., Sato, N., Hunter, S., et al. (2003). Fcgamma receptor transmembrane domains: Role in cell surface expression, gamma chain interaction, and phagocytosis. *Blood*, *101*(11), 4479–4484.

Kim, W. M., & Sigalov, A. B. (2008). Viral pathogenesis, modulation of immune receptor signaling and treatment. *Advances in Experimental Medicine and Biology*, *640*, 325–349. https://doi.org/10.1007/978-0-387-09789-3_22.

Kimbrell, D. A., & Beutler, B. (2001). The evolution and genetics of innate immunity. *Nature Reviews. Genetics*, *2*(4), 256–267. https://doi.org/10.1038/35066006.

Kinne, R. W., Brauer, R., Stuhlmuller, B., Palombo-Kinne, E., & Burmester, G. R. (2000). Macrophages in rheumatoid arthritis. *Arthritis Research*, *2*(3), 189–202.

Kinne, R. W., Stuhlmuller, B., & Burmester, G. R. (2007). Cells of the synovium in rheumatoid arthritis. Macrophages. *Arthritis Research & Therapy*, *9*(6), 224. https://doi.org/10.1186/ar2333.

Klemm, J. D., Schreiber, S. L., & Crabtree, G. R. (1998). Dimerization as a regulatory mechanism in signal transduction. *Annual Review of Immunology*, *16*, 569–592.

Klesney-Tait, J., Turnbull, I. R., & Colonna, M. (2006). The TREM receptor family and signal integration. *Nature Immunology*, *7*(12), 1266–1273.

Kuai, J., Gregory, B., Hill, A., Pittman, D. D., Feldman, J. L., Brown, T., et al. (2009). TREM-1 expression is increased in the synovium of rheumatoid arthritis patients and induces the expression of pro-inflammatory cytokines. *Rheumatology (Oxford)*, *48*(11), 1352–1358. https://doi.org/10.1093/rheumatology/kep235.

Kubota, Y., Takubo, K., Shimizu, T., Ohno, H., Kishi, K., Shibuya, M., et al. (2009). M-CSF inhibition selectively targets pathological angiogenesis and lymphangiogenesis. *The Journal of Experimental Medicine*, *206*(5), 1089–1102. https://doi.org/10.1084/jem.20081605.

Kukar, M., Petryna, O., & Efthimiou, P. (2009). Biological targets in the treatment of rheumatoid arthritis: A comprehensive review of current and in-development biological disease modifying anti-rheumatic drugs. *Biologics*, *3*, 443–457.

Kurosaka, N., Ali, M., Byth, K., & Manolios, N. (2007a). The mode of anti-arthritic peptide delivery impacts on the severity and outcome of adjuvant induced arthritis. *APLAR Journal of Rheumatology*, *10*(3), 198–203.

Kurosaka, N., Bolte, A., Ali, M., & Manolios, N. (2007b). T-cell antigen receptor assembly and cell surface expression is not affected by treatment with T-cell antigen receptor-alpha chain transmembrane peptide. *Protein and Peptide Letters*, *14*(3), 299–303.

Kuwahara, K., Sasaki, T., Kuwada, Y., Murakami, M., Yamasaki, S., & Chayama, K. (2003). Expressions of angiogenic factors in pancreatic ductal carcinoma: A correlative study with clinicopathologic parameters and patient survival. *Pancreas*, *26*(4), 344–349.

Lagler, H., Sharif, O., Haslinger, I., Matt, U., Stich, K., Furtner, T., et al. (2009). TREM-1 activation alters the dynamics of pulmonary IRAK-M expression in vivo and improves host defense during pneumococcal pneumonia. *Journal of Immunology*, *183*(3), 2027–2036. https://doi.org/10.4049/jimmunol.0803862.

Laouri, M., Chen, E., Looman, M., & Gallagher, M. (2011). The burden of disease of retinal vein occlusion: Review of the literature. *Eye (Lond)*, *25*(8), 981–988. https://doi.org/10.1038/eye.2011.92.

Lee, H. K., Dunzendorfer, S., & Tobias, P. S. (2004). Cytoplasmic domain-mediated dimerizations of toll-like receptor 4 observed by beta-lactamase enzyme fragment complementation. *The Journal of Biological Chemistry*, *279*(11), 10564–10574.

Leslie, M. (2010). Cell biology. Beyond clotting: The powers of platelets. *Science*, *328*(5978), 562–564.

Li, E., & Hristova, K. (2006). Role of receptor tyrosine kinase transmembrane domains in cell signaling and human pathologies. *Biochemistry*, *45*(20), 6241–6251.

Li, J., Hsu, H. C., & Mountz, J. D. (2012). Managing macrophages in rheumatoid arthritis by reform or removal. *Current Rheumatology Reports*, *14*(5), 445–454. https://doi.org/10.1007/s11926-012-0272-4.

Li, H., Lockyer, S., Concepcion, A., Gong, X., Takizawa, H., Guertin, M., et al. (2007). The fab fragment of a novel anti-GPVI monoclonal antibody, OM4, reduces in vivo thrombosis without bleeding risk in rats. *Arteriosclerosis, Thrombosis, and Vascular Biology*, *27*(5), 1199–1205.

Lin, E. Y., Li, J. F., Gnatovskiy, L., Deng, Y., Zhu, L., Grzesik, D. A., et al. (2006). Macrophages regulate the angiogenic switch in a mouse model of breast cancer. *Cancer Research*, *66*(23), 11238–11246. https://doi.org/10.1158/0008-5472.CAN-06-1278.

Litman, G. W., & Cooper, M. D. (2007). Why study the evolution of immunity? *Nature Immunology*, *8*(6), 547–548. https://doi.org/10.1038/ni0607-547.

Liu, J., Copland, D. A., Horie, S., Wu, W. K., Chen, M., Xu, Y., et al. (2013). Myeloid cells expressing VEGF and arginase-1 following uptake of damaged retinal pigment epithelium suggests potential mechanism that drives the onset of choroidal angiogenesis in mice. *PLoS One*, *8*(8), e72935. https://doi.org/10.1371/journal.pone.0072935.

Liu, W., Xu, G. Z., Jiang, C. H., & Da, C. D. (2009). Expression of macrophage colony-stimulating factor (M-CSF) and its receptor in streptozotocin-induced diabetic rats. *Current Eye Research*, *34*(2), 123–133. https://doi.org/10.1080/02713680802650369.

Livnah, O., Johnson, D. L., Stura, E. A., Farrell, F. X., Barbone, F. P., You, Y., et al. (1998). An antagonist peptide–EPO receptor complex suggests that receptor dimerization is not sufficient for activation. *Nature Structural Biology*, *5*(11), 993–1004.

Lockyer, S., Okuyama, K., Begum, S., Le, S., Sun, B., Watanabe, T., et al. (2006). GPVI-deficient mice lack collagen responses and are protected against experimentally induced pulmonary thromboembolism. *Thrombosis Research*, *118*(3), 371–380.

Lorenz, H. M. (2003). T-cell-activation inhibitors in rheumatoid arthritis. *BioDrugs*, *17*(4), 263–270.

Maharaj, A. S., Walshe, T. E., Saint-Geniez, M., Venkatesha, S., Maldonado, A. E., Himes, N. C., et al. (2008). VEGF and TGF-beta are required for the maintenance of the choroid plexus and ependyma. *The Journal of Experimental Medicine*, *205*(2), 491–501. https://doi.org/10.1084/jem.20072041.

Mancuso, M. R., Davis, R., Norberg, S. M., O'Brien, S., Sennino, B., Nakahara, T., et al. (2006). Rapid vascular regrowth in tumors after reversal of VEGF inhibition. *The Journal of Clinical Investigation*, *116*(10), 2610–2621. https://doi.org/10.1172/JCI24612.

Manolios, N., Bonifacino, J. S., & Klausner, R. D. (1990). Transmembrane helical interactions and the assembly of the T cell receptor complex. *Science*, *249*(4966), 274–277.

Manolios, N., Collier, S., Taylor, J., Pollard, J., Harrison, L. C., & Bender, V. (1997). T-cell antigen receptor transmembrane peptides modulate T-cell function and T cell-mediated disease. *Nature Medicine*, *3*(1), 84–88.

Manolios, N., Huynh, N. T., & Collier, S. (2002). Peptides in the treatment of inflammatory skin disease. *The Australasian Journal of Dermatology*, *43*(3), 226–227.

Martin, G. S., Mannino, D. M., Eaton, S., & Moss, M. (2003). The epidemiology of sepsis in the United States from 1979 through 2000. *The New England Journal of Medicine*, *348*(16), 1546–1554. https://doi.org/10.1056/NEJMoa022139.

Metzger, H. (2004). The high affinity receptor for IgE, FcepsilonRI. *Novartis Foundation Symposium*, *257*(59), 51. discussion 59–64, 98–100, 276–185.

Miura, Y., Takahashi, T., Jung, S. M., & Moroi, M. (2002). Analysis of the interaction of platelet collagen receptor glycoprotein VI (GPVI) with collagen. A dimeric form of GPVI, but not the monomeric form, shows affinity to fibrous collagen. *The Journal of Biological Chemistry*, *277*(48), 46197–46204.

Molloy, P. E., Sewell, A. K., & Jakobsen, B. K. (2005). Soluble T cell receptors: Novel immunotherapies. *Current Opinion in Pharmacology*, *5*(4), 438–443.

Moriki, T., Maruyama, H., & Maruyama, I. N. (2001). Activation of preformed EGF receptor dimers by ligand-induced rotation of the transmembrane domain. *Journal of Molecular Biology*, *311*(5), 1011–1026.

Moroi, M., & Jung, S. M. (2004). Platelet glycoprotein VI: Its structure and function. *Thrombosis Research*, *114*(4), 221–233.

Mulherin, D., Fitzgerald, O., & Bresnihan, B. (1996). Synovial tissue macrophage populations and articular damage in rheumatoid arthritis. *Arthritis and Rheumatism, 39*(1), 115–124.

Murakami, Y., Akahoshi, T., Aoki, N., Toyomoto, M., Miyasaka, N., & Kohsaka, H. (2009). Intervention of an inflammation amplifier, triggering receptor expressed on myeloid cells 1, for treatment of autoimmune arthritis. *Arthritis and Rheumatism, 60*(6), 1615–1623.

Mutlu, F. M., & Sarici, S. U. (2013). Treatment of retinopathy of prematurity: A review of conventional and promising new therapeutic options. *International Journal of Ophthalmology, 6*(2), 228–236. https://doi.org/10.3980/j.issn.2222-3959.2013.02.23.

Nagai, A., Nakagawa, E., Hatori, K., Choi, H. B., McLarnon, J. G., Lee, M. A., et al. (2001). Generation and characterization of immortalized human microglial cell lines: Expression of cytokines and chemokines. *Neurobiology of Disease, 8*(6), 1057–1068. https://doi.org/10.1006/nbdi.2001.0437.

Nieswandt, B., Brakebusch, C., Bergmeier, W., Schulte, V., Bouvard, D., Mokhtari-Nejad,-R., et al. (2001). Glycoprotein VI but not alpha2beta1 integrin is essential for platelet interaction with collagen. *The EMBO Journal, 20*(9), 2120–2130.

Niu, J., Azfer, A., Zhelyabovska, O., Fatma, S., & Kolattukudy, P. E. (2008). Monocyte chemotactic protein (MCP)-1 promotes angiogenesis via a novel transcription factor, MCP-1-induced protein (MCPIP). *The Journal of Biological Chemistry, 283*(21), 14542–14551. https://doi.org/10.1074/jbc.M802139200.

Nomura, T., & Kabashima, K. (2016). Advances in atopic dermatitis in 2015. *The Journal of Allergy and Clinical Immunology, 138*(6), 1548–1555. https://doi.org/10.1016/j.jaci.2016.10.004.

Nourse, A., & Mittag, T. (2014). The cytoplasmic domain of the T-cell receptor zeta subunit does not form disordered dimers. *Journal of Molecular Biology, 426*(1), 62–70. https://doi.org/10.1016/j.jmb.2013.09.036.

Novak, N. (2009). New insights into the mechanism and management of allergic diseases: Atopic dermatitis. *Allergy, 64*(2), 265–275.

Oren, H., Duman, N., Abacioglu, H., Ozkan, H., & Irken, G. (2001). Association between serum macrophage colony-stimulating factor levels and monocyte and thrombocyte counts in healthy, hypoxic, and septic term neonates. *Pediatrics, 108*(2), 329–332.

Ortega, E. (1995). How do multichain immune recognition receptors signal? A structural hypothesis. *Molecular Immunology, 32*(13), 941–945.

Ortega, E., Schweitzer-Stenner, R., & Pecht, I. (1988). Possible orientational constraints determine secretory signals induced by aggregation of IgE receptors on mast cells. *The EMBO Journal*, 7(13), 4101–4109.

Pagliaro, L., Felding, J., Audouze, K., Nielsen, S. J., Terry, R. B., Krog-Jensen, C., et al. (2004). Emerging classes of protein–protein interaction inhibitors and new tools for their development. *Current Opinion in Chemical Biology, 8*(4), 442–449.

Pelham, C. J., & Agrawal, D. K. (2014). Emerging roles for triggering receptor expressed on myeloid cells receptor family signaling in inflammatory diseases. *Expert Review of Clinical Immunology, 10*(2), 243–256. https://doi.org/10.1586/1744666X.2014.866519.

Penfold, P. L., Madigan, M. C., Gillies, M. C., & Provis, J. M. (2001). Immunological and aetiological aspects of macular degeneration. *Progress in Retinal and Eye Research, 20*(3), 385–414.

Pieramici, D. J., & Rabena, M. D. (2008). Anti-VEGF therapy: Comparison of current and future agents. *Eye (Lond), 22*(10), 1330–1336. https://doi.org/10.1038/eye.2008.88.

Pike, K. A., Baig, E., & Ratcliffe, M. J. (2004). The avian B-cell receptor complex: Distinct roles of Igalpha and Igbeta in B-cell development. *Immunological Reviews, 197*, 10–25.

Pitcher, L. A., & van Oers, N. S. (2003). T-cell receptor signal transmission: Who gives an ITAM? *Trends in Immunology, 24*(10), 554–560.

Posner, R. G., Savage, P. B., Peters, A. S., Macias, A., DelGado, J., Zwartz, G., et al. (2002). A quantitative approach for studying IgE-FcepsilonRI aggregation. *Molecular Immunology, 38*(16–18), 1221–1228.
Pribluda, V. S., Pribluda, C., & Metzger, H. (1994). Transphosphorylation as the mechanism by which the high-affinity receptor for IgE is phosphorylated upon aggregation. *Proceedings of the National Academy of Sciences of the United States of America, 91*(23), 11246–11250.
Puffer, E. B., Pontrello, J. K., Hollenbeck, J. J., Kink, J. A., & Kiessling, L. L. (2007). Activating B cell signaling with defined multivalent ligands. *ACS Chemical Biology, 2*(4), 252–262.
Quintana, F. J., Gerber, D., Kent, S. C., Cohen, I. R., & Shai, Y. (2005). HIV-1 fusion peptide targets the TCR and inhibits antigen-specific T cell activation. *The Journal of Clinical Investigation, 115*(8), 2149–2158.
Radaev, S., & Sun, P. D. (2003). Structure and function of natural killer cell surface receptors. *Annual Review of Biophysics and Biomolecular Structure, 32*, 93–114.
Read, C. B., Kuijper, J. L., Hjorth, S. A., Heipel, M. D., Tang, X., Fleetwood, A. J., et al. (2015). Cutting Edge: Identification of neutrophil PGLYRP1 as a ligand for TREM-1. *Journal of Immunology, 194*(4), 1417–1421. https://doi.org/10.4049/jimmunol.1402303.
Reth, M. (1989). Antigen receptor tail clue. *Nature, 338*(6214), 383–384.
Reth, M., & Wienands, J. (1997). Initiation and processing of signals from the B cell antigen receptor. *Annual Review of Immunology, 15*, 453–479.
Robinson, J. A. (2009). Design of protein–protein interaction inhibitors based on protein epitope mimetics. *ChemBioChem, 10*(6), 971–973.
Rudd, C. E. (2006). Disabled receptor signaling and new primary immunodeficiency disorders. *The New England Journal of Medicine, 354*(18), 1874–1877. https://doi.org/10.1056/NEJMp068062.
Ryan, D. P., & Matthews, J. M. (2005). Protein–protein interactions in human disease. *Current Opinion in Structural Biology, 15*(4), 441–446.
Sabroe, I., Read, R. C., Whyte, M. K., Dockrell, D. H., Vogel, S. N., & Dower, S. K. (2003). Toll-like receptors in health and disease: Complex questions remain. *Journal of Immunology, 171*(4), 1630–1635.
Sakaguchi, S., Benham, H., Cope, A. P., & Thomas, R. (2012). T-cell receptor signaling and the pathogenesis of autoimmune arthritis: Insights from mouse and man. *Immunology and Cell Biology, 90*(3), 277–287. https://doi.org/10.1038/icb.2012.4.
Sanchez-Mejorada, G., & Rosales, C. (1998). Signal transduction by immunoglobulin fc receptors. *Journal of Leukocyte Biology, 63*(5), 521–533.
Schamel, W. W., Arechaga, I., Risueno, R. M., van Santen, H. M., Cabezas, P., Risco, C., et al. (2005). Coexistence of multivalent and monovalent TCRs explains high sensitivity and wide range of response. *The Journal of Experimental Medicine, 202*(4), 493–503.
Schenk, M., Bouchon, A., Seibold, F., & Mueller, C. (2007). TREM-1—Expressing intestinal macrophages crucially amplify chronic inflammation in experimental colitis and inflammatory bowel diseases. *The Journal of Clinical Investigation, 117*(10), 3097–3106.
Schneider, G., Siveke, J. T., Eckel, F., & Schmid, R. M. (2005). Pancreatic cancer: Basic and clinical aspects. *Gastroenterology, 128*(6), 1606–1625. https://doi.org/S0016508505006323 [pii].
Schweitzer-Stenner, R., Tamir, I., & Pecht, I. (1997). Analysis of Fc(epsilon)RI-mediated mast cell stimulation by surface-carried antigens. *Biophysical Journal, 72*(6), 2470–2478.
Shen, Z. T., & Sigalov, A. B. (2016). SARS coronavirus fusion peptide-derived sequence suppresses collagen-induced arthritis in DBA/1J mice. *Scientific Reports, 6*, 28672. https://doi.org/10.1038/srep28672.

Shen, Z. T., & Sigalov, A. B. (2017). Rationally designed ligand-independent peptide inhibitors of TREM-1 ameliorate collagen-induced arthritis. *J Cell Mol Med*. https://doi.org/10.1111/jcmm.13173.

Shih, J.-Y., Yuan, A., Chen, J. J.-W., & Yang, P.-C. (2006). Tumor-associated macrophage: Its role in cancer invasion and metastasis. *Journal of Cancer Molecules*, *2*(3), 101–106.

Siegel, R. M., Muppidi, J. R., Sarker, M., Lobito, A., Jen, M., Martin, D., et al. (2004). SPOTS: Signaling protein oligomeric transduction structures are early mediators of death receptor-induced apoptosis at the plasma membrane. *The Journal of Cell Biology*, *167*(4), 735–744.

Sigalov, A. B. (2004). Multichain immune recognition receptor signaling: Different players, same game? *Trends in Immunology*, *25*(11), 583–589.

Sigalov, A. (2005). Multi-chain immune recognition receptors: Spatial organization and signal transduction. *Seminars in Immunology*, *17*(1), 51–64.

Sigalov, A. B. (2006). Immune cell signaling: A novel mechanistic model reveals new therapeutic targets. *Trends in Pharmacological Sciences*, *27*(10), 518–524.

Sigalov, A. B., inventor; University of Massachusetts, Boston, MA (US), assignee. (2007a). Inhibiting Collagen-induced Platelet Aggregation and Activation with Peptide Variants. US Patent US 8,278,271. 2012 Oct 2 (12/001,258).

Sigalov, A. B. (2007b). Interaction between HIV gp41 fusion peptide and T cell receptor: Putting the puzzle pieces back together. *The FASEB Journal*, *21*(8), 1633–1634. author reply 1635 https://doi.org/10.1096/fj.07-0503ltr.

Sigalov, A. B. (2007c). More on: Glycoprotein VI oligomerization: A novel concept of platelet inhibition. *Journal of Thrombosis and Haemostasis*, *5*(11), 2310–2312.

Sigalov, A. B. (Ed.), (2008a). *Multichain immune recognition receptor signaling: From spatiotemporal organization to human disease*. New York: Springer-Verlag.

Sigalov, A. B. (2008b). Novel mechanistic concept of platelet inhibition. *Expert Opinion on Therapeutic Targets*, *12*(6), 677–692.

Sigalov, A. B. (2008c). SCHOOL model and new targeting strategies. *Advances in Experimental Medicine and Biology*, *640*, 268–311. https://doi.org/10.1007/978-0-387-09789-3_20.

Sigalov, A. B. (2009). Novel mechanistic insights into viral modulation of immune receptor signaling. *PLoS Pathogens*, *5*(7), e1000404. https://doi.org/10.1371/journal.ppat.1000404.

Sigalov, A. B. (2010a). New therapeutic strategies targeting transmembrane signal transduction in the immune system. *Cell Adhesion & Migration*, *4*(2), 255–267.

Sigalov, A. B. (2010b). Protein intrinsic disorder and oligomericity in cell signaling. *Molecular BioSystems*, *6*(3), 451–461. https://doi.org/10.1039/B916030M.

Sigalov, A. B. (2010c). The SCHOOL of nature. I. Transmembrane signaling. *Self Nonself*, *1*(1), 4–39. https://doi.org/10.4161/self.1.1.10832.

Sigalov, A. B. (2010d). The SCHOOL of nature: II. Protein order, disorder and oligomericity in transmembrane signaling. *Self Nonself*, *1*(2), 89–102. https://doi.org/10.4161/self.1.2.11590.

Sigalov, A. B. (2010e). The SCHOOL of nature: III. From mechanistic understanding to novel therapies. *Self Nonself*, *1*(3), 192–224. https://doi.org/10.4161/self.1.3.12794.

Sigalov, A. B. (2010f). The SCHOOL of nature: IV. Learning from viruses. *Self Nonself*, *1*(4), 282–298. https://doi.org/10.4161/self.1.4.13279.

Sigalov, A. B. (2010g). Transmembrane signaling: From understanding to action. *Self Nonself*, *1*(4), 341–342.

Sigalov, A. B. (2011a). Cells diversify transmembrane signaling through the controlled chaos of protein disorder. *Self Nonself*, *2*(2), 75–79. https://doi.org/10.4161/self.2.2.15756.

Sigalov, A. B. (2011b). Cells diversify transmembrane signaling through the controlled chaos of protein disorder. *Self Nonself*, *2*(2).

Sigalov, A. B. (2011c). Uncoupled binding and folding of immune signaling-related intrinsically disordered proteins. *Progress in Biophysics and Molecular Biology, 106*(3), 525–536. https://doi.org/10.1016/j.pbiomolbio.2011.08.005.
Sigalov, A. B. (2012a). Evolution of immunity: No development without risk. *Immunologic Research, 52*(3), 176–181. https://doi.org/10.1007/s12026-011-8256-4.
Sigalov, A. B. (2012b). Interplay between protein order, disorder and oligomericity in receptor signaling. *Advances in Experimental Medicine and Biology, 725*, 50–73. https://doi.org/10.1007/978-1-4614-0659-4_4.
Sigalov, A. B. (2014). A novel ligand-independent peptide inhibitor of TREM-1 suppresses tumor growth in human lung cancer xenografts and prolongs survival of mice with lipopolysaccharide-induced septic shock. *International Immunopharmacology, 21*(1), 208–219. https://doi.org/10.1016/j.intimp.2014.05.001.
Sigalov, A. B. (2016). Structural biology of intrinsically disordered proteins: Revisiting unsolved mysteries. *Biochimie, 125*, 112–118. https://doi.org/10.1016/j.biochi.2016.03.006.
Sigalov, A., Aivazian, D., & Stern, L. (2004). Homooligomerization of the cytoplasmic domain of the T cell receptor zeta chain and of other proteins containing the immunoreceptor tyrosine-based activation motif. *Biochemistry, 43*(7), 2049–2061.
Sigalov, A. B., Aivazian, D. A., Uversky, V. N., & Stern, L. J. (2006). Lipid-binding activity of intrinsically unstructured cytoplasmic domains of multichain immune recognition receptor signaling subunits. *Biochemistry, 45*(51), 15731–15739. https://doi.org/10.1021/bi061108f.
Sigalov, A. B., Zhuravleva, A. V., & Orekhov, V. Y. (2007). Binding of intrinsically disordered proteins is not necessarily accompanied by a structural transition to a folded form. *Biochimie, 89*(3), 419–421. https://doi.org/10.1016/j.biochi.2006.11.003.
Sillerud, L. O., & Larson, R. S. (2005). Design and structure of peptide and peptidomimetic antagonists of protein–protein interaction. *Current Protein & Peptide Science, 6*(2), 151–169.
Smith, S. O., Smith, C., Shekar, S., Peersen, O., Ziliox, M., & Aimoto, S. (2002). Transmembrane interactions in the activation of the Neu receptor tyrosine kinase. *Biochemistry, 41*(30), 9321–9332.
Solinas, G., Germano, G., Mantovani, A., & Allavena, P. (2009). Tumor-associated macrophages (TAM) as major players of the cancer-related inflammation. *Journal of Leukocyte Biology, 86*(5), 1065–1073. https://doi.org/10.1189/jlb.0609385.
Sun, J. C., & Lanier, L. L. (2009). Natural killer cells remember: An evolutionary bridge between innate and adaptive immunity? *European Journal of Immunology, 39*(8), 2059–2064. https://doi.org/10.1002/eji.200939435.
Syed, R. S., Reid, S. W., Li, C., Cheetham, J. C., Aoki, K. H., Liu, B., et al. (1998). Efficiency of signalling through cytokine receptors depends critically on receptor orientation. *Nature, 395*(6701), 511–516.
Tamir, I., & Cambier, J. C. (1998). Antigen receptor signaling: Integration of protein tyrosine kinase functions. *Oncogene, 17*(11 Reviews), 1353–1364.
Tammaro, A., Derive, M., Gibot, S., Leemans, J. C., Florquin, S., & Dessing, M. C. (2017). TREM-1 and its potential ligands in non-infectious diseases: From biology to clinical perspectives. *Pharmacology & Therapeutics*. https://doi.org/10.1016/j.pharmthera.2017.02.043.
Tarasova, N. I., Rice, W. G., & Michejda, C. J. (1999). Inhibition of G-protein-coupled receptor function by disruption of transmembrane domain interactions. *The Journal of Biological Chemistry, 274*(49), 34911–34915.
Tjomsland, V., Spangeus, A., Valila, J., Sandstrom, P., Borch, K., Druid, H., et al. (2011). Interleukin 1alpha sustains the expression of inflammatory factors in human pancreatic

cancer microenvironment by targeting cancer-associated fibroblasts. *Neoplasia, 13*(8), 664–675.
Toogood, P. L. (2002). Inhibition of protein–protein association by small molecules: Approaches and progress. *Journal of Medicinal Chemistry, 45*(8), 1543–1558.
Uversky, V. N., & Dunker, A. K. (2013). The case for intrinsically disordered proteins playing contributory roles in molecular recognition without a stable 3D structure. *F1000 Biology Reports, 5*, 1. https://doi.org/10.3410/B5-1 [1 [pii].
van Bremen, T., Dromann, D., Luitjens, K., Dodt, C., Dalhoff, K., Goldmann, T., et al. (2013). Triggering receptor expressed on myeloid cells-1 (Trem-1) on blood neutrophils is associated with cytokine inducibility in human *E. coli* sepsis. *Diagnostic Pathology, 8*, 24. https://doi.org/10.1186/1746-1596-8-24.
Verheul, H. M., & Pinedo, H. M. (2007). Possible molecular mechanisms involved in the toxicity of angiogenesis inhibition. *Nature Reviews. Cancer*, 7(6), 475–485. https://doi.org/10.1038/nrc2152.
Veselovsky, A. V., & Archakov, A. I. (2007). Inhibitors of protein–protein interactions as potential drugs. *Current Computer-Aided Drug Design, 3*(1), 51–58.
Walker, E. J., & Ko, A. H. (2014). Beyond first-line chemotherapy for advanced pancreatic cancer: An expanding array of therapeutic options? *World Journal of Gastroenterology, 20*(9), 2224–2236. https://doi.org/10.3748/wjg.v20.i9.2224.
Wang, X. M., Djordjevic, J. T., Bender, V., & Manolios, N. (2002a). T cell antigen receptor (TCR) transmembrane peptides colocalize with TCR, not lipid rafts, in surface membranes. *Cellular Immunology, 215*(1), 12–19.
Wang, X. M., Djordjevic, J. T., Kurosaka, N., Schibeci, S., Lee, L., Williamson, P., et al. (2002b). T-cell antigen receptor peptides inhibit signal transduction within the membrane bilayer. *Clinical Immunology, 105*(2), 199–207.
Wang, F., Liu, S., Wu, S., Zhu, Q., Ou, G., Liu, C., et al. (2012). Blocking TREM-1 signaling prolongs survival of mice with Pseudomonas aeruginosa induced sepsis. *Cellular Immunology, 272*(2), 251–258. https://doi.org/10.1016/j.cellimm.2011.10.006.
Weber, B., Schuster, S., Zysset, D., Rihs, S., Dickgreber, N., Schurch, C., et al. (2014). TREM-1 deficiency can attenuate disease severity without affecting pathogen clearance. *PLoS Pathogens, 10*(1), e1003900. https://doi.org/10.1371/journal.ppat.1003900.
Welsh, T. J., Green, R. H., Richardson, D., Waller, D. A., O'Byrne, K. J., & Bradding, P. (2005). Macrophage and mast-cell invasion of tumor cell islets confers a marked survival advantage in non-small-cell lung cancer. *Journal of Clinical Oncology, 23*(35), 8959–8967.
Won, J., & Lee, G. H. (2008). T-cell-targeted signaling inhibitors. *International Reviews of Immunology, 27*(1–2), 19–41. https://doi.org/10.1080/08830180701798976.
Wu, J., Cherwinski, H., Spies, T., Phillips, J. H., & Lanier, L. L. (2000). DAP10 and DAP12 form distinct, but functionally cooperative, receptor complexes in natural killer cells. *The Journal of Experimental Medicine, 192*(7), 1059–1068.
Yako, Y. Y., Kruger, D., Smith, M., & Brand, M. (2016). Cytokines as biomarkers of pancreatic ductal adenocarcinoma: A systematic review. *PLoS One, 11*(5), e0154016. https://doi.org/10.1371/journal.pone.0154016.
Yamasaki, S., Ishikawa, E., Kohno, M., & Saito, T. (2004). The quantity and duration of FcRgamma signals determine mast cell degranulation and survival. *Blood, 103*(8), 3093–3101.
Yoshida, S., Kobayashi, Y., Nakama, T., Zhou, Y., Ishikawa, K., Arita, R., et al. (2015). Increased expression of M-CSF and IL-13 in vitreous of patients with proliferative diabetic retinopathy: Implications for M2 macrophage-involving fibrovascular membrane formation. *The British Journal of Ophthalmology, 99*(5), 629–634. https://doi.org/10.1136/bjophthalmol-2014-305860.

Yoshida, S., Yoshida, A., Ishibashi, T., Elner, S. G., & Elner, V. M. (2003). Role of MCP-1 and MIP-1alpha in retinal neovascularization during postischemic inflammation in a mouse model of retinal neovascularization. *Journal of Leukocyte Biology*, *73*(1), 137–144.

Zhu, Y., Knolhoff, B. L., Meyer, M. A., Nywening, T. M., West, B. L., Luo, J., et al. (2014). CSF1/CSF1R blockade reprograms tumor-infiltrating macrophages and improves response to T-cell checkpoint immunotherapy in pancreatic cancer models. *Cancer Research*, *74*(18), 5057–5069. https://doi.org/10.1158/0008-5472.CAN-13-3723.

Zinzalla, G., & Thurston, D. E. (2009). Targeting protein–protein interactions for therapeutic intervention: A challenge for the future. *Future Medicinal Chemistry*, *1*(1), 65–93. https://doi.org/10.4155/fmc.09.12.

FURTHER READING

Sigalov, A. B. (2007d). Transmembrane interactions as immunotherapeutic targets: Lessons from viral pathogenesis. *Advances in Experimental Medicine and Biology*, *601*, 335–344.

Sigalov, A. B. (2008d). Signaling chain homooligomerization (SCHOOL) model. *Advances in Experimental Medicine and Biology*, *640*, 121–163.

CHAPTER THREE

Targeting the Architecture of Deregulated Protein Complexes in Cancer

Eduard Stefan*[,1], Jakob Troppmair†, Klaus Bister*
*Institute of Biochemistry and Center for Molecular Biosciences Innsbruck, University of Innsbruck, Innsbruck, Austria
†Daniel Swarovski Research Laboratory, Medical University of Innsbruck, Innsbruck, Austria
[1]Corresponding author: e-mail address: eduard.stefan@uibk.ac.at

Contents

Abstract

The architectures of central signaling hubs are precisely organized by static and dynamic protein–protein interactions (PPIs). Upon deregulation, these PPI platforms are capable to propagate or initiate pathophysiological signaling events. This causes the acquisition of molecular features contributing to the etiology or progression of many diseases, including cancer, where deregulated molecular interactions of signaling proteins have been best studied. The reasons for PPI-dependent reprogramming of cancer-initiating cells are manifold; in many cases, mutations perturb PPIs, enzyme

Advances in Protein Chemistry and Structural Biology, Volume 111
ISSN 1876-1623
https://doi.org/10.1016/bs.apcsb.2017.07.001
© 2018 Elsevier Inc.
All rights reserved.

activities, protein abundance, or protein localization. Consequently, the pharmaceutical targeting of PPIs promises to be of remarkable therapeutic value. For this review we have selected three key players of oncogenic signaling which are differently affected by PPI deregulation: two (the small G proteins of the RAS family and the transcription factor MYC) are considered "undruggable" using classical drug discovery approaches and in the case of the third protein discussed here, PKA, standard kinase inhibitors, may be unsuitable in the clinic. These circumstances require alternative strategies, which may lie in pharmaceutical drug interference of critical PPIs accountable for oncogenic signaling.

1. INTRODUCTION

Cells are dynamic systems which depend on an assortment of differently organized protein complexes to control the external and internal state. These are involved not only in the extracellular sensing of environmental cues by membrane-localized receptor complexes but also in converting and relaying the incoming signal through intracellular signaling circuits (Brandman & Meyer, 2008; Fisher & White, 2004; Neves, Ram, & Iyengar, 2002). In the cytoplasm, scaffolding proteins organize signaling proteins for the accurate propagation of the signal to cytoplasmic and nuclear effectors, mainly members of various transcription factor families (Good, Zalatan, & Lim, 2011; Langeberg & Scott, 2015; Scott & Pawson, 2009). From these considerations it is evident that protein–protein interactions (PPIs) are involved in practically every aspect of biological function. Flexible PPIs are the framework for signal transmission. It is the spatiotemporal communication within and between protein complexes which is essential for the proper functioning of the cell. Oncogenic signaling largely relies on existing macromolecular protein complexes and PPIs, but due to mutations in individual components differences are seen (i) in the reliance upon upstream events, (ii) in the duration of PPIs, and (iii) eventually in altered preferences regarding interaction partners. Deregulated protein complexes have the potential to promote uncontrolled cell proliferation (Ivanov, Khuri, & Fu, 2013) and PPIs thus have recently become targets for small molecule inhibitors. However, PPIs are still considered as challenging drug targets due to missing cavities for small molecule binding. In some cases, they are even considered as essentially undruggable. Nevertheless, several PPI inhibitors have reached clinical trials (Jubb, Higueruelo, Winter, & Blundell, 2012; Scott, Bayly, Abell, & Skidmore, 2016). Before we discuss specific examples of deregulated PPIs localized to different cell compartments and their targeting strategies, we would like to list basic PPI features. These characteristics need to be considered for drug discovery efforts before listing possible reasons of deregulation. First, PPIs cannot be considered as general new types of drug targets like kinases or G-protein-coupled receptor pathways with defined domains and binding pockets (Fleuren, Zhang, Wu, & Daly, 2016; O'Hayre, Degese, & Gutkind, 2014). PPIs come in many flavors, with different shapes and sizes. So far, most of clinical stage inhibitors target PPIs with small binding pockets (Arkin, Tang, & Wells, 2014). Typically, these inhibitors disrupt the interaction of a globular protein and a single interaction site in the partnering protein (Scott et al., 2016). Second, in this respect structural features are critical for the development of PPI inhibitors. PPIs that contain single secondary structures for interaction have been shown to be amenable for small molecule inhibition (Arkin et al., 2014; Jubb et al., 2012). Third, the mode of PPI formation needs to be examined. In general, two types of PPI complexes can be discriminated. On the one hand, stable complexes which organize macromolecular complexes, and on the other hand transient interactions of signaling proteins. The fourth point is related to the latter feature; one central key characteristic of PPIs are their affinities. Affinities of proteins span from picomolar to micromolar range (Scott et al., 2016; Smith & Gestwicki, 2012). Now, the question arises which are possible reasons for the deconstruction of critical PPIs. In Fig. 1 we illustrate three scenarios. First, the protein abundance is a critical factor. Changes of gene expression patterns, regulator or effector protein stability, protein:protein ratios may alter PPIs and thereby affect the downstream signaling. Second, mutations of regulator or effector proteins uncouple downstream signaling from regulatory signaling circuits and enhance signal propagation. Third yet importantly, the compartmentalization of signaling units or macromolecular protein machineries affects signal transmission. These feasible means of PPI deconstruction endorse drug discovery efforts to target protein complex formation specifically. Mechanistically, PPI modulators can be classified into PPI disruptors and stabilizers. Further, the possibility of direct (competitive) or indirect (allosteric) interference of complex formation needs to be examined (Fischer, Rossmann, & Hyvonen, 2015). Dependent on the type of PPI, alternatives to small molecule inhibitors should be considered. Chemical analyses of small molecule PPI inhibitors revealed that they differ from classical drug properties in bigger size, higher hydrophobicity, less hydrogen bonds, and lower ligand efficiency values (London, Raveh, &

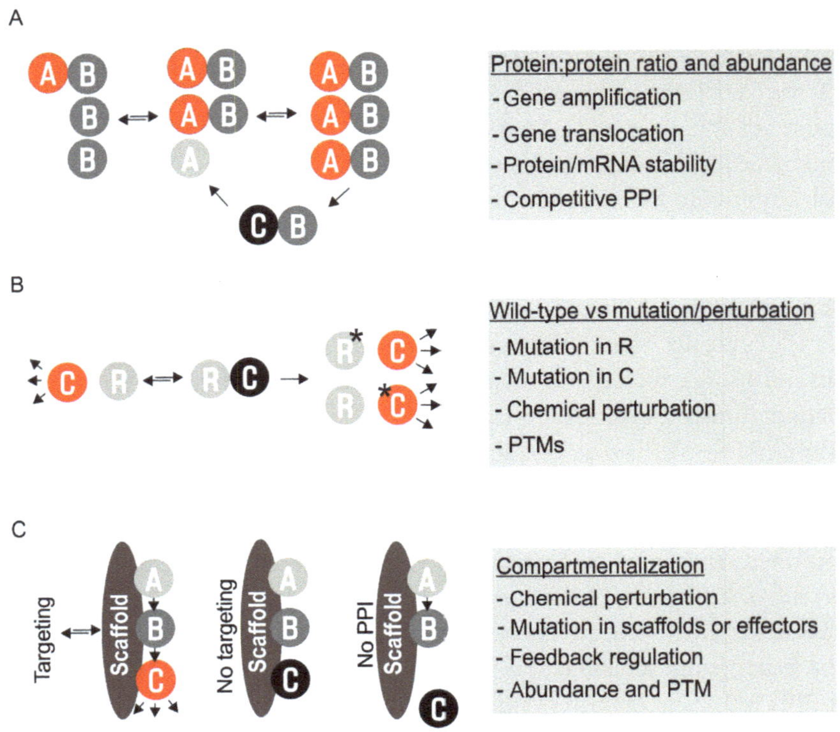

Fig. 1 Deregulated architecture of protein complexes. Shown are three examples how deregulated protein complexes may lead to elevated pathological downstream signaling. (A) Alterations of gene expression, gene translocations, protein abundance, or competitive PPIs (e.g., B with protein C) lead to changes of activities of oncogenic protein units A:B. (B) Mutations in regulatory (R) or catalytic (C) protein subunits or their chemical perturbation lead to R-uncoupled and constitutive C signaling; * indicates mutation. (C) Compartmentalization of signaling complexes is relevant for cascading signal transmission. Deconstruction of specific PPIs affects signaling as well (through, e.g., mutations, competitive PPI, or chemical perturbation); posttranslational modifications (PTMs).

Schueler-Furman, 2013; Morelli, Bourgeas, & Roche, 2011). In this review, we exemplarily present three functionally diverse protein complexes displaying different modes of interaction, which are unfavorably deregulated in distinct types of cancers. Following a short description of the physiological and pathological functions, we survey and compare the pharmaceutical strategies to interfere with deregulated complex formation of the GTPase RAS with the kinase RAF, the prototypical kinase protein kinase A (PKA), and the cancer driver and transcription factor MYC using small molecules or peptides.

2. RAS:RAF SIGNALOSOME AND NETWORK

2.1 Background

Both RAS GTPases and RAF family serine/threonine kinases are integral components of an evolutionary conserved signaling cascade, which together with the RAF downstream kinases MEK1/2 and their effectors ERK1/2 link the activation of receptor tyrosine kinases (RTK) at the cell membrane to the activation of cytoplasmic and nuclear effectors (Fig. 2). A peculiarity of this signaling scheme is the positioning of RAS proteins between RTKs and RAF, whereby the signal, rather than directly passing from the upstream RTK to a three-tiered cytoplasmic signaling cascades (RAF–MEK–ERK), is diverted to a guanine nucleotide exchange factor (GEF) for the small G protein RAS which functions as a binary switch in the signal propagation. RAS activation results in structural changes which allow binding of target proteins to the effector domain, among which RAF kinases and the p110 subunit of PI3K have been best studied (Fig. 2). RAS and RAF have become preferred targets for therapeutic interference because of their frequent involvement in human cancers. While mutant

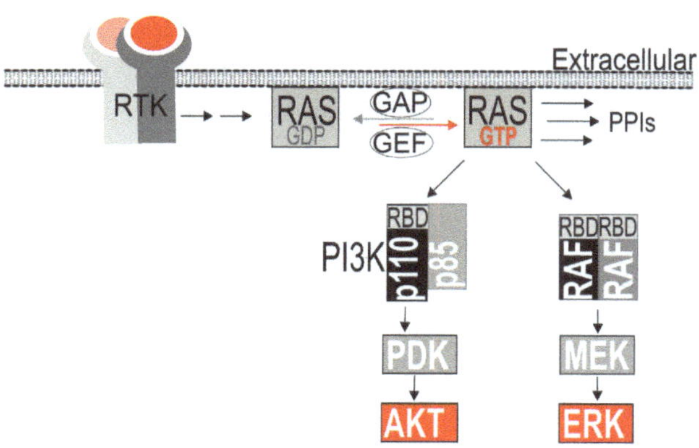

Fig. 2 Schematic diagram of the RAS activation cascade/cycle. Growth factor-mediated activation of receptor tyrosine kinases (RTK) triggers recruitment of adaptor proteins and nucleotide exchange factors that enhance GDP/GTP exchange leading to RAS activation (RAS-GTP; under physiological conditions reversible through GTPase activating proteins, GAP). RAF or PI3K are activated through a multistep process. In the case of RAF this includes RAS binding, recruitment to the membrane, conformational change, phosphorylation, and dimerization. Activated RAF phosphorylates and activates the gatekeeper kinase MEK which in turn phosphorylates ERK. Also, the PI3K pathway (linked to PDK and AKT kinase activities) and other PPI-dependent pathways involved in the control of cellular growth, proliferation, and survival are activated.

RAF kinases have been amenable to therapeutic targeting by small molecule kinase inhibitors (Holderfield, Deuker, McCormick, & McMahon, 2014), the situation is much more complicated in the case of RAS (McCormick, 2016; Stephen, Esposito, Bagni, & McCormick, 2014), and FDA-approved drugs for the treatment of RAS-driven tumors have yet to reach clinics.

2.2 Small G Proteins of the RAS Family and RAF Family Serine–Threonine Kinases

RAS oncogenes were originally discovered in murine retroviruses and by direct transfection of human tumor cell DNA (Parada, Tabin, Shih, & Weinberg, 1982; Shih, Weeks, Young, & Scolnick, 1979). RAS proteins are small, membrane-bound guanine nucleotide-binding proteins that act as molecular switches by cycling between an inactive GDP-bound and an active GTP-bound conformation (Karnoub & Weinberg, 2008). RAS activation is regulated through GEFs, e.g., SOS, while inactivation results from the low intrinsic RAS GTPase activity, which is considerable enhanced through the action of GTPase-activating proteins (GAPs). RAS proteins have a crucial role in the regulation of cell proliferation, differentiation, and survival by signaling through a number of important pathways, including the RAF–MEK–ERK (RAF–MAPK/ERK kinase–extracellular signal-regulated kinase), PI3K–AKT–mTOR, and RALGDS–RAL (RAL guanine nucleotide-dissociation stimulator–RAL) pathways, among others. Three genes (*HRAS*, *NRAS*, and *KRAS*) give rise to four 21 kDa proteins (HRAS, NRAS, KRAS4A, and KRAS4B). While the amino-terminal residues 1–165 share 92%–98% sequence identity, the remaining 23–24 differ substantially in sequence (hypervariable region, HVR). The HVR contains the membrane anchor sequence including the terminal CAAX box (C, cysteine; A, aliphatic amino acid; X, any amino acid), which is subject to post-translational modification important for localization of RAS proteins to the inner cell membrane. The cysteine in HRAS-<u>C</u>AAX is modified by farnesyltransferase, whereas the remaining three proteins can undergo alternative prenylation at this site by geranylgeranyltransferase. Palmitoylation of a second cysteine in the HVR (with exemption of KRAS4B lacking this residue) creates an additional hydrophobic anchor. The first 166–168 residues of RAS proteins form a single structured domain (the G domain), containing the phosphate-binding loop (P-loop, residues 10–17), switch I (residues 30–38), switch II (residues 60–76), and the base-binding loops (residues 116–120 and 145–147). GTP binding enables several residues, primarily in the switch I and the switch II region, to adopt a conformation that permits RAS effector proteins to bind (Hall, Bar-Sagi, & Nassar, 2002).

Like RAS G proteins, the RAF serine/threonine kinases were initially discovered as protein products of retroviral oncogenes, which caused tumors in rodents and fowl (Jansen et al., 1984; Jansen, Ruckert, Lurz, & Bister, 1983; Rapp et al., 1983). Three paralogous genes have been identified in humans and rodents, *ARAF*, *BRAF*, and *RAF1* (Lavoie & Therrien, 2015; Stefan & Bister, 2017; Wellbrock, Karasarides, & Marais, 2004; Zebisch & Troppmair, 2006). Structurally, the RAF family proteins have two functional domains, an N-terminal regulatory domain and the C-terminal kinase domain, which are conserved in all three isoforms. For the interaction with RAS proteins, the N-terminal region is responsible, while phosphorylation sites are located in both domains. RAS binding to RAF results in the localization of the kinases to the inner cell membrane, where further phosphorylations occur, which are required for full activation (Lavoie & Therrien, 2015; Stefan & Bister, 2017; Wellbrock et al., 2004).

2.3 Physiological and Pathophysiological Functions

RAS and RAF are components of an evolutionary highly conserved signaling pathway with essential functions in the embryonic development demonstrated for RAF1, BRAF, and KRAS through mouse knockout studies (Castellano & Santos, 2011; Matallanas et al., 2011). In humans, mutations in components of this signaling pathway contribute to a range of developmental malfunctions referred to as rasopathies (Aoki, Niihori, Inoue, & Matsubara, 2016; Halaban & Krauthammer, 2016; Niemeyer, 2014; Rauen, 2013).

RAS and *RAF* family genes have been stringently classified as cancer driver genes (Tokheim, Papadopoulos, Kinzler, Vogelstein, & Karchin, 2016). Genes from both families are frequently mutated in human cancer (30% or 10% of all tumors, respectively), and their protein products have thus attracted considerable attention as possible targets for therapy. Mutations of *RAS* genes are commonly found in numerous malignancies, including tumors of pancreas (90%), colon (45%), and lung (35%) (Prior, Lewis, & Mattos, 2012). Many tumor types have been shown to be dependent on continued expression of oncogenic RAS proteins in cell and animal models (McCormick, 2011). RAS mutations are not evenly distributed among the various isoforms: KRAS is mutated in 86% of RAS-driven tumors, followed by NRAS (11%) and finally HRAS (3%) and differences are also seen with regard to the tumor entities primarily affected (COSMIC) (Cox, Fesik, Kimmelman, Luo, & Der, 2014) Most of these mutations have been shown to decrease GAP-catalyzed hydrolysis, intrinsic GTPase activity, or both, leading to an increase in the active GTP-bound RAS population (Cox et al., 2014). Oncogenic mutations in RAF kinases, like the frequent BRAF (V600E) mutation in melanoma, cause decoupled phosphorylation events leading to tumorigenesis (Holderfield et al., 2014; Stefan & Bister, 2017; Wellbrock et al., 2004; Zebisch et al., 2007; Zebisch & Troppmair, 2006).

2.4 Pharmacological Targeting of RAS:RAF Complexes

Given the high frequency of RAS mutations in human tumors, RAS has become a target for therapeutic intervention early on. However, still no drugs have been approved by the FDA directly inhibiting the growth of RAS-driven tumors (Keeton, Salter, & Piazza, 2017), while progress has been made in the successful targeting of downstream components using kinase inhibitors for the RAF/MEK/ERK or the PI3K/AKT pathways.

The development of effective inhibitors of RAS signaling has been complicated by the fact that RAS activates downstream signaling through direct PPIs with its effectors and that oncogenic hyperactivation in cancer results from impaired RAS enzymatic activity. RAS inhibitors would therefore need to block the productive interaction between RAS proteins and their effectors. This has been initially attempted by preventing RAS localization to the inner cell membrane, which is necessary for the biological function of the wild-type and oncogenic RAS. Inhibitors of farnesyltransferase to prevent lipid modification of RAS proteins during maturation, however, failed to efficiently block RAS membrane localization due to possibility of alternative prenylation by geranylgeranyltransferase I. A more recent approach involved a series of inhibitors developed to target the noncatalytic δ-subunit of the cyclic GMP phosphodiesterase (PDE) 6 isozyme, which functions as a chaperone, binding the prenyl group and shuttling RAS to the plasma membrane (Dharmaiah et al., 2016; Zimmermann et al., 2013). A small molecule inhibitor of PDEδ (deltarasin) inhibited KRAS signaling and pancreatic tumor cell growth in vitro and in vivo. The clinical utility remains to be proven, and concerns have been raised about the nonspecific inhibition of other prenylated proteins (Keeton et al., 2017), which also includes other members of the RAS superfamily (Wennerberg, Rossman, & Der, 2005).

Real progress came from the development of inhibitors of PPI once proof was obtained that RAS proteins harbor sufficiently large and deep hydrophobic pocket for small molecule binding (Cox et al., 2014;

Keeton et al., 2017; Papke & Der, 2017). Extensive structural studies combined with mutagenesis provided the insight required to efficiently target RAS output signaling. Most intensively the interaction of HRAS with RAF1 has been studied. This interaction is mainly affected by mutations in switch I (Moodie et al., 1995; Shirouzu et al., 1994), which results in decreased protein binding and simultaneously also decreased oncogenic potential of mutant RAS-G12V (Shirouzu et al., 1994). Switch II mutations seem to have a greater effect on other RAS interactions, such as PI3K, neurofibromin, and exchange factors (Moodie et al., 1995). First evidence that disrupting complexes of activated RAS with effectors may result in a therapeutic strategy was shown using a membrane-directed, single immunoglobulin antibody domain (iDab#6-memb), which can bind to the switch I and II domains of GTP-bound HRAS-G12V in a way predicted to be mutually exclusive of the amino-terminal RAS-binding domain of RAF1 (RAF-RBD), RAL guanine nucleotide dissociation stimulator (RALGDS), and PI3K. Phosphorylation of downstream signaling was robustly inhibited, which modestly suppressed the growth of colon tumor cells in vitro. In a transgenic mouse model of KRAS-driven lung cancer, tumor initiation was dramatically reduced by expressing the antibody fragment. However, the tumor cells remained viable and resumed growth if antibody expression was halted (Tanaka & Rabbitts, 2010). Early small compounds identified on their ability to impair RAS-driven transformation included the nonsteroidal antiinflammatory drug sulindac sulfide (Muller et al., 2004; Waldmann et al., 2004), as well as putative RAS:RAF interaction inhibitors identified in a yeast two-hybrid screen (Gonzalez-Perez et al., 2010; Kato-Stankiewicz et al., 2002). However, it remains unclear whether their mode of action really involves binding to RAS proteins to block their interaction with downstream effectors. This also applies to the RAS-binding compounds, Kobe0065 and Kobe2602, identified through an in silico screening approach (Shima et al., 2013). Rigosertib, another compound currently in clinical trials, works as an indirect inhibitor of RAS by binding to the RBDs of RAS effectors and preventing interaction with RAS (Athuluri-Divakar et al., 2016). However, recent experimental evidence proposed a different indirect mode of action (Ritt et al., 2016).

Other approaches involving high-throughput screening have been conducted to identify compounds that disrupt the HRAS:RAF1 PPI. Such efforts resulted in the demonstration of a series of compounds (e.g., MCP1) capable of inhibiting mutant or wild-type HRAS-driven promoter/reporter constructs, HRAS-driven RAF1 activation, and mutant HRAS- or NRAS-dependent tumor phenotypes without interfering with RAS-induced activation of AKT (Kato-Stankiewicz et al., 2002). An important aspect in the search for inhibitors of the RAS effector domain is the preferential targeting of mutant over wild-type protein to avoid substantial side effects due to the requirement of RAS signaling in normal tissues. Inhibitors that target only the mutant protein may be achieved by attaching the inhibitor to the mutant residue, preferably a cysteine, which is the most reactive amino acid, as exemplified by the FDA-approved Bruton tyrosine kinase inhibitor Ibrutinib (Byrd et al., 2013; Wang et al., 2013). Such a strategy is applicable to KRAS-G12C, which is one of the three most common RAS mutants in cancer. A series of such compounds has been described that irreversibly target KRAS-G12C in a fragment-based screen paired with crystallographic studies and molecular dynamics simulation (Ostrem, Peters, Sos, Wells, & Shokat, 2013). These compounds bind in an allosteric pocket beneath switch II impairing SOS activity, decreasing the affinity of RAS for GTP relative to GDP, RAF binding and the viability of KRAS-G12C-transformed cells. Follow-up work resulted in the KRAS-G12C cell-specific active inhibitor, ARS-853 (Patricelli et al., 2016). A separate group developed GDP analogues, which also covalently bind the cysteine of KRAS (G12C) (Lim et al., 2014). Both classes should result in an inactive RAS–nucleotide complex, thus blocking downstream signal transduction. However, the relatively low frequency of the G12C mutation may limit the utility of G12C-selective compounds for cancers other than those of lung cancer, where this mutation is more prevalent than other KRAS mutations.

A more recent development was the search for a RAS inhibitor targeting multiple sides on the protein and inhibiting all RAS isoforms (pan RAS inhibitor) (Welsch et al., 2017). Compound 3144 was found to bind to RAS proteins using microscale thermophoresis, nuclear magnetic resonance spectroscopy, and isothermal titration calorimetry and to exhibit lethality in cells partially dependent on expression of RAS proteins. Compound 3144 also displayed antitumor activity in xenograft mouse cancer models.

3. PKA SIGNALOSOME AND NETWORK

3.1 Background

The cAMP-dependent PKA is a serine/threonine kinase that is evolutionary conserved in all eukaryotes. PKA is ubiquitously expressed and regulates central cellular processes that are relevant for metabolism, development, memory formation, and proliferation. The mechanism of PKA activation and inactivation is one of the best studied examples for protein allostery and small molecule:protein interactions. PKA is the prototypical cellular effector protein for the canonical second messenger molecule 3′,5′-cyclic adenosine monophosphate (cAMP) (Langeberg & Scott, 2015; Tasken & Aandahl, 2004; Taylor, Ilouz, Zhang, & Kornev, 2012; Taylor & Kornev, 2011). The activation of multiple receptor-controlled signaling pathways leads to oscillations of the diffusible intracellular second messenger molecule cAMP. The list of hormone or neurotransmitter-controlled signaling circuits that utilize the cAMP–PKA signaling system is expansive. Deregulation of several components of the GPCR–cAMP–PKA signaling axis contributes to a selection of human diseases. Mutations in PKA subunits affect spatiotemporal-controlled PKA activities and contribute to specific disease patterns. Recently, a collection of different activating mutations in the cAMP-sensing PKA have been identified which contribute to carcinogenesis or degenerative diseases (Beuschlein et al., 2014; Cao et al., 2014; Cheung et al., 2015; Espiard, Ragazzon, & Bertherat, 2014; Honeyman et al., 2014; Sato et al., 2014; Stratakis, 2013; Wong et al., 2014; Zilbermint & Stratakis, 2015). In contrast, kinase-inactivating mutations cause hormone resistance (Assie, 2012; Linglart et al., 2011; Silve, Clauser, & Linglart, 2012). It is of interest that dependent on the cell type, cAMP mobilization either boosts proliferation or inhibits cell growth (Bachmann et al., 2013; Dumaz & Marais, 2005; Gerits, Kostenko, Shiryaev, Johannessen, & Moens, 2008; Stork & Schmitt, 2002; Thaker et al., 2006). The question rises how to interfere with deregulated PKA functions. One pharmacological strategy would be the direct targeting of PKA activities. An alternative and presumably more specific approach is to tackle molecular interactions of PKA either with variants of the small molecule cAMP or by targeting selected regulatory proteins.

3.2 Macromolecular PKA Complexes

The PKA holoenzyme is a heterotetramer consisting of two catalytic (C, PKAc) subunits that bind a dimer of identical regulatory (R) subunits. The physical interaction of the cAMP-sensing R subunit dimer with the C subunits locks the enzyme in its inactive state. C subunits have a bilobal subdomain organization, consisting of a small and a large lobe. The N-terminal small lobe is separated from the C-terminal large lobe by a nonconserved α-helix, which spans both lobes. The active site of the kinase contains critical and conserved residues for catalysis and is located in the cleft between both lobes. Each lobe contains two hydrophobic motifs, which are classified as regulatory (R spine) and catalytic spines (C spine), respectively. The relative position of the lobes to each other and the ordering of R and C spines contribute to the opening and closing state of the active site cleft and are general features of catalytic kinase process (Hu et al., 2015; Smith, Radzio-Andzelm, Madhusudan, Akamine, & Taylor, 1999; Taylor et al., 2012; Taylor & Kornev, 2011). PKA R subunits simultaneously associate with C subunits and specific scaffolding proteins (A-kinase anchoring proteins [AKAPs]). Four different R subunits have been identified and shown to be functionally nonredundant. R subunits contain a conserved domain organization. At the N-terminus the dimerization and docking domain (D/D domain) is essential for the forming the R subunit dimer and required for the interaction with AKAPs (Banky, Huang, & Taylor, 1998; Kinderman et al., 2006). A flexible linker region is located between the N-terminal D/D domain and two C-terminally located cAMP-binding domains and contains the inhibitor site. The inhibitor site is comprised of five residues of the PKA substrate recognition sequence which acts as (pseudo)-substrate sequence for binding to the active site of C subunits (Diller, Madhusudan, Xuong, & Taylor, 2001; Taylor et al., 2012). The structures of PKA type I and type II holoenzyme complexes differ considerably in relation to the overall tetrameric holoenzyme structure, Mg^{2+} ions and ATP binding, and varying dissociation constants for the C subunit (Herberg, Taylor, & Dostmann, 1996; Kim, Xuong, & Taylor, 2005; Solberg et al., 1994; Taylor et al., 2012; Wu, Brown, von Daake, & Taylor, 2007; Zawadzki & Taylor, 2004). cAMP binding to the cAMP-binding domains causes a conformational change and leads to PKA holoenzyme disassembly. The cAMP–PKA system is involved in numerous cellular responses. This fact requires that on the cellular level the specificity, duration, and intensity of PKA responses are spatially and temporally restricted. Indeed, the receptor-sensed input signals are relayed through macromolecular PKA complexes which are organized by scaffolding proteins, the AKAPs. The AKAP family consists of functionally related scaffolding proteins with a common interaction domain but, in other respect, with different targeting domains (Langeberg & Scott, 2015; Scott, Dessauer, & Tasken, 2013; Skroblin, Grossmann, Schafer, Rosenthal, & Klussmann, 2010; Wong & Scott, 2004). AKAPs directly bind PKA R subunits and coordinate second messenger responsive events by tethering the kinase holoenzyme to distinct subcellular compartments. There, these PKA-centered signaling complexes ensure the controlled phosphorylation of possibly more than 250 colocalized and AKAP-organized PKA substrates (Greenwald & Saucerman, 2011). Different mechanisms are involved in controlling cAMP fluxes and PKA activities which involves, among others, physical PPIs of PKA subunits and AKAPs with PDE, the ubiquitin–proteasome system (UPS), the protein kinase inhibitor peptide (PKI), and/or other PKA interactors or substrates (Fig. 3).

3.3 Pathological PKA Functions

Activation of cAMP synthesis by different GPCR agonists produces distinct physiological outputs. PKA acts as key effector of cellular cAMP elevations which affects essential cellular processes such as metabolism, immune function, growth, differentiation, gene expression, and proliferation. Given its major functions in normal physiology, defects in cAMP signaling have been observed, e.g., in the cardiac setting (Lissandron & Zaccolo, 2006). However, recent sequencing efforts have identified mutations in PKA R and C subunits which are associated with different disease states. In Cushing's

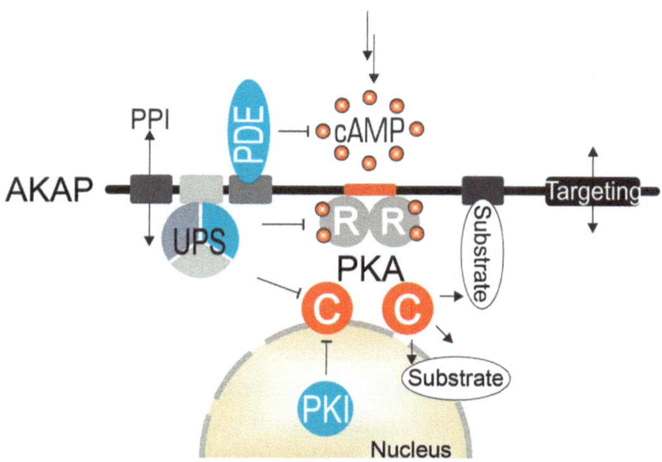

Fig. 3 General mechanism of transient activation and inactivation of compartmentalized PKA signalosomes. Diffusible cAMP fluxes are sensed by AKAP-tethered PKA complexes leading to PKA subunit (regulatory, R; catalytic, C) activation and compartmentalized substrate phosphorylation. The domain organization of a hypothetical AKAP with modular interaction domains is shown. The carboxy-terminal targeting domain mediates PKA compartmentalization. PPI sites for binary complex formation with substrates and/or additional interactors such as PDEs or the ubiquitin–proteasome system (UPS) are indicated. Compartmentalized PDE activities or UPS participate to spatiotemporally control cAMP-dependent and PKA-mediated signal propagation. PKI is shown as an example for a physical interaction partner of PKA subunits which desensitizes PKA activities.

syndrome, a disease which is characterized by excess glucocorticoid production from tumors and hyperplasia of the adrenal cortex, C subunit mutations such as L205R have been identified. It has been shown that this mutation leads to higher basal phosphotransferase activity of C subunits. Several studies underline the concept that disturbance of the C subunit structure leads to deregulated phosphotransferase activities (Beuschlein et al., 2014; Cao et al., 2014; Sato et al., 2014). In addition, also gene amplifications of the C subunit gene have been identified in Cushing's syndrome patients (Lodish & Stratakis, 2016). Another disease of the adrenal gland is called Carney complex. Carney complex patients are at increased risk for developing adrenal gland tumors or benign tumors (so-called myxomas) in the heart. Sequencing efforts have identified various inactivating mutations in the gene for RIα subunits. The consequence is that a decrease in functional RIα subunits leads to upregulations of C subunit phosphotransferase activities (Espiard et al., 2014; Horvath et al., 2010; Turnham & Scott, 2016). A further involvement of PKA in cancer stems from a direct involvement of C subunit fusions in fibrolamellar hepatocellular carcinoma. This rare liver tumor results in a fusion of the molecular chaperone DnaJ with the gene of the Cα subunit. It results in a fusion of exon 1 of DnaJ with exon 2 of Cα. It has been postulated that these mutations affect the complex formation with inhibitory R subunits (Cheung et al., 2015; Honeyman et al., 2014). These findings underline that alterations of C subunit transcripts, deregulations of C subunit and R subunit emanating PPIs, and changes of the cAMP-sensing properties contribute to a variety of cancer-related disorders. In this aspect, also the perturbance of kinase compartmentalization through deregulated AKAP scaffolds needs to be considered. It should not be forgotten that controlling PKA activities depends among others mainly on the mobilization of cAMP levels. cAMP levels can also be perturbed in different disease settings, for example, downstream of deregulated GPCR pathways (Bachmann et al., 2016; Briscoe & Therond, 2013; Dorsam & Gutkind, 2007; Mukhopadhyay et al., 2013; O'Hayre et al., 2013; Pierce, Premont, & Lefkowitz, 2002). Exemplarily, we just would like to mention activating and inactivating mutations of the gene for the stimulating G protein alpha, Gαs. Activating mutations in Gαs genes promote aberrant growth of human thyroid and pituitary tumors (Landis et al., 1989; Weinstein et al., 1991). In a different cellular setting, the Gαs–PKA axis is part of a tumor suppressive pathway in skin stem cells (Iglesias-Bartolome et al., 2015).

3.4 Pharmacological Targeting of PKA Emanating PPIs

Protein kinases have become one of the most effective families of drug targets next to the GPCR superfamily. So far, a collection of ATP-competitive and allosteric kinase inhibitors have been approved, and hundreds of kinase inhibitors are currently in different stages of clinical development (Roskoski, 2015). However, the phosphotransferase activities of PKA have not been considered to become the target of major drug discovery efforts due to the broad involvement of PKA in major physiological signaling events. The recent implications of PKA in specific disease patterns might change this (Berthon, Szarek, & Stratakis, 2015; Calebiro, Bathon, & Weigand, 2017; Cheung et al., 2015; Stratakis, 2013; Turnham & Scott, 2016). ATP-competitive inhibitors of C subunit activities are frequently used to study cell-based PKA functions. Though, the lack of kinase specificity lowers the enthusiasm to apply the small molecule inhibitors KT5720 and H89 for targeting oncogenic PKA functions (Murray, 2008). Another strategy to interfere with PKA functions are cyclic nucleotide analogs, such as variants of Rp-cAMPS, which act cAMP antagonistic and prevent PKA activation (Kennedy & Scott, 2015). Active site-directed PKA inhibitors (ATP-competitive) or cAMP-bindings site blockers have the potential to cause off-target effects. With this in mind, tackling PPIs of macromolecular PKA complexes seems to be a more target-oriented alternative approach for interfering with compartmentalized PKA functions. We assume that the pharmaceutical targeting of binary PKA subunit interactions may show more efficacy and specificity. One physiological strategy to shut down PKA activities is through binary PPIs of the C subunit with the endogenously present PKA inhibitor peptide (PKI; Fig. 3) (Scott, Fischer, Demaille, & Krebs, 1985). The inactivation mechanism of PKA by PKI is based on the PKA pseudosubstrate sequence and inhibits nuclear PKA functions (Dalton & Dewey, 2006; Dalton, Smith, Smith, & Dewey, 2005). Peptides derived from PKI are very selective PKA inhibitors and frequently used for analyzing PKA phosphotransferase functions in experimental settings by blocking substrate binding (Dalton & Dewey, 2006). Recently, ARHGAP36, a member of the Rho GTPase-activating protein (RhoGAP) family, has been identified to directly bind to C subunits in a similar binding mode as PKI. Upon binding of ARHGAP36 to C subunits, ubiquitin-mediated lysosomal degradation is initiated, and PKA signaling is suppressed (Eccles et al., 2016). The development of peptides or peptide mimetics which are based on C:PKI or C:ARHGAP36 interface structures might be a feasible strategy to interfere with deregulated PKA functions which have been decoupled from R subunit interactions through mutations. Such peptides should function as substrate mimics for blocking PKA phosphotransferase activities with higher efficacy and selectivity. Further, also the physical interaction with the UPS is relevant for controlling PKA signal propagation. Controlling either R or C subunit stability through proteolysis will affect R:PKAc ratios thereby modulating PKA substrate phosphorylation (Lignitto et al., 2013, 2011; Rinaldi, Sepe, Donne, & Feliciello, 2015; Yu, Chen, Tagle, & Cai, 2002). Hereby, the identification of the ubiquitin ligase targeting C subunits is of major interest (Kandel, 2001). The pharmaceutical targeting would involve peptide mimetic compounds which have been developed to disrupt the linear ubiquitin ligase complex and, dependent on the target, block prooncogenic signals in specific cell systems (Popovic, Vucic, & Dikic, 2014).

One alternative strategy to interfere with compartmentalized PKA function is either to displace or modify enzyme activities from the AKAP signaling platform. Typically, AKAPs coordinate GTPases, kinases, PDEs, and phosphatases by binding, for controlling the spatiotemporal cAMP signaling response. The mechanistic details and the functional relevance for several AKAPs have been revealed in recent years and have been reviewed in detail (Langeberg & Scott, 2015; Lee, Maurice, & Baillie, 2013; Scott et al., 2013; Skroblin et al., 2010; Tasken & Aandahl, 2004; Torres-Quesada, Mayrhofer, & Stefan, 2017; Wong & Scott, 2004). The growing family of canonical AKAPs shares a short characteristic motif for R subunit interaction. Additional binding sites in AKAPs are diverse and responsible for subcellular targeting or other PPIs (Fig. 3). The R subunit-binding motif of AKAPs is an amphipathic α-helix with commonly 14–18 residues. Through dimerization of the R-subunit dimer and docking domains, an antiparallel X-type four-helix bundle is formed thereby creating a hydrophobic surface, for AKAP binding. Hydrophobic residues of the amphipathic α-helix of AKAPs dock diagonally into this groove. All four R subunits contain a conserved binding interface for AKAP interactions with, dependent on the scaffold, nano- or micromolar affinities (Alto et al., 2003; Burgers, van der Heyden, Kok, Heck, & Scholten, 2015; Dema, Perets, Schulz, Deak, & Klussmann, 2015; Kinderman et al., 2006; Skroblin et al., 2010). The common binding interface of AKAP: PKA motivated research efforts for the identification of disrupting molecules to elucidate the individual roles of these compartmentalized kinase signalosomes. A variety of therapeutic strategies have been proposed to interfere with PPI within the AKAP compartmentalized cAMP–PKA signaling system (Calejo & Tasken, 2015; Dema et al., 2015; Kennedy & Scott, 2015; Lee et al., 2013). The first AKAP:PKA disrupting peptide was named Ht31 and was derived from the PKA anchoring domain of AKAP-Lbc (Carr, Hausken, Fraser, Stofko-Hahn, & Scott, 1992). Besides other type I or type II PKA:AKAP complex disrupting peptides which have been developed, Ht31 is still used to characterize and identify new AKAPs and their functions. Other examples for peptides are AKAPis and RIAD which show different specificities for disrupting AKAP:PKA complexes (Alto et al., 2003; Carlson et al., 2006). To avoid some limitations of these peptides, recently stapled AKAP disruptor peptides (STAD peptides) have been developed showing increased binding affinity and improved pharmacological performance (Calejo & Tasken, 2015; Kennedy & Scott, 2015; Wang et al., 2015). Besides peptidomimetics, bioactive small molecules are favorable alternatives to peptides. In a high-throughput drug screening, one allosterically acting small molecule inhibitor of specific AKAP:PKA complexes has been identified (Christian et al., 2011). AKAPs compartmentalize PKA phosphotransferase activities by acting as a physical linker between distinct pharmaceutically relevant receptor and kinase pathways to control signal transmission spatially and temporally. In a pathological cell setting, PPIs of PKA subunits are deregulated. That is the reason why targeting of selected PKA PPIs using peptides or small molecules seems to be a promising therapeutic strategy to interfere with subcellular PKA functions.

4. MYC:MAX SIGNALOSOME AND NETWORK

4.1 The MYC:MAX Network

The *MYC* oncogene was originally discovered as a transduced allele of the cellular *MYC* protooncogene in the retroviral genome of avian acute leukemia virus MC29 (Duesberg, Bister, & Vogt, 1977; Stefan & Bister, 2017; Vogt, 2012). The discovery of specific chromosomal translocations of the human ortholog of the *MYC* protooncogene in Burkitt lymphoma was the first indication of the involvement of *MYC* oncogenes in human tumorigenesis (Dalla-Favera et al., 1982). Today, deregulation of *MYC* expression by amplification, chromosomal rearrangements, or defects in upstream signaling is a hallmark and important driving force in most, if not all, human cancers (Dang, 2012; Stefan & Bister, 2017; Stine, Walton, Altman, Hsieh, & Dang, 2015; Tokheim et al., 2016; Vogt, 2012). In addition to *MYC*, the human genome contains two paralogous genes, *MYCN* and

MYCL, that are also critically involved in the genesis of specific tumor types (Dang, 2012; Stefan & Bister, 2017).

The MYC protein product of the *MYC* protooncogene is a member of a family of proteins that share a characteristic hallmark, a dimerization and DNA-binding domain (bHLH-LZ) containing a basic region (b) as DNA-binding surface, and a helix–loop–helix (HLH) and a leucine repeat (zipper; LZ) region as consecutive PPI domains. MYC with its C-terminal bHLH-LZ domain preferentially binds to MAX, another member of the bHLH-LZ protein family, and the stable MYC:MAX heterodimer is a ubiquitous transcriptional regulator with a broad functional spectrum (Blackwood & Eisenman, 1991; Conacci-Sorrell, McFerrin, & Eisenman, 2014; Eisenman, 2001; Stefan & Bister, 2017). At physiological concentrations and in the absence of the MAX bHLH-LZ-binding partner, MYC is monomeric, with the typical properties of an intrinsically disordered protein (Fieber et al., 2001). In contrast, MAX forms homodimers, however, with lower stability than MYC:MAX heterodimers. X-ray structure determination of the bHLH-LZ domains in MYC:MAX heterodimers in complex with DNA revealed the structural details of the parallel protein chains arranged in a four-helix bundle (Fig. 4) (Nair & Burley, 2003). MYC:MAX heterodimers specifically bind to DNA elements (*E*-boxes) with the preferred nucleotide sequence 5′-CACGTG-3′ by forming specific contacts between the α-helical basic regions and the major groove of DNA (Fig. 4) (Blackwood & Eisenman, 1991; Nair & Burley, 2003). MYC:MAX heterodimers induce transcription of a vast number of target genes by binding to E-boxes in promoter or enhancer regions and by interaction with a variety of other protein factors. MYC proteins contain several conserved regions (MYC boxes) in their N-terminal half that are essential for transcriptional transactivation, for posttranslational modifications and turnover of these proteins, and for interactions with a multitude of other regulatory proteins (Fig. 4) (Stefan & Bister, 2017).

The MYC:MAX heterodimer holds an apex position within a large network of specifically interacting bHLH-LZ proteins (Conacci-Sorrell et al., 2014; Eisenman, 2001; Stefan & Bister, 2017). Whereas MYC does not form homodimers and forms heterodimers only with MAX, MAX undergoes homodimerization and also forms heterodimers with other bHLH-LZ proteins like MXD family proteins or MNT and MGA (Fig. 4). On the other hand, MXD and MNT form heterodimers with MLX which again interacts with transcriptional regulators of the Mondo protein family (Conacci-Sorrell et al., 2014; Stefan & Bister, 2017). MYC:MAX heterodimers

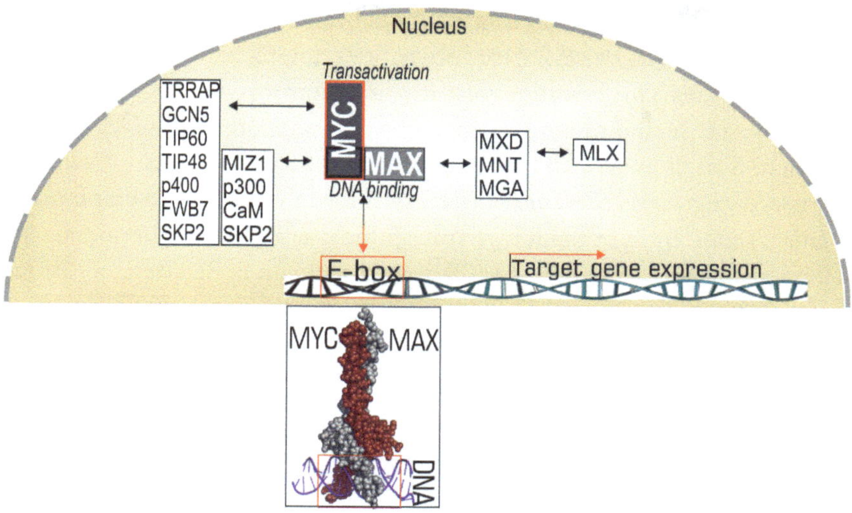

Fig. 4 Interactions of the MYC:MAX network with proteins and DNA elements. The oncoprotein MYC and its dimerization partner MAX bind to specific DNA motifs (E-Box) and control the expression of a vast array of target genes. Shown are the structural details of basic/helix–loop–helix/leucine zipper (bHLH-LZ) interactions of MYC: MAX (Nair & Burley, 2003). Elevated levels of MYC:MAX complexes drive cell proliferation and tumorigenesis. The indicated interacting proteins bind to the transactivation domain or to bHLH-LZ domains of MYC or MAX, respectively. TRRAP (adaptor protein), TIP60 and GCN5 (histone acetyltransferases), TIP48 (part of an ATPase complex), p300 (coactivator, acetyltransferase), p400 (histone exchange factor), FWB7, SKP2 (E3 ubiquitin ligase complexes), MIZ1 (zinc finger transcription factor), CaM (calcium sensor calmodulin), MXD, MNT, MGA, MLX (bHLH-LZ proteins).

typically activate transcription, whereas the antagonistic MXD proteins in heterodimeric complexes with MAX suppress transcription.

4.2 Physiological and Oncogenic Functions

The broad impact of MYC on virtually all major physiological processes is dependent on dimerization with MAX and mediated by the MYC:MAX transcriptional regulator complex, although MAX-independent functions and even cytoplasmic functions of MYC cleavage products have also been reported (Conacci-Sorrell et al., 2014; Eilers & Eisenman, 2008; Eisenman, 2001; Stefan & Bister, 2017). As a true master regulator, MYC controls an overwhelmingly broad spectrum of fundamental cellular processes, including cell proliferation, cell cycle regulation, cellular growth, biosyntheses, energy metabolism, differentiation, stem cell regulation, and apoptosis (Dang, 2012, 2013; Stefan & Bister, 2017; Stine et al., 2015). This is reflected by an equally broad range of transcriptional regulation: MYC affects transcription conducted by all RNA polymerases, interacts with histone modifiers and other chromatin regulators, and controls virtually the entire transcriptome including miRNAs and long noncoding RNAs (Fig. 4) (Conacci-Sorrell et al., 2014; Croce, 2009; Hart, Roberts, Weinberg, Morris, & Vogt, 2014; Stefan & Bister, 2017). MYC regulates a vast number of target genes, most of them by transcriptional activation, but repression of specific genes was also observed (Eilers & Eisenman, 2008; Stefan & Bister, 2017; Zeller et al., 2006). MYC may function both as specific activator of transcriptional initiation by binding to promoters or enhancers of selected target genes, but also as a general transcriptional amplifier by a broad nonspecific stimulation of transcriptional elongation, thereby primarily enhancing preexisting transcriptional programs (Dang, 2013; Rahl & Young, 2014; Stine et al., 2015). One of the striking areas of cell physiology controlled by MYC is energy metabolism, including glycolysis and glutaminolysis. This leads to nutrient addiction—glucose or glutamine dependency—of tumor cells with deregulated *MYC* expression, offering possible targets for therapeutic intervention (Dang, 2013; Shim et al., 1997; Stine et al., 2015). Contrary to its distinctive role in cell proliferation, MYC is also involved in cell death. In the absence of survival factors, deregulated *MYC* expression leads to apoptosis (Evan et al., 1992; Shortt & Johnstone, 2012). Hence, MYC-induced tumorigenesis is frequently accompanied by the loss of apoptotic checkpoints (Dang, 2012; Stefan & Bister, 2017; Stine et al., 2015).

Deregulated *MYC* alleles are an important driving force in most human cancers. Accordingly, *MYC* is an obvious key target for cancer research and development of therapeutic strategies (Dang, 2012; Stine et al., 2015; Vogt, 2012). All *MYC* family members were listed in the census of human cancer genes (Futreal et al., 2004), and *MYC* and *MYCN* were included in a list of cancer driver genes recently compiled by a stringent reevaluation of cancer gene classifications (Tokheim et al., 2016). Commonly, oncogenic activation of *MYC* family protooncogenes does not require mutations in the protein-coding region, rather deregulation of gene expression, and hence decoupled levels of MYC proteins are at the roots of *MYC* oncogenic function. The most important mechanisms underlying decoupling of *MYC* family gene expression from normal cellular control in human tumorigenesis are gene amplification and, less frequently, chromosomal translocations. Such genetic duplications and rearrangements of *MYC* family genes and the subsequent deregulation of gene expression have been observed in a broad range of practically all human tumors (Vita & Henriksson, 2006). In addition to these direct genetic alterations of the *MYC* loci, disturbances of upstream signaling or downstream control elements including negative feedback pathways can also lead to deregulation of gene expression (Stefan & Bister, 2017).

4.3 Pharmacological Interference

In view of the pivotal role of *MYC* genes in human tumorigenesis, the development of small molecule inhibitors of MYC function with suitable pharmacokinetic properties is an obvious and urgent goal. For oncogenic kinases, drug development targeted at the enzymatic activities of these proteins has shown stunning clinical success, although frequent occurrence of drug resistance is still an enormous challenge (Holderfield et al., 2014). However, MYC is not an enzyme and is mainly engaged in macromolecular interactions, in particular in PPI with its dimerization partner MAX but also in PPI with additional protein factors (Fig. 4) (Stefan & Bister, 2017). In contrast to structurally well-defined catalytic centers and clefts of enzymes, protein surfaces involved in PPI are much more difficult to target with small molecules. Nonetheless, several small molecule inhibitors of MYC:MAX dimerization and of MYC-induced cell transformation were isolated from chemical libraries and provided proof of principle for targeted MYC inhibition (Berg et al., 2002; Yin, Giap, Lazo, & Prochownik, 2003). Small molecule inhibitors of MYC:MAX dimerization with pharmacokinetic properties suitable for possible drug development were isolated recently from a Kröhnke pyridine library (Hart, Garner, et al., 2014; Raffeiner et al., 2014). These inhibitors, termed KJ-Pyr-9 and KJ-Pyr-10, efficiently inhibited MYC:MAX dimerization, the MYC-driven transcriptome, and, most importantly, MYC-induced cell transformation and even tumor growth in vivo (Hart, Garner, et al., 2014; Raffeiner et al., 2014). In view of the intrinsic difficulties of direct inhibition of MYC proteins, indirect inhibitory strategies were also developed, in particular targeted at *MYC* transcription. G-quadruplex regulatory regions in the *MYC* locus were targeted, and also BET bromodomain proteins that facilitate transcriptional activation of *MYC* (Balasubramanian, Hurley, & Neidle, 2011; Shu & Polyak, 2017). The striking nutrient addiction of *MYC*-driven cancer cells may offer additional opportunities for pharmacological inhibition, in particular targeting glycolytic or glutaminolytic enzymes or transporters (Altman, Stine, & Dang, 2016; Dang, 2012; Stine et al., 2015). Furthermore, the pharmacological inhibition of genes that are required for survival of *MYC*-driven tumors, i.e., the systematic screening for synthetic lethality may be promising approaches to target *MYC* oncogenicity (Stefan & Bister, 2017; Varmus, Unni, & Lockwood, 2017).

5. CONCLUSIONS

We have discussed three different protein complexes which are deregulated in different cancer types due to variable molecular mechanisms. Their frequent involvement in this devastating disease called traditional drug discovery efforts into action. In the last two decades, efficient targeting of the cancer drivers MYC and RAS has been more than challenging and led to their classification as "undruggable". That is the reason why there is a growing interest to advance alternative approaches through intervention of binary molecular interactions. Recent efforts in computational methods, fragment-based drug discovery, cell-based systems measuring PPIs, and the availability of structural information enables the precise elucidation of PPI-binding interfaces. Targeting these more sophisticated interaction sites requires the combination and improvement of small molecule and peptide-based strategies of interference. First, direct and indirect modulators of PPIs need to be taken into account. Second, polypharmacology approaches using PPI modulators and enzyme inhibitors should be considered. Third, accumulating data on PPI inhibitor structures will lead to customized chemical libraries specialized for small molecules:PPI screenings. Fourth, new means to analyze and quantify PPIs and small molecule efficacies in vitro and in vivo will be relevant. With the achievement of these goals we envision perturbations of PPI networks to uncouple deregulated signaling pathways from pathological signaling responses. Integration and combination of different means of direct or indirect PPI inhibition along with traditional drug discovery efforts will pave the road for personalized cancer therapy and precision medicine.

ACKNOWLEDGMENTS

Work by the authors has been supported by Grants from the Austrian Science Fund (FWF; P27606, P23652, SFB-F44).

REFERENCES

Altman, B. J., Stine, Z. E., & Dang, C. V. (2016). From Krebs to clinic: Glutamine metabolism to cancer therapy. *Nature Reviews. Cancer*, *16*(10), 619–634. https://doi.org/10.1038/nrc.2016.71.

Alto, N. M., Soderling, S. H., Hoshi, N., Langeberg, L. K., Fayos, R., Jennings, P. A., et al. (2003). Bioinformatic design of A-kinase anchoring protein-in silico: A potent and selective peptide antagonist of type II protein kinase A anchoring. *Proceedings of the National Academy of Sciences of the United States of America*, *100*(8), 4445–4450. https://doi.org/10.1073/pnas.0330734100.

Aoki, Y., Niihori, T., Inoue, S., & Matsubara, Y. (2016). Recent advances in RASopathies. *Journal of Human Genetics*, *61*(1), 33–39. https://doi.org/10.1038/jhg.2015.114.

Arkin, M. R., Tang, Y., & Wells, J. A. (2014). Small-molecule inhibitors of protein-protein interactions: Progressing toward the reality. *Chemistry & Biology*, *21*(9), 1102–1114. https://doi.org/10.1016/j.chembiol.2014.09.001.

Assie, G. (2012). One single signaling pathway for so many different biological functions: Lessons from the cyclic adenosine monophosphate/protein kinase A pathway-related diseases. *The Journal of Clinical Endocrinology and Metabolism*, *97*(12), 4355–4357. https://doi.org/10.1210/jc.2012-3659.

Athuluri-Divakar, S. K., Vasquez-Del Carpio, R., Dutta, K., Baker, S. J., Cosenza, S. C., Basu, I., et al. (2016). A small molecule RAS-mimetic disrupts RAS association with effector proteins to block signaling. *Cell*, *165*(3), 643–655. https://doi.org/10.1016/j.cell.2016.03.045.

Bachmann, V. A., Mayrhofer, J. E., Ilouz, R., Tschaikner, P., Raffeiner, P., Rock, R., et al. (2016). Gpr161 anchoring of PKA consolidates GPCR and cAMP signaling. *Proceedings of the National Academy of Sciences of the United States of America*, *113*, 7786–7791. https://doi.org/10.1073/pnas.1608061113.

Bachmann, V. A., Riml, A., Huber, R. G., Baillie, G. S., Liedl, K. R., Valovka, T., et al. (2013). Reciprocal regulation of PKA and Rac signaling. *Proceedings of the National Academy of Sciences of the United States of America*, *110*(21), 8531–8536. https://doi.org/10.1073/pnas.1215902110.

Balasubramanian, S., Hurley, L. H., & Neidle, S. (2011). Targeting G-quadruplexes in gene promoters: A novel anticancer strategy? *Nature Reviews. Drug Discovery*, *10*(4), 261–275. https://doi.org/10.1038/nrd3428.

Banky, P., Huang, L. J., & Taylor, S. S. (1998). Dimerization/docking domain of the type Ialpha regulatory subunit of cAMP-dependent protein kinase. Requirements for dimerization and docking are distinct but overlapping. *The Journal of Biological Chemistry*, *273*(52), 35048–35055.

Berg, T., Cohen, S. B., Desharnais, J., Sonderegger, C., Maslyar, D. J., Goldberg, J., et al. (2002). Small-molecule antagonists of Myc/Max dimerization inhibit Myc-induced transformation of chicken embryo fibroblasts. *Proceedings of the National Academy of Sciences of the United States of America*, *99*(6), 3830–3835. https://doi.org/10.1073/pnas.062036999.

Berthon, A. S., Szarek, E., & Stratakis, C. A. (2015). PRKACA: The catalytic subunit of protein kinase A and adrenocortical tumors. *Frontiers in Cell and Development Biology*, *3*, 26. https://doi.org/10.3389/fcell.2015.00026.

Beuschlein, F., Fassnacht, M., Assie, G., Calebiro, D., Stratakis, C. A., Osswald, A., et al. (2014). Constitutive activation of PKA catalytic subunit in adrenal Cushing's syndrome. *The New England Journal of Medicine*, *370*(11), 1019–1028. https://doi.org/10.1056/NEJMoa1310359.

Blackwood, E. M., & Eisenman, R. N. (1991). Max: A helix-loop-helix zipper protein that forms a sequence-specific DNA-binding complex with Myc. *Science*, *251*(4998), 1211–1217.

Brandman, O., & Meyer, T. (2008). Feedback loops shape cellular signals in space and time. *Science*, *322*(5900), 390–395. https://doi.org/10.1126/science.1160617.

Briscoe, J., & Therond, P. P. (2013). The mechanisms of Hedgehog signalling and its roles in development and disease. *Nature Reviews. Molecular Cell Biology*, *14*(7), 416–429. https://doi.org/10.1038/nrm3598.

Burgers, P. P., van der Heyden, M. A., Kok, B., Heck, A. J., & Scholten, A. (2015). A systematic evaluation of protein kinase A-A-kinase anchoring protein interaction motifs. *Biochemistry*, *54*(1), 11–21. https://doi.org/10.1021/bi500721a.

Byrd, J. C., Furman, R. R., Coutre, S. E., Flinn, I. W., Burger, J. A., Blum, K. A., et al. (2013). Targeting BTK with ibrutinib in relapsed chronic lymphocytic leukemia. *The New England Journal of Medicine*, *369*(1), 32–42. https://doi.org/10.1056/NEJMoa1215637.

Calebiro, D., Bathon, K., & Weigand, I. (2017). Mechanisms of aberrant PKA activation by calpha subunit mutations. *Hormone and Metabolic Research*, *49*(4), 307–314. https://doi.org/10.1055/s-0042-112817.

Calejo, A. I., & Tasken, K. (2015). Targeting protein-protein interactions in complexes organized by A kinase anchoring proteins. *Frontiers in Pharmacology*, *6*, 192. https://doi.org/10.3389/fphar.2015.00192.

Cao, Y., He, M., Gao, Z., Peng, Y., Li, Y., Li, L., et al. (2014). Activating hotspot L205R mutation in PRKACA and adrenal Cushing's syndrome. *Science*, *344*(6186), 913–917. https://doi.org/10.1126/science.1249480.

Carlson, C. R., Lygren, B., Berge, T., Hoshi, N., Wong, W., Tasken, K., et al. (2006). Delineation of type I protein kinase A-selective signaling events using an RI anchoring disruptor. *The Journal of Biological Chemistry*, *281*(30), 21535–21545. https://doi.org/10.1074/jbc.M603223200.

Carr, D. W., Hausken, Z. E., Fraser, I. D., Stofko-Hahn, R. E., & Scott, J. D. (1992). Association of the type II cAMP-dependent protein kinase with a human thyroid RII-anchoring protein. Cloning and characterization of the RII-binding domain. *The Journal of Biological Chemistry*, *267*(19), 13376–13382.

Castellano, E., & Santos, E. (2011). Functional specificity of ras isoforms: So similar but so different. *Genes & Cancer*, *2*(3), 216–231. https://doi.org/10.1177/1947601911408081.

Cheung, J., Ginter, C., Cassidy, M., Franklin, M. C., Rudolph, M. J., Robine, N., et al. (2015). Structural insights into mis-regulation of protein kinase A in human tumors. *Proceedings of the National Academy of Sciences of the United States of America*, *112*(5), 1374–1379. https://doi.org/10.1073/pnas.1424206112.

Christian, F., Szaszak, M., Friedl, S., Drewianka, S., Lorenz, D., Goncalves, A., et al. (2011). Small molecule AKAP-protein kinase A (PKA) interaction disruptors that activate PKA interfere with compartmentalized cAMP signaling in cardiac myocytes. *The Journal of Biological Chemistry*, *286*(11), 9079–9096. https://doi.org/10.1074/jbc.M110.160614.

Conacci-Sorrell, M., McFerrin, L., & Eisenman, R. N. (2014). An overview of MYC and its interactome. *Cold Spring Harbor Perspectives in Medicine*, *4*(1), a014357. https://doi.org/10.1101/cshperspect.a014357.

Cox, A. D., Fesik, S. W., Kimmelman, A. C., Luo, J., & Der, C. J. (2014). Drugging the undruggable RAS: Mission possible? *Nature Reviews. Drug Discovery*, *13*(11), 828–851. https://doi.org/10.1038/nrd4389.

Croce, C. M. (2009). Causes and consequences of microRNA dysregulation in cancer. *Nature Reviews. Genetics*, *10*(10), 704–714. https://doi.org/10.1038/nrg2634.

Dalla-Favera, R., Bregni, M., Erikson, J., Patterson, D., Gallo, R. C., & Croce, C. M. (1982). Human c-myc onc gene is located on the region of chromosome 8 that is translocated in Burkitt lymphoma cells. *Proceedings of the National Academy of Sciences of the United States of America*, *79*(24), 7824–7827.

Dalton, G. D., & Dewey, W. L. (2006). Protein kinase inhibitor peptide (PKI): A family of endogenous neuropeptides that modulate neuronal cAMP-dependent protein kinase function. *Neuropeptides*, *40*(1), 23–34. https://doi.org/10.1016/j.npep.2005.10.002.

Dalton, G. D., Smith, F. L., Smith, P. A., & Dewey, W. L. (2005). Alterations in brain protein kinase A activity and reversal of morphine tolerance by two fragments of native protein kinase A inhibitor peptide (PKI). *Neuropharmacology*, *48*(5), 648–657. https://doi.org/10.1016/j.neuropharm.2004.12.006.

Dang, C. V. (2012). MYC on the path to cancer. *Cell, 149*(1), 22–35. https://doi.org/10.1016/j.cell.2012.03.003.
Dang, C. V. (2013). MYC, metabolism, cell growth, and tumorigenesis. *Cold Spring Harbor Perspectives in Medicine, 3*(8). https://doi.org/10.1101/cshperspect.a014217.
Dema, A., Perets, E., Schulz, M. S., Deak, V. A., & Klussmann, E. (2015). Pharmacological targeting of AKAP-directed compartmentalized cAMP signalling. *Cellular Signalling, 27*(12), 2474–2487. https://doi.org/10.1016/j.cellsig.2015.09.008.
Dharmaiah, S., Bindu, L., Tran, T. H., Gillette, W. K., Frank, P. H., Ghirlando, R., et al. (2016). Structural basis of recognition of farnesylated and methylated KRAS4b by PDEdelta. *Proceedings of the National Academy of Sciences of the United States of America, 113*(44), E6766–E6775. https://doi.org/10.1073/pnas.1615316113.
Diller, T. C., Madhusudan, Xuong, N. H., & Taylor, S. S. (2001). Molecular basis for regulatory subunit diversity in cAMP-dependent protein kinase: Crystal structure of the type II beta regulatory subunit. *Structure, 9*(1), 73–82 [doi:S0969212600005566 [pii]].
Dorsam, R. T., & Gutkind, J. S. (2007). G-protein-coupled receptors and cancer. *Nature Reviews. Cancer,* 7(2), 79–94. https://doi.org/10.1038/nrc2069.
Duesberg, P. H., Bister, K., & Vogt, P. K. (1977). The RNA of avian acute leukemia virus MC29. *Proceedings of the National Academy of Sciences of the United States of America, 74*(10), 4320–4324.
Dumaz, N., & Marais, R. (2005). Integrating signals between cAMP and the RAS/RAF/MEK/ERK signalling pathways. Based on the anniversary prize of the Gesellschaft fur Biochemie und Molekularbiologie Lecture delivered on 5 July 2003 at the Special FEBS Meeting in Brussels. *The FEBS Journal, 272*(14), 3491–3504. https://doi.org/10.1111/j.1742-4658.2005.04763.x.
Eccles, R. L., Czajkowski, M. T., Barth, C., Muller, P. M., McShane, E., Grunwald, S., et al. (2016). Bimodal antagonism of PKA signalling by ARHGAP36. *Nature Communications,* 7, 12963. https://doi.org/10.1038/ncomms12963.
Eilers, M., & Eisenman, R. N. (2008). Myc's broad reach. *Genes & Development, 22*(20), 2755–2766. https://doi.org/10.1101/gad.1712408.
Eisenman, R. N. (2001). Deconstructing myc. *Genes & Development, 15*(16), 2023–2030. https://doi.org/10.1101/gad928101.
Espiard, S., Ragazzon, B., & Bertherat, J. (2014). Protein kinase a alterations in adrenocortical tumors. *Hormone and Metabolic Research, 46*(12), 869–875. https://doi.org/10.1055/s-0034-1385908.
Evan, G. I., Wyllie, A. H., Gilbert, C. S., Littlewood, T. D., Land, H., Brooks, M., et al. (1992). Induction of apoptosis in fibroblasts by c-myc protein. *Cell, 69*(1), 119–128.
Fieber, W., Schneider, M. L., Matt, T., Krautler, B., Konrat, R., & Bister, K. (2001). Structure, function, and dynamics of the dimerization and DNA-binding domain of oncogenic transcription factor v-Myc. *Journal of Molecular Biology, 307*(5), 1395–1410. https://doi.org/10.1006/jmbi.2001.4537.
Fischer, G., Rossmann, M., & Hyvonen, M. (2015). Alternative modulation of protein-protein interactions by small molecules. *Current Opinion in Biotechnology, 35*, 78–85. https://doi.org/10.1016/j.copbio.2015.04.006.
Fisher, T. L., & White, M. F. (2004). Signaling pathways: The benefits of good communication. *Current Biology, 14*(23), R1005–1007.
Fleuren, E. D., Zhang, L., Wu, J., & Daly, R. J. (2016). The kinome 'at large' in cancer. *Nature Reviews. Cancer, 16*(2), 83–98. https://doi.org/10.1038/nrc.2015.18.
Futreal, P. A., Coin, L., Marshall, M., Down, T., Hubbard, T., Wooster, R., et al. (2004). A census of human cancer genes. *Nature Reviews. Cancer, 4*(3), 177–183. https://doi.org/10.1038/nrc1299.
Gerits, N., Kostenko, S., Shiryaev, A., Johannessen, M., & Moens, U. (2008). Relations between the mitogen-activated protein kinase and the cAMP-dependent protein kinase

pathways: Comradeship and hostility. *Cellular Signalling*, *20*(9), 1592–1607. https://doi.org/10.1016/j.cellsig.2008.02.022.

Gonzalez-Perez, V., Reiner, D. J., Alan, J. K., Mitchell, C., Edwards, L. J., Khazak, V., et al. (2010). Genetic and functional characterization of putative Ras/Raf interaction inhibitors in *C. elegans* and mammalian cells. *Journal of Molecular Signaling*, *5*, 2. https://doi.org/10.1186/1750-2187-5-2.

Good, M. C., Zalatan, J. G., & Lim, W. A. (2011). Scaffold proteins: Hubs for controlling the flow of cellular information. *Science*, *332*(6030), 680–686. https://doi.org/10.1126/science.1198701.

Greenwald, E. C., & Saucerman, J. J. (2011). Bigger, better, faster: Principles and models of AKAP anchoring protein signaling. *Journal of Cardiovascular Pharmacology*, *58*(5), 462–469. https://doi.org/10.1097/FJC.0b013e31822001e3.

Halaban, R., & Krauthammer, M. (2016). RASopathy gene mutations in melanoma. *The Journal of Investigative Dermatology*, *136*(9), 1755–1759. https://doi.org/10.1016/j.jid.2016.05.095.

Hall, B. E., Bar-Sagi, D., & Nassar, N. (2002). The structural basis for the transition from Ras-GTP to Ras-GDP. *Proceedings of the National Academy of Sciences of the United States of America*, *99*(19), 12138–12142. https://doi.org/10.1073/pnas.192453199.

Hart, J. R., Garner, A. L., Yu, J., Ito, Y., Sun, M., Ueno, L., et al. (2014). Inhibitor of MYC identified in a Krohnke pyridine library. *Proceedings of the National Academy of Sciences of the United States of America*, *111*(34), 12556–12561. https://doi.org/10.1073/pnas.1319488111.

Hart, J. R., Roberts, T. C., Weinberg, M. S., Morris, K. V., & Vogt, P. K. (2014). MYC regulates the non-coding transcriptome. *Oncotarget*, *5*(24), 12543–12554. https://doi.org/10.18632/oncotarget.3033.

Herberg, F. W., Taylor, S. S., & Dostmann, W. R. (1996). Active site mutations define the pathway for the cooperative activation of cAMP-dependent protein kinase. *Biochemistry*, *35*(9), 2934–2942. https://doi.org/10.1021/bi951647c.

Holderfield, M., Deuker, M. M., McCormick, F., & McMahon, M. (2014). Targeting RAF kinases for cancer therapy: BRAF-mutated melanoma and beyond. *Nature Reviews. Cancer*, *14*(7), 455–467. https://doi.org/10.1038/nrc3760.

Honeyman, J. N., Simon, E. P., Robine, N., Chiaroni-Clarke, R., Darcy, D. G., Lim, I. I., et al. (2014). Detection of a recurrent DNAJB1-PRKACA chimeric transcript in fibrolamellar hepatocellular carcinoma. *Science*, *343*(6174), 1010–1014. https://doi.org/10.1126/science.1249484.

Horvath, A., Bertherat, J., Groussin, L., Guillaud-Bataille, M., Tsang, K., Cazabat, L., et al. (2010). Mutations and polymorphisms in the gene encoding regulatory subunit type 1-alpha of protein kinase A (PRKAR1A): An update. *Human Mutation*, *31*(4), 369–379. https://doi.org/10.1002/humu.21178.

Hu, J., Ahuja, L. G., Meharena, H. S., Kannan, N., Kornev, A. P., Taylor, S. S., et al. (2015). Kinase regulation by hydrophobic spine assembly in cancer. *Molecular and Cellular Biology*, *35*(1), 264–276. https://doi.org/10.1128/MCB.00943-14.

Iglesias-Bartolome, R., Torres, D., Marone, R., Feng, X., Martin, D., Simaan, M., et al. (2015). Inactivation of a Galpha(s)-PKA tumour suppressor pathway in skin stem cells initiates basal-cell carcinogenesis. *Nature Cell Biology*, *17*(6), 793–803. https://doi.org/10.1038/ncb3164.

Ivanov, A. A., Khuri, F. R., & Fu, H. (2013). Targeting protein-protein interactions as an anticancer strategy. *Trends in Pharmacological Sciences*, *34*(7), 393–400. https://doi.org/10.1016/j.tips.2013.04.007.

Jansen, H. W., Lurz, R., Bister, K., Bonner, T. I., Mark, G. E., & Rapp, U. R. (1984). Homologous cell-derived oncogenes in avian carcinoma virus MH2 and murine sarcoma virus 3611. *Nature*, *307*(5948), 281–284.

Jansen, H. W., Ruckert, B., Lurz, R., & Bister, K. (1983). Two unrelated cell-derived sequences in the genome of avian leukemia and carcinoma inducing retrovirus MH2. *The EMBO Journal*, *2*(11), 1969–1975.

Jubb, H., Higueruelo, A. P., Winter, A., & Blundell, T. L. (2012). Structural biology and drug discovery for protein-protein interactions. *Trends in Pharmacological Sciences*, *33*(5), 241–248. https://doi.org/10.1016/j.tips.2012.03.006.

Kandel, E. R. (2001). The molecular biology of memory storage: A dialogue between genes and synapses. *Science*, *294*(5544), 1030–1038. https://doi.org/10.1126/science.1067020.

Karnoub, A. E., & Weinberg, R. A. (2008). Ras oncogenes: Split personalities. *Nature Reviews. Molecular Cell Biology*, *9*(7), 517–531. https://doi.org/10.1038/nrm2438.

Kato-Stankiewicz, J., Hakimi, I., Zhi, G., Zhang, J., Serebriiskii, I., Guo, L., et al. (2002). Inhibitors of Ras/Raf-1 interaction identified by two-hybrid screening revert Ras-dependent transformation phenotypes in human cancer cells. *Proceedings of the National Academy of Sciences of the United States of America*, *99*(22), 14398–14403. https://doi.org/10.1073/pnas.222222699.

Keeton, A. B., Salter, E. A., & Piazza, G. A. (2017). The RAS-effector interaction as a drug target. *Cancer Research*, *77*(2), 221–226. https://doi.org/10.1158/0008-5472.CAN-16-0938.

Kennedy, E. J., & Scott, J. D. (2015). Selective disruption of the AKAP signaling complexes. *Methods in Molecular Biology*, *1294*, 137–150. https://doi.org/10.1007/978-1-4939-2537-7_11.

Kim, C., Xuong, N. H., & Taylor, S. S. (2005). Crystal structure of a complex between the catalytic and regulatory (RIalpha) subunits of PKA. *Science*, *307*(5710), 690–696. https://doi.org/10.1126/science.1104607.

Kinderman, F. S., Kim, C., von Daake, S., Ma, Y., Pham, B. Q., Spraggon, G., et al. (2006). A dynamic mechanism for AKAP binding to RII isoforms of cAMP-dependent protein kinase. *Molecular Cell*, *24*(3), 397–408. https://doi.org/10.1016/j.molcel.2006.09.015.

Landis, C. A., Masters, S. B., Spada, A., Pace, A. M., Bourne, H. R., & Vallar, L. (1989). GTPase inhibiting mutations activate the alpha chain of Gs and stimulate adenylyl cyclase in human pituitary tumours. *Nature*, *340*(6236), 692–696. https://doi.org/10.1038/340692a0.

Langeberg, L. K., & Scott, J. D. (2015). Signalling scaffolds and local organization of cellular behaviour. *Nature Reviews. Molecular Cell Biology*, *16*(4), 232–244. https://doi.org/10.1038/nrm3966.

Lavoie, H., & Therrien, M. (2015). Regulation of RAF protein kinases in ERK signalling. *Nature Reviews. Molecular Cell Biology*, *16*(5), 281–298. https://doi.org/10.1038/nrm3979.

Lee, L. C., Maurice, D. H., & Baillie, G. S. (2013). Targeting protein-protein interactions within the cyclic AMP signaling system as a therapeutic strategy for cardiovascular disease. *Future Medicinal Chemistry*, *5*(4), 451–464. https://doi.org/10.4155/fmc.12.216.

Lignitto, L., Arcella, A., Sepe, M., Rinaldi, L., Delle Donne, R., Gallo, A., et al. (2013). Proteolysis of MOB1 by the ubiquitin ligase praja2 attenuates hippo signalling and supports glioblastoma growth. *Nature Communications*, *4*, 1822. https://doi.org/10.1038/ncomms2791.

Lignitto, L., Carlucci, A., Sepe, M., Stefan, E., Cuomo, O., Nistico, R., et al. (2011). Control of PKA stability and signalling by the RING ligase praja2. *Nature Cell Biology*, *13*(4), 412–422. https://doi.org/10.1038/ncb2209.

Lim, S. M., Westover, K. D., Ficarro, S. B., Harrison, R. A., Choi, H. G., Pacold, M. E., et al. (2014). Therapeutic targeting of oncogenic K-Ras by a covalent catalytic site inhibitor. *Angewandte Chemie (International Ed. in English)*, *53*(1), 199–204. https://doi.org/10.1002/anie.201307387.

Linglart, A., Menguy, C., Couvineau, A., Auzan, C., Gunes, Y., Cancel, M., et al. (2011). Recurrent PRKAR1A mutation in acrodysostosis with hormone resistance. *The New England Journal of Medicine, 364*(23), 2218–2226. https://doi.org/10.1056/NEJMoa1012717.

Lissandron, V., & Zaccolo, M. (2006). Compartmentalized cAMP/PKA signalling regulates cardiac excitation-contraction coupling. *Journal of Muscle Research and Cell Motility, 27*(5–7), 399–403. https://doi.org/10.1007/s10974-006-9077-2.

Lodish, M., & Stratakis, C. A. (2016). A genetic and molecular update on adrenocortical causes of cushing syndrome. *Nature Reviews. Endocrinology, 12*(5), 255–262. https://doi.org/10.1038/nrendo.2016.24.

London, N., Raveh, B., & Schueler-Furman, O. (2013). Druggable protein-protein interactions—From hot spots to hot segments. *Current Opinion in Chemical Biology, 17*(6), 952–959. https://doi.org/10.1016/j.cbpa.2013.10.011.

Matallanas, D., Birtwistle, M., Romano, D., Zebisch, A., Rauch, J., von Kriegsheim, A., et al. (2011). Raf family kinases: Old dogs have learned new tricks. *Genes & Cancer, 2*(3), 232–260. https://doi.org/10.1177/1947601911407323.

McCormick, F. (2011). Cancer therapy based on oncogene addiction. *Journal of Surgical Oncology, 103*(6), 464–467. https://doi.org/10.1002/jso.21749.

McCormick, F. (2016). K-Ras protein as a drug target. *Journal of Molecular Medicine (Berlin), 94*(3), 253–258. https://doi.org/10.1007/s00109-016-1382-7.

Moodie, S. A., Paris, M., Villafranca, E., Kirshmeier, P., Willumsen, B. M., & Wolfman, A. (1995). Different structural requirements within the switch II region of the Ras protein for interactions with specific downstream targets. *Oncogene, 11*(3), 447–454.

Morelli, X., Bourgeas, R., & Roche, P. (2011). Chemical and structural lessons from recent successes in protein-protein interaction inhibition (2P2I). *Current Opinion in Chemical Biology, 15*(4), 475–481. https://doi.org/10.1016/j.cbpa.2011.05.024.

Mukhopadhyay, S., Wen, X., Ratti, N., Loktev, A., Rangell, L., Scales, S. J., et al. (2013). The ciliary G-protein-coupled receptor Gpr161 negatively regulates the sonic hedgehog pathway via cAMP signaling. *Cell, 152*(1–2), 210–223. https://doi.org/10.1016/j.cell.2012.12.026.

Muller, O., Gourzoulidou, E., Carpintero, M., Karaguni, I. M., Langerak, A., Herrmann, C., et al. (2004). Identification of potent Ras signaling inhibitors by pathway-selective phenotype-based screening. *Angewandte Chemie (International Ed. in English), 43*(4), 450–454. https://doi.org/10.1002/anie.200352587.

Murray, A. J. (2008). Pharmacological PKA inhibition: All may not be what it seems. *Science Signaling, 1*(22), re4. https://doi.org/10.1126/scisignal.122re4.

Nair, S. K., & Burley, S. K. (2003). X-ray structures of Myc-Max and Mad-Max recognizing DNA. Molecular bases of regulation by proto-oncogenic transcription factors. *Cell, 112*(2), 193–205.

Neves, S. R., Ram, P. T., & Iyengar, R. (2002). G protein pathways. *Science, 296*(5573), 1636–1639. https://doi.org/10.1126/science.1071550, 296/5573/1636 [pii].

Niemeyer, C. M. (2014). RAS diseases in children. *Haematologica, 99*(11), 1653–1662. https://doi.org/10.3324/haematol.2014.114595.

O'Hayre, M., Degese, M. S., & Gutkind, J. S. (2014). Novel insights into G protein and G protein-coupled receptor signaling in cancer. *Current Opinion in Cell Biology, 27*, 126–135. https://doi.org/10.1016/j.ceb.2014.01.005.

O'Hayre, M., Vazquez-Prado, J., Kufareva, I., Stawiski, E. W., Handel, T. M., Seshagiri, S., et al. (2013). The emerging mutational landscape of G proteins and G-protein-coupled receptors in cancer. *Nature Reviews. Cancer, 13*(6), 412–424. https://doi.org/10.1038/nrc3521.

Ostrem, J. M., Peters, U., Sos, M. L., Wells, J. A., & Shokat, K. M. (2013). K-Ras(G12C) inhibitors allosterically control GTP affinity and effector interactions. *Nature, 503*(7477), 548–551. https://doi.org/10.1038/nature12796.

Papke, B., & Der, C. J. (2017). Drugging RAS: Know the enemy. *Science, 355*(6330), 1158–1163. https://doi.org/10.1126/science.aam7622.

Parada, L. F., Tabin, C. J., Shih, C., & Weinberg, R. A. (1982). Human EJ bladder carcinoma oncogene is homologue of Harvey sarcoma virus ras gene. *Nature, 297*(5866), 474–478.

Patricelli, M. P., Janes, M. R., Li, L. S., Hansen, R., Peters, U., Kessler, L. V., et al. (2016). Selective inhibition of oncogenic KRAS output with small molecules targeting the inactive state. *Cancer Discovery, 6*(3), 316–329. https://doi.org/10.1158/2159-8290.CD-15-1105.

Pierce, K. L., Premont, R. T., & Lefkowitz, R. J. (2002). Seven-transmembrane receptors. *Nature Reviews. Molecular Cell Biology, 3*(9), 639–650. https://doi.org/10.1038/nrm908, nrm908 [pii].

Popovic, D., Vucic, D., & Dikic, I. (2014). Ubiquitination in disease pathogenesis and treatment. *Nature Medicine, 20*(11), 1242–1253. https://doi.org/10.1038/nm.3739.

Prior, I. A., Lewis, P. D., & Mattos, C. (2012). A comprehensive survey of Ras mutations in cancer. *Cancer Research, 72*(10), 2457–2467. https://doi.org/10.1158/0008-5472.CAN-11-2612.

Raffeiner, P., Rock, R., Schraffl, A., Hartl, M., Hart, J. R., Janda, K. D., et al. (2014). In vivo quantification and perturbation of Myc-Max interactions and the impact on oncogenic potential. *Oncotarget, 5*(19), 8869–8878.

Rahl, P. B., & Young, R. A. (2014). MYC and transcription elongation. *Cold Spring Harbor Perspectives in Medicine, 4*(1), a020990. https://doi.org/10.1101/cshperspect.a020990.

Rapp, U. R., Goldsborough, M. D., Mark, G. E., Bonner, T. I., Groffen, J., Reynolds, F. H., Jr., et al. (1983). Structure and biological activity of v-raf, a unique oncogene transduced by a retrovirus. *Proceedings of the National Academy of Sciences of the United States of America, 80*(14), 4218–4222.

Rauen, K. A. (2013). The RASopathies. *Annual Review of Genomics and Human Genetics, 14*, 355–369. https://doi.org/10.1146/annurev-genom-091212-153523.

Rinaldi, L., Sepe, M., Donne, R. D., & Feliciello, A. (2015). A dynamic interface between ubiquitylation and cAMP signaling. *Frontiers in Pharmacology, 6*, 177. https://doi.org/10.3389/fphar.2015.00177.

Ritt, D. A., Abreu-Blanco, M. T., Bindu, L., Durrant, D. E., Zhou, M., Specht, S. I., et al. (2016). Inhibition of Ras/Raf/MEK/ERK pathway signaling by a stress-induced phospho-regulatory circuit. *Molecular Cell, 64*(5), 875–887. https://doi.org/10.1016/j.molcel.2016.10.029.

Roskoski, R., Jr. (2015). A historical overview of protein kinases and their targeted small molecule inhibitors. *Pharmacological Research, 100*, 1–23. https://doi.org/10.1016/j.phrs.2015.07.010.

Sato, Y., Maekawa, S., Ishii, R., Sanada, M., Morikawa, T., Shiraishi, Y., et al. (2014). Recurrent somatic mutations underlie corticotropin-independent Cushing's syndrome. *Science, 344*(6186), 917–920. https://doi.org/10.1126/science.1252328.

Scott, D. E., Bayly, A. R., Abell, C., & Skidmore, J. (2016). Small molecules, big targets: Drug discovery faces the protein-protein interaction challenge. *Nature Reviews. Drug Discovery, 15*(8), 533–550. https://doi.org/10.1038/nrd.2016.29.

Scott, J. D., Dessauer, C. W., & Tasken, K. (2013). Creating order from chaos: Cellular regulation by kinase anchoring. *Annual Review of Pharmacology and Toxicology, 53*, 187–210. https://doi.org/10.1146/annurev-pharmtox-011112-140204.

Scott, J. D., Fischer, E. H., Demaille, J. G., & Krebs, E. G. (1985). Identification of an inhibitory region of the heat-stable protein inhibitor of the cAMP-dependent protein kinase. *Proceedings of the National Academy of Sciences of the United States of America, 82*(13), 4379–4383.

Scott, J. D., & Pawson, T. (2009). Cell signaling in space and time: Where proteins come together and when they're apart. *Science, 326*(5957), 1220–1224. https://doi.org/10.1126/science.1175668.

Shih, T. Y., Weeks, M. O., Young, H. A., & Scolnick, E. M. (1979). Identification of a sarcoma virus-coded phosphoprotein in nonproducer cells transformed by Kirsten or Harvey murine sarcoma virus. *Virology*, *96*(1), 64–79.

Shim, H., Dolde, C., Lewis, B. C., Wu, C. S., Dang, G., Jungmann, R. A., et al. (1997). C-Myc transactivation of LDH-A: Implications for tumor metabolism and growth. *Proceedings of the National Academy of Sciences of the United States of America*, *94*(13), 6658–6663.

Shima, F., Yoshikawa, Y., Ye, M., Araki, M., Matsumoto, S., Liao, J., et al. (2013). In silico discovery of small-molecule Ras inhibitors that display antitumor activity by blocking the Ras-effector interaction. *Proceedings of the National Academy of Sciences of the United States of America*, *110*(20), 8182–8187. https://doi.org/10.1073/pnas.1217730110.

Shirouzu, M., Koide, H., Fujita-Yoshigaki, J., Oshio, H., Toyama, Y., Yamasaki, K., et al. (1994). Mutations that abolish the ability of Ha-Ras to associate with Raf-1. *Oncogene*, *9*(8), 2153–2157.

Shortt, J., & Johnstone, R. W. (2012). Oncogenes in cell survival and cell death. *Cold Spring Harbor Perspectives in Biology*, *4*(12). https://doi.org/10.1101/cshperspect.a009829.

Shu, S., & Polyak, K. (2017). BET Bromodomain proteins as cancer therapeutic targets. *Cold Spring Harbor Symposia on Quantitative Biology*, *81*, 123–129. https://doi.org/10.1101/sqb.2016.81.030908.

Silve, C., Clauser, E., & Linglart, A. (2012). Acrodysostosis. *Hormone and Metabolic Research*, *44*(10), 749–758. https://doi.org/10.1055/s-0032-1316330.

Skroblin, P., Grossmann, S., Schafer, G., Rosenthal, W., & Klussmann, E. (2010). Mechanisms of protein kinase A anchoring. *International Review of Cell and Molecular Biology*, *283*, 235–330. https://doi.org/10.1016/S1937-6448(10)83005-9.

Smith, M. C., & Gestwicki, J. E. (2012). Features of protein-protein interactions that translate into potent inhibitors: Topology, surface area and affinity. *Expert Reviews in Molecular Medicine*, *14*, e16. https://doi.org/10.1017/erm.2012.10.

Smith, C. M., Radzio-Andzelm, E., Madhusudan, Akamine, P., & Taylor, S. S. (1999). The catalytic subunit of cAMP-dependent protein kinase: Prototype for an extended network of communication. *Progress in Biophysics and Molecular Biology*, *71*(3–4), 313–341.

Solberg, R., Tasken, K., Wen, W., Coghlan, V. M., Meinkoth, J. L., Scott, J. D., et al. (1994). Human regulatory subunit RI beta of cAMP-dependent protein kinases: Expression, holoenzyme formation and microinjection into living cells. *Experimental Cell Research*, *214*(2), 595–605. https://doi.org/10.1006/excr.1994.1297.

Stefan, E., & Bister, K. (2017). MYC and RAF: Key effectors in cellular signaling and major drivers in human cancer. *Current Topics in Microbiology and Immunology*, *407*, 117–151. https://doi.org/10.1007/82_2017_4.

Stephen, A. G., Esposito, D., Bagni, R. K., & McCormick, F. (2014). Dragging ras back in the ring. *Cancer Cell*, *25*(3), 272–281. https://doi.org/10.1016/j.ccr.2014.02.017.

Stine, Z. E., Walton, Z. E., Altman, B. J., Hsieh, A. L., & Dang, C. V. (2015). MYC, metabolism, and cancer. *Cancer Discovery*, *5*(10), 1024–1039. https://doi.org/10.1158/2159-8290.CD-15-0507.

Stork, P. J., & Schmitt, J. M. (2002). Crosstalk between cAMP and MAP kinase signaling in the regulation of cell proliferation. *Trends in Cell Biology*, *12*(6), 258–266. https://doi.org/10.1016/S0962-8924(02)02294-8.

Stratakis, C. A. (2013). cAMP/PKA signaling defects in tumors: Genetics and tissue-specific pluripotential cell-derived lesions in human and mouse. *Molecular and Cellular Endocrinology*, *371*(1–2), 208–220. https://doi.org/10.1016/j.mce.2013.01.015.

Tanaka, T., & Rabbitts, T. H. (2010). Interfering with RAS-effector protein interactions prevent RAS-dependent tumour initiation and causes stop-start control of cancer growth. *Oncogene*, *29*(45), 6064–6070. https://doi.org/10.1038/onc.2010.346.

Tasken, K., & Aandahl, E. M. (2004). Localized effects of cAMP mediated by distinct routes of protein kinase A. *Physiological Reviews*, *84*(1), 137–167. https://doi.org/10.1152/physrev.00021.2003.

Taylor, S. S., Ilouz, R., Zhang, P., & Kornev, A. P. (2012). Assembly of allosteric macromolecular switches: Lessons from PKA. *Nature Reviews. Molecular Cell Biology*, *13*(10), 646–658. https://doi.org/10.1038/nrm3432.

Taylor, S. S., & Kornev, A. P. (2011). Protein kinases: Evolution of dynamic regulatory proteins. *Trends in Biochemical Sciences*, *36*(2), 65–77. https://doi.org/10.1016/j.tibs.2010.09.006.

Thaker, P. H., Han, L. Y., Kamat, A. A., Arevalo, J. M., Takahashi, R., Lu, C., et al. (2006). Chronic stress promotes tumor growth and angiogenesis in a mouse model of ovarian carcinoma. *Nature Medicine*, *12*(8), 939–944.

Tokheim, C. J., Papadopoulos, N., Kinzler, K. W., Vogelstein, B., & Karchin, R. (2016). Evaluating the evaluation of cancer driver genes. *Proceedings of the National Academy of Sciences of the United States of America*, *113*(50), 14330–14335. https://doi.org/10.1073/pnas.1616440113.

Torres-Quesada, O., Mayrhofer, J. E., & Stefan, E. (2017). The many faces of compartmentalized PKA signalosomes. *Cellular Signalling*, *37*, 1–11. https://doi.org/10.1016/j.cellsig.2017.05.012.

Turnham, R. E., & Scott, J. D. (2016). Protein kinase A catalytic subunit isoform PRKACA; History, function and physiology. *Gene*, *577*(2), 101–108. https://doi.org/10.1016/j.gene.2015.11.052.

Varmus, H., Unni, A. M., & Lockwood, W. W. (2017). How cancer genomics drives cancer biology: Does synthetic lethality explain mutually exclusive oncogenic mutations? *Cold Spring Harbor Symposia on Quantitative Biology*, *81*, 247–255. https://doi.org/10.1101/sqb.2016.81.030866.

Vita, M., & Henriksson, M. (2006). The Myc oncoprotein as a therapeutic target for human cancer. *Seminars in Cancer Biology*, *16*(4), 318–330. https://doi.org/10.1016/j.semcancer.2006.07.015.

Vogt, P. K. (2012). Retroviral oncogenes: A historical primer. *Nature Reviews. Cancer*, *12*(9), 639–648. https://doi.org/10.1038/nrc3320.

Waldmann, H., Karaguni, I. M., Carpintero, M., Gourzoulidou, E., Herrmann, C., Brockmann, C., et al. (2004). Sulindac-derived Ras pathway inhibitors target the Ras-Raf interaction and downstream effectors in the Ras pathway. *Angewandte Chemie (International Ed. in English)*, *43*(4), 454–458. https://doi.org/10.1002/anie.200353089.

Wang, Y., Ho, T. G., Franz, E., Hermann, J. S., Smith, F. D., Hehnly, H., et al. (2015). PKA-type I selective constrained peptide disruptors of AKAP complexes. *ACS Chemical Biology*, *10*(6), 1502–1510. https://doi.org/10.1021/acschembio.5b00009.

Wang, M. L., Rule, S., Martin, P., Goy, A., Auer, R., Kahl, B. S., et al. (2013). Targeting BTK with ibrutinib in relapsed or refractory mantle-cell lymphoma. *The New England Journal of Medicine*, *369*(6), 507–516. https://doi.org/10.1056/NEJMoa1306220.

Weinstein, L. S., Shenker, A., Gejman, P. V., Merino, M. J., Friedman, E., & Spiegel, A. M. (1991). Activating mutations of the stimulatory G protein in the McCune-Albright syndrome. *The New England Journal of Medicine*, *325*(24), 1688–1695. https://doi.org/10.1056/NEJM199112123252403.

Wellbrock, C., Karasarides, M., & Marais, R. (2004). The RAF proteins take centre stage. *Nature Reviews. Molecular Cell Biology*, *5*(11), 875–885.

Welsch, M. E., Kaplan, A., Chambers, J. M., Stokes, M. E., Bos, P. H., Zask, A., et al. (2017). Multivalent small-molecule Pan-RAS inhibitors. *Cell*, *168*(5), 878–889. e829. https://doi.org/10.1016/j.cell.2017.02.006.

Wennerberg, K., Rossman, K. L., & Der, C. J. (2005). The Ras superfamily at a glance. *Journal of Cell Science*, *118*(Pt. 5), 843–846. https://doi.org/10.1242/jcs.01660.
Wong, T. H., Chiu, W. Z., Breedveld, G. J., Li, K. W., Verkerk, A. J., Hondius, D., et al. (2014). PRKAR1B mutation associated with a new neurodegenerative disorder with unique pathology. *Brain*, *137*(Pt. 5), 1361–1373. https://doi.org/10.1093/brain/awu067.
Wong, W., & Scott, J. D. (2004). AKAP signalling complexes: Focal points in space and time. *Nature Reviews. Molecular Cell Biology*, *5*(12), 959–970.
Wu, J., Brown, S. H., von Daake, S., & Taylor, S. S. (2007). PKA type IIalpha holoenzyme reveals a combinatorial strategy for isoform diversity. *Science*, *318*(5848), 274–279. https://doi.org/10.1126/science.1146447.
Yin, X., Giap, C., Lazo, J. S., & Prochownik, E. V. (2003). Low molecular weight inhibitors of Myc-Max interaction and function. *Oncogene*, *22*(40), 6151–6159. https://doi.org/10.1038/sj.onc.1206641.
Yu, P., Chen, Y., Tagle, D. A., & Cai, T. (2002). PJA1, encoding a RING-H2 finger ubiquitin ligase, is a novel human X chromosome gene abundantly expressed in brain. *Genomics*, *79*(6), 869–874. https://doi.org/10.1006/geno.2002.6770.
Zawadzki, K. M., & Taylor, S. S. (2004). cAMP-dependent protein kinase regulatory subunit type IIbeta: Active site mutations define an isoform-specific network for allosteric signaling by cAMP. *The Journal of Biological Chemistry*, *279*(8), 7029–7036. https://doi.org/10.1074/jbc.M310804200.
Zebisch, A., Czernilofsky, A. P., Keri, G., Smigelskaite, J., Sill, H., & Troppmair, J. (2007). Signaling through RAS-RAF-MEK-ERK: From basics to bedside. *Current Medicinal Chemistry*, *14*(5), 601–623.
Zebisch, A., & Troppmair, J. (2006). Back to the roots: The remarkable RAF oncogene story. *Cellular and Molecular Life Sciences*, *63*(11), 1314–1330.
Zeller, K. I., Zhao, X., Lee, C. W., Chiu, K. P., Yao, F., Yustein, J. T., et al. (2006). Global mapping of c-Myc binding sites and target gene networks in human B cells. *Proceedings of the National Academy of Sciences of the United States of America*, *103*(47), 17834–17839. https://doi.org/10.1073/pnas.0604129103.
Zilbermint, M., & Stratakis, C. A. (2015). Protein kinase A defects and cortisol-producing adrenal tumors. *Current Opinion in Endocrinology, Diabetes, and Obesity*, *22*(3), 157–162. https://doi.org/10.1097/MED.0000000000000149.
Zimmermann, G., Papke, B., Ismail, S., Vartak, N., Chandra, A., Hoffmann, M., et al. (2013). Small molecule inhibition of the KRAS-PDEdelta interaction impairs oncogenic KRAS signalling. *Nature*, *497*(7451), 638–642. https://doi.org/10.1038/nature12205.

CHAPTER FOUR

Multifaceted Nucleolin Protein and Its Molecular Partners in Oncogenesis

Iva Ugrinova*[,1], Maria Petrova*, Mounira Chalabi-Dchar[†], Philippe Bouvet[†,‡]

*"Roumen Tsanev" Institute of Molecular Biology, Bulgarian Academy of Sciences, Sofia, Bulgaria
[†]Université de Lyon, Centre de Recherche en Cancérologie de Lyon, Centre Léon Bérard, Lyon, France
[‡]Université de Lyon, Ecole Normale Supérieure de Lyon, Lyon, France
[1]Corresponding author: e-mail address: ugryiva@gmail.com

Contents

Abstract

Discovered in 1973, nucleolin is one of the most abundant phosphoproteins of the nucleolus. The ability of nucleolin to be involved in many cellular processes is probably related to its structural organization and its capability to form many different interactions with other proteins. Many functions of nucleolin affect cellular processes involved in oncogenesis—for instance: in ribosome biogenesis; in DNA repair, remodeling, and genome stability; in cell division and cell survival; in chemokine and growth factor signaling pathways; in angiogenesis and lymphangiogenesis; in epithelial–mesenchymal transition; and in stemness. In this review, we will describe the different functions of nucleolin in oncogenesis through its interaction with other proteins.

Advances in Protein Chemistry and Structural Biology, Volume 111
ISSN 1876-1623
https://doi.org/10.1016/bs.apcsb.2017.08.001
© 2018 Elsevier Inc.
All rights reserved.

1. INTRODUCTION

Discovered in 1973, nucleolin (or C23) is one of the most abundant phosphoproteins of the nucleolus (Orrick, Olson, & Busch, 1973). It is also found in the nucleoplasm and the cytoplasm as well as on the cell membrane to control cellular homeostasis by modulating the proliferation, survival, and apoptosis of cells. Growing evidences have demonstrated that the elevated nucleolin expression is associated with a poor prognosis of patients with different cancers, including gastric cancer, hepatocellular carcinoma (HCC), acute myeloid leukemia, non-small-cell lung cancer, and pancreatic ductal adenocarcinoma (Guo et al., 2014; Marcel et al., 2017; Peng et al., 2010; Qiu et al., 2013; Zhao et al., 2013). The oncogenic effect of nucleolin appears to be multifactorial, which probably reflects the multiple functions of this protein (Abdelmohsen & Gorospe, 2012).

In this review, we will describe the different functions of nucleolin in oncogenesis through the interaction of nucleolin with other proteins.

2. STRUCTURE OF NUCLEOLIN

The ability of nucleolin to be involved in many cellular processes is probably related to its structural organization and its capability to form many different interactions with other proteins. The primary sequence of nucleolin is organized into three main structural domains (Lapeyre & Amalric, 1985):

The N-terminal domain of nucleolin (~300 residues) is rich in acidic glutamate/aspartate sequence repeats whose number varies depending on the species. This region contains multiple posttranslational modification including acetylation (Das et al., 2013) and phosphorylation sites of kinases like CDK1 and CK2 (Belenguer, Caizergues-Ferrer, Labbe, Doree, & Amalric, 1990). This domain is also the site of multiple protein–protein interactions such as interaction with histone H1 (for chromatin decondensation) (Erard, Belenguer, Caizergues-Ferrer, Pantaloni, & Amalric, 1988) and nucleophosmin, another abundant nucleolar protein (Lischwe, Smetana, Olson, & Busch, 1979).

The central domain of nucleolin contains four RNA-binding domains (RBDs, also known as RRMs) that allow the specific interactions of nucleolin with several nucleic acid species including mRNAs and pre-rRNA (Ghisolfi-Nieto, Joseph, Puvion-Dutilleul, Amalric, & Bouvet,

1996). The number of RBDs varies from yeast to human. However, knockout experiment suggests that these domains harbor redundant functions and that specific RNA-binding activity may not be required for the vital function of nucleolin (Storck, Thiry, & Bouvet, 2009).

The C-terminal region called GAR (glycine- and arginine-rich) or RGG (Arg-Gly-Gly) domain is rich in glycine, arginine, and phenylalanine residues. This domain from pea nucleolin shows ATP-dependent DNA helicase activities (Nasirudin, Ehtesham, Tuteja, Sopory, & Tuteja, 2005). It is also a protein–protein interaction domain (Bouvet, Diaz, Kindbeiter, Madjar, & Amalric, 1998; Ginisty, Sicard, Roger, & Bouvet, 1999) and is subject to posttranslational methylation modifications affecting nucleolin interaction with nucleic acids (Lischwe, Roberts, Yeoman, & Busch, 1982). Although the length of this region is variable among different species, it is relatively well conserved. Until now, it has not been possible to get a 3D structure of the full protein or of the four RBDs. However, different nuclear magnetic resonance studies have succeeded in the resolution of the structure of different individual RBDs alone or in complex with nucleic acid sequences of hamster and human nucleolin (Allain, Bouvet, Dieckmann, & Feigon, 2000; Arumugam et al., 2010; Bouvet, Allain, Finger, Dieckmann, & Feigon, 2001).

Nucleolin can interact with both nucleic acids and proteins. Using a Tap-tag approach, nuclear proteins that interact with nucleolin have been identified (Salvetti et al., 2016). About 140 proteins copurified with nucleolin, and for most of them, this association was independent of RNA. In addition to many proteins involved in ribosome biogenesis, proteins implicated in pre-mRNA splicing, transport, or stability were identified, suggesting that nuclear nucleolin is also involved in these pathways. It would be interesting to perform the same interactomic approach for cytoplasmic and cell surface nucleolin to identify the full interactome of nucleolin.

This global proteomic approach (Salvetti et al., 2016) suggests that nucleolin may form different complexes that could also explain its numerous functions and subcellular localization.

3. ROLE OF NUCLEOLIN IN RIBOSOME BIOGENESIS AND GENE EXPRESSION IN CANCER CELLS

Ribosome biogenesis is crucial to support cell growth and proliferation. In cancer cells, an increase in metabolism and proliferation requires a higher level of protein synthesis (Marcel, Catez, & Diaz, 2015; Ruggero, 2013). It is therefore expected that the synthesis of ribosome is also increased in these cells. Indeed, the nucleolar structure is usually strongly modified during cell transformation, which can be used by pathologists for diagnosis (Pich, Chiusa, & Margaria, 2000). Expression of several nucleolar proteins (AgNOR proteins) gives a strong nucleolar labeling in cancer cells (Derenzini et al., 2000). Nucleoli structures are results of the different steps of ribosome biogenesis and in particular of the transcription of ribosomal DNA (rDNA). An imbalance of rDNA transcription is closely correlated with tumor progression (Liao et al., 2014). Interestingly, different drugs that target the production of ribosomal RNA have been developed (Pickard & Bierbach, 2013). Among them, CX-3543 selectively disrupts the interaction of nucleolin with rDNA G-quadruplex in nucleoli resulting in the inhibition of RNAPI transcription (Bywater et al., 2012). Additional compounds AS1411 (G-quartet aptamer) (Bates et al., 2017) and HB19/N6L (pseudopeptide) (Destouches et al., 2012) that target nucleolin show also some activity in reducing tumor growth, but it is not clear if this antitumoral activity is through the regulation of nucleolar nucleolin. Nucleolin is associated with the activation of transcription of rDNA genes. Nucleolin interacts with rDNA chromatin and overexpression of nucleolin in human cells is associated with a decrease of heterochromatin marks on rDNA, while euchromatin marks and the rate of RNAPI transcription elongation increase (Cong, Das, & Bouvet, 2011; Cong et al., 2012). In addition, nucleolin possesses a histone chaperone activity and facilitates the transcription through nucleosomal structures (FACT activity) (Angelov et al., 2006; Mongelard & Bouvet, 2007) that probably also contribute to this high transcription rate, without any change in rDNA methylation or in the number of active rDNA genes. Therefore, the increased expression of nucleolin in cancer cells may contribute to the higher production of ribosomal RNA and the production of ribosomes to support the increased metabolism of cancer cells.

In addition to the increased level of nucleolin, its posttranslational modifications may also contribute to the regulation of its activity in RNAPI transcription regulation. PTEN-β, a new isoform of PTEN, a critical tumor suppressor that is frequently mutated in human cancer (Liang et al., 2017), localizes in the nucleolus and interacts with nucleolin. It acts as a nucleolin phosphatase (at Thr84 site) and it is proposed that PTEN-β negatively regulates rDNA transcription and cell proliferation through nucleolin dephosphorylation. In cancer, mutation of PTEN-β may therefore contribute to the deregulated RNAPI transcription function of nucleolin in these cells.

Several reports describe also nucleolin as a key factor for regulation of transcription of RNA polymerase II genes. Although a global approach (ChIP-seq) has not been yet performed, nucleolin was found associated with many promoter regions and to participate to the regulation of transcription. Grinstein and coworkers showed that nucleolin is a key activator of HPV18 oncogene transcription and might be directly linked to tumor development and clinical outcome (Grinstein et al., 2002). Human papilloma viruses (HPVs) play a central role in the development of cervical cancer (zur Hausen & de Villiers, 1994). Nucleolin is expressed in HPV18+ squamous cell carcinomas and adenocarcinomas of the cervix uteri, and its intranuclear localization is abnormal compared with normal cervical epithelial cells (Grinstein et al., 2002). Nucleolin binds to an HPV16 genomic region between nucleotides 604 and 614 in a sequence-dependent manner and might be involved in the dynamics of the HPV genome in cells (Sato et al., 2009).

A transcriptomic analysis of cervical cancer cells (HeLa) depleted in nucleolin was performed to determine which genes and biological pathways could be affected by a deregulation of nucleolin expression (Kumar et al., 2017). It was found that nucleolin expression was required for the expression of many genes involved metabolism and several signaling pathways. In particular, the cholesterol synthesis and fatty acid degradation pathways were greatly affected by the downregulation of nucleolin. These data suggest that nucleolin could be a key factor to connect ribosome biogenesis in the nucleoli with cell metabolism and signaling pathways to support cell proliferation.

4. FUNCTIONS OF NUCLEOLIN IN DNA REPAIR AND GENOME STABILITY

Genomic instability is a characteristic of almost all human cancers. It arises from many different pathways, such as DNA damage from endogenous and exogenous sources, telomere damage, centrosome amplification, and epigenetic modifications (Ferguson et al., 2015). Hanahan and Weinberg have described six functional capabilities of cancers that they called hallmarks of cancer (Hanahan & Weinberg, 2000). Genomic instability was considered separately from them because it is not a functional capability of cancer per se but a property that enables the acquisition of the hallmarks (Hanahan & Weinberg, 2000). Nowadays the expansion of the concept of hallmarks leads to the idea that genomic instability is present in all stages of cancer, from precancerous lesions, even before TP53 mutations are acquired, to advanced cancers (Negrini, Gorgoulis, & Halazonetis, 2010).

Under nonstress conditions, nucleolin is a predominantly nucleolar protein, but in response to exposure to genotoxic stress the protein exhibits rapid, reversible, and posttranslationally regulated nucleoplasmic translocation. For instance, it has been shown that nucleolin translocates from the nucleolus to the nucleoplasm following heat shock (Daniely & Borowiec, 2000). Its relocalization was found to be accompanied by an increase in complex formation between nucleolin and replication protein A (RPA), the primary eukaryotic single-stranded DNA-binding protein (Iftode, Daniely, & Borowiec, 1999). A combination of in vitro and in vivo data demonstrated that nucleolin–RPA complex formation is strongly inhibitory to DNA replication, likely by preventing RPA from active participation in the initiation of DNA synthesis. Nucleolin has also been found to be a stimulatory factor for DNA annealing, suggesting that it may play a role in heteroduplex-mediated DNA repair reactions (Hanakahi, Bu, & Maizels, 2000). Its mobilization also occurs following exposure to ionizing radiation (IR) and treatment with the radiomimetic agent camptothecin (CPT). It turns out that nucleolin mobilization is dependent on the activation of p53 by cellular stress but is independent of the ability of p53 to activate transcription. Based on these data, Daniely and coworkers have proposed a novel p53-dependent mechanism in which cell stress mobilizes nucleolin for transient replication inhibition and DNA repair by interacting with activated p53 (Daniely, Dimitrova, & Borowiec, 2002). Together, these data indicate that nucleolin mobilization occurs to globally modulate DNA metabolism following stress. Nucleolin has been shown to associate with various DNA repair, recombination, and replication proteins and possesses DNA helicase, strand annealing, and strand pairing activities. Examination of nuclear protein extracts from human somatic cells revealed that nucleolin and Rad51 coimmunoprecipitate (De et al., 2006). Furthermore, recombinant Rad51 associates with in vitro translated nucleolin. Electroporation-mediated introduction of antinucleolin antibody resulted in a 10- to 20-fold reduction in intraplasmid homologous recombination (HR) activity in human fibrosarcoma cells. Additionally, introduction of antinucleolin antibody sensitized cells to death induced by the topoisomerase II inhibitor, amsacrine, and also reduced intraplasmid HR activity and induced hypersensitivity to amsacrine-induced cell death. The authors propose the intriguing possibility that nucleolin binding to Rad51 may function to regulate homologous recombinational repair of chromosomal DNA (De et al., 2006). Two different groups have reported that both nucleolin and Werner protein (WRNp), a member of the RecQ helicase family, bind to Rad51 (De et al., 2006;

Otterlei et al., 2006). Indeed, it has been shown that DNA-damaging agents affect the interaction between WRNp and nucleolin. For instance, treatment of U2OS cells with topoisomerase I inhibitor, CPT, causes the dissociation of the nucleolin–Werner complex in the nucleolus, followed by partial reassociation in the nucleoplasm. Other DNA-damaging agents, such as hydroxyurea, mitomycin C, and aphidicolin, did not have these effects. These data suggest that nucleolin may regulate DNA unwinding by WRNp, possibly in response to certain DNA-damaging agents (Indig et al., 2012).

Nucleolin has been also implicated in the repair of double-stranded DNA breaks (DSBs). Both single-stranded DNA gaps (SSGs) and DSBs are created every cell cycle as a consequence of DNA replication through imperfect DNA templates (Kowalczykowski, 2000). DSBs are the most dangerous and potential lethal DNA lesions. Often, they lead to chromosome rearrangements, a crucial step in carcinogenesis. DNA nonhomologous end-joining (NHEJ) is the major DSB repair process, which involves direct juxtaposition and relegation of free DNA ends, providing an efficient, but error-prone process (Ochi, Wu, & Blundell, 2014). Being a compact process that requires a little change to the chromatin in the double-strand break vicinity NHEJ occurs in all cell cycle stages. In mammalian cells, HR requires extensive resection, repositioning of damage response proteins, and engagement of the sister chromatid and functions only in late S/G2 phase HR (Kakarougkas & Jeggo, 2014; Kowalczykowski, 2015).

Studying the components of the H2AX complex, which forms at DSB, Kobayashy et al. identified nucleolin as a candidate of H2AX-interacting protein. Nucleolin was found to accumulate into DSB damage sites in a γH2AX (a phosphorylated form of histone H2AX)-dependent manner. Furthermore, the depletion of nucleolin by siRNA reduced accumulation of several DNA double-strand repair (DDR) proteins and resulted in reduction of ataxia telangiectasia mutated (ATM) kinase-dependent cell cycle checkpoint and DSB repair. The authors suggest that nucleolin could be responsible for MDC1 (mediator of DNA damage checkpoint 1)-dependent histone ubiquitination and other MDC1-related responses, as MDC1 formed a DSB damage-dependent complex with nucleolin. Thus, nucleolin could be important for H2AX/MDC1-related DNA damage responses (Kobayashi et al., 2012). Indeed, several groups also reported that in IR- or CPT-treated cells, nucleolin forms nucleoplasmic foci that colocalize with sites of DSB repair, and returns to the nucleolus upon their clearance (Daniely et al., 2002; Goldstein, Derheimer, Tait-Mulder, & Kastan, 2013; Kobayashi et al., 2012), supporting the hypothesis that nucleolin plays a direct role in the repair and clearance of DSBs. As nucleolin possesses chaperone and FACT-like activity (facilitation of transcription through the nucleosomes thanks to partial unfolding of the nucleosome structure and loss of H2A/H2B dimers) (Angelov et al., 2006; Mongelard & Bouvet, 2007), the γH2AX-dependent recruitment of nucleolin to DNA damage sites could facilitate MDC1-related DNA damage response including HR repair and ATM-mediating cell cycle checkpoint (Kobayashi et al., 2012; Mongelard & Bouvet, 2007). The NHEJ repair pathway, the major DSB repair mechanism in mammalian cells (Goodarzi, Jeggo, & Lobrich, 2010), is dependent on nucleolin-mediated partial nucleosome disruption, whereas a nucleolin-dependent complete nucleosome disruption occurs during the S and G2 phases and is associated with DNA resection, a process that is required for homologous strand invasion and HR (Goldstein et al., 2013).

Growing evidence has begun to provide detailed insights into the potential role of nucleolin at DSBs in vivo. Initial steps in the cellular response to DSBs involve an early recruitment of the MRE11–RAD50–NBS1 (MRN) complex to the DSB site and activation and recruitment of ATM kinase resulting in the activation of multiple signaling pathways (Bakkenist & Kastan, 2003; Ciccia & Elledge, 2010). Interestingly, depletion of nucleolin, either by antibody microinjection or by siRNA knockdown, did not inhibit recruitment of early DSB repair factors such as the MRN complex or γH2AX, but did cause a failure to repair DSBs by both HR and NHEJ (Goldstein et al., 2013; Kobayashi et al., 2012; Scott & Oeffinger, 2016).

Using a ChIP-based assay on a site-specific, NHEJ-targeted DSB, Goldstein et al. (2013) demonstrated that nucleolin is recruited to sites of DSB repair through an interaction of the RGG domain of nucleolin with the RAD50 subunit of the MRN complex, and that while this recruitment has no effect on the loading of DSB repair factors such as γH2AX, ATM, MDC1, RNF8, and RNF168, it did coincide with the rapid depletion of H2A–H2B, but not H3–H4, dimers within a few kilobases of the DSB site. Kobayashi and colleagues similarly observed reduced H2A–H2B mobilization from chromatin to the nucleoplasm upon depletion of nucleolin (Kobayashi et al., 2012). Collectively, these data argue for a model in which nucleolin recruitment to DSBs via the MRN complex facilitates the proximal nucleosome destabilization by chromatin-remodeling complexes, possibly through the actions of MDC1 and its targets RNF8/RNF168, and ultimately facilitates NHEJ- and HR-mediated DSB resolution (Goldstein et al., 2013).

Using GST-pull-down assays and coimmunoprecipitation, Yang and coworkers have shown that recombinant nucleolin interacts with the proliferating cell nuclear antigen (PCNA). This nucleolin–PCNA interaction occurs in intact cells and this interaction increases after exposure of colon carcinoma RKO cells to UV radiation. Furthermore, the overexpression of nucleolin decreases the repair efficiency of UV-damaged plasmid DNA in RKO cells and vice versa, the reduction of the protein levels increases DNA repair efficiency in RKO and CHO cells and consequently increases cell survival. These data indicate that the direct interaction of nucleolin with PCNA inhibits NER efficiency of UV-damaged DNA. This effect could contribute to oncogenesis and aging in cells overexpressing nucleolin (Yang, Kim, Chakravarty, Indig, & Carrier, 2009). However, it is still obscure if the nucleolin–PCNA interaction itself is responsible for nucleolin's modulation of NER or nucleolin regulates NER indirectly through a pathway dependent on PCNA, or through regulation of other PCNA-regulatory NER components such as RPA (Daniely & Borowiec, 2000; Kim et al., 2005). Considerable further work will be required to establish the precise mechanistic roles, if any, of nucleolin in NER.

Despite all these data that suggest direct interaction between nucleolin and different partners participating in DDR cited earlier, the possibility that nucleolin may regulate various DNA repair processes mainly through its ability to promote chromatin remodeling remains open.

5. FUNCTIONS OF NUCLEOLIN IN CELL DIVISION

Nucleolin depletion leads to cell cycle arrest (G2/M in HeLa and G1 in DT40 cells) (Ma et al., 2007; Storck, et al., 2009; Ugrinova et al., 2007). In addition to numerous nuclear alterations, including the presence of micronuclei, multiple nuclei, or large nuclei, a defect in the control of centrosome duplication is observed (Ugrinova et al., 2007). Nearly a century ago, Boveri (1914) proposed that tumors develop as a consequence of chromosomal imbalances, suggesting that centrosome aberrations could be the cause of such imbalances (Boveri, 1914). Nowadays, many human tumors have been described to carry extensive centrosome aberrations, and a strong correlation between the extent of these aberrations and the clinical aggressiveness of tumors has been reported (Lingle, Lukasiewicz, & Salisbury, 2005; Nigg, 2002; Zyss & Gergely, 2009). Bipolar mitotic spindle is important for proper chromatid separation during mitosis. It is worth mentioning that many cells use centrosome-independent spindle assembly mechanisms to cluster extra centrosomes and to allow the formation of bipolar spindles (Gadde & Heald, 2004; Kwon et al., 2008; Quintyne, Reing, Hoffelder, Gollin, & Saunders, 2005; Ring, Hubble, & Kirschner, 1982). Nevertheless, numerical centriole aberrations and centrosome aneuploidy are important sources of chromosomal instability in tumor cells (Nigg, Cajanek, & Arquint, 2014; Pihan, 2013). It has been reported that nucleolin is involved in centrosome misregulation (Gaume et al., 2015; Ugrinova et al., 2007). The knocking down of nucleolin by RNA interference leads to abnormal centrosome amplification and to the formation of multipolar mitotic spindles (Ma et al., 2007; Ugrinova et al., 2007). Several groups have reported experiments suggesting the implication of nucleolin in centrosome regulation. Nucleolin has been associated to human centrosomes (Andersen et al., 2003) and mitotic spindles (Sauer et al., 2005) in different proteomic studies. By immunofluorescence, a phosphorylated form of nucleolin has been detected in mitosis at the spindle poles (Ma et al., 2007). Interestingly, in interphase, nucleolin was found to colocalize only with the mature centrioles and its depletion to induce an amplification of immature centriole markers that were therefore unable to nucleate microtubules (Gaume et al., 2015). Furthermore, nucleolin was coimmunoprecipitated with centrosomal proteins—γ-tubulin and ninein, a protein of the subdistal appendages involved in microtubule anchoring (Gaume et al., 2015). In addition, during microtubule growth phases, nucleolin affects microtubule stabilization by modifying both the speed and lifetime of polymerization and by reducing catastrophe frequency (Gaume et al., 2016).

Altogether, these data suggest that nucleolin is involved in microtubules anchoring to centrosomes in interphase cells and microtubules polymerization (Gaume et al., 2016, 2015). Although the centrosome misregulation is associated with tumorigenicity, the potential effects of centrosomal nucleolin on cancer cell apparition or progression have never been addressed (Berger, Gaume, & Bouvet, 2015).

6. FUNCTIONS OF NUCLEOLIN IN CELL SURVIVAL

The ability of transforming cells to overcome cell death and to evade apoptosis is an important step for cancerogenesis. Nucleolin interacts with diverse mRNAs (Abdelmohsen et al., 2011; Ishimaru et al., 2010; Sengupta, Bandyopadhyay, Fernandes, & Spicer, 2004; Zhang, Tsaprailis, & Bowden, 2008), many of which are involved either directly or indirectly in the regulation of apoptosis and cell survival (Berger et al., 2015).

Overexpression of nucleolin in chronic lymphocytic leukemia cells induces stabilization of Bcl-2 mRNA and Bcl-2 overexpression (Otake et al., 2007; Sengupta et al., 2004). Bcl-2 protein functions as an inhibitor of programmed cell death or apoptosis. It is therefore a prosurvival protein that has an essential function in normal immunity and whose constitutive expression leads to the development of lymphomas. Although transcriptional control of Bcl-2 has been reported, increasing evidences suggest that an important component of Bcl-2 regulation is mediated through posttranscriptional regulation (Willimott & Wagner, 2010). The 5′ and 3′ untranslated region (UTR) of Bcl-2 mRNA contains regulatory elements like the AU-rich elements (ARE) that function in regulating Bcl-2 stability. The first two RBDs of nucleolin were found to interact with a region close to the Bcl-2 ARE region, stabilizing Bcl-2 mRNA, thereby increasing resistance to apoptosis in chronic lymphocytic leukemia cells (Ishimaru et al., 2010; Sengupta et al., 2004).

RNA-binding protein AUF1 has been shown to recruit Bcl-2 mRNA to exosomes for its degradation, while nucleolin seems to protect Bcl-2 ARE from the exosomal decay (Ishimaru et al., 2010). The authors suggest that in normal cells, Bcl-2 mRNA is rendered intrinsically unstable through the action of AUF1 which binds to the ARE and recruits the mRNA to the exosome, thus preventing aberrant overexpression of Bcl-2 protein allowing cells to enter apoptosis in response to stresses. HuR, another RNA-binding protein, binds also the Bcl-2 ARE (different recognition site that nucleolin). Nucleolin and HuR seem to be present in the same ribonucleoparticles and have a synergistic protective effect on the decay of Bcl-2 ARE RNA (Ishimaru et al., 2009).

In addition, several predicted or experimentally proven microRNAs, like miR-15a and miR-16-1, have been found to inhibit Bcl-2 expression in B cells (Lin et al., 2014). Interestingly, nucleolin expression affects also miR-15a expression (Kumar et al., 2017). An upregulation of miR-15a is observed in nucleolin-depleted cells, which probably contributes to the apoptosis observed in these cells that express low level of nucleolin (Ugrinova et al., 2007).

Nucleolin was reported to stabilize mRNA of one other member of the Bcl-2 family—Bcl-xl mRNA also by specifically recognizing the ARE in the 3′ UTR of Bcl-xl mRNA. As a consequence of this interaction, Bcl-xl mRNA is protected from nuclease degradation (Zhang et al., 2008).

The Syk protein, a protein tyrosine kinase, involved in cell survival and that is known to regulate immune cell function interacts with nucleolin (Wang, Childress, & Geahlen, 2014). This interaction induces the phosphorylation of nucleolin on a tyrosine which was shown to enhance the binding of nucleolin for the Bcl-xl mRNA, Interestingly the authors found that the reduction of the level of nucleolin destabilizes the Bcl-xl mRNA and inhibits the ability of Syk to protect cells from apoptosis (Wang et al., 2014). Other evidences for the role of nucleolin as a Bcl-2 mRNA-stabilizing factor came from the reports, showing that different treatments like *trans*-retinoic acid, taxol, or okadaic acid causing nucleolin degradation also promoted Bcl-2 mRNA decay (Otake, Sengupta, Bandyopadhyay, Spicer, & Fernandes, 2004; Sengupta et al., 2004; Wang et al., 2014).

Nucleolin was reported to enhance translation of the prosurvival proteins Akt1 and cyclin-I (Abdelmohsen & Gorospe, 2012) by interacting with G-rich elements present in the coding region and 5′–3′ UTRs (Abdelmohsen et al., 2011). Akt is one of the most active kinases in cancer cells that promote cell survival via the negative regulation of the proapoptotic Bcl-2 homology domain (BH3)-only proteins, which carry out their function by binding to and inactivating the prosurvival Bcl-2 family members (Griffin, Olivier, Pippin, Roberts, & Shankland, 2006; Hers, Vincent, & Tavare, 2011; Manning, 2004). Cyclin-I protein belongs to the highly conserved cyclin family and shows the highest similarity with cyclin G. The precise function of this cyclin has not yet been determined despite that it is considered as antiapoptotic protein (Nakamura et al., 1995). The protein was found highly expressed in breast cancer and its increased level in serum is associated with pancreatic cancer (Landberg et al., 2005; Sun et al., 2007).

Nucleolin is involved in the regulation of p53 synthesis through several mechanisms. This control of p53 protein expression allows tumor cells to escape from apoptosis and promotes survival (Takagi, Absalon, McLure, & Kastan, 2005). First, nucleolin interacts with RPL26 and binds to the 5′ and 3′ UTR of p53 mRNA. This results in the inhibition of translation of p53 after DNA damage (Chen, Guo, & Kastan, 2012; Takagi et al., 2005). The second mechanism of the control of p53 protein is through the regulation of p53 ubiquitination and subsequent degradation. Human double minute 2 (Hdm2) is an E3 ubiquitin ligase that targets the p53 tumor suppressor protein for proteasomal degradation. After cellular stress, like DNA damage, the activity of Hdm2 is reduced, raising p53 levels, which cause growth arrest and the induction of apoptosis or senescence. Interaction of nucleolin with Hdm2 inhibits both p53 ubiquitination and Hdm2 autoubiquitination (Bhatt, d'Avout, Kane, Borowiec, & Saxena, 2012; Saxena, Rorie, Dimitrova, Daniely, & Borowiec, 2006). The function of nucleolin in this process seems to be regulated by phosphorylation, as a phosphorylation-deficient nucleolin-mutant, 6/S*A, where all the six CK2 consensus serine sites residing in the N-terminus of nucleolin were mutated to alanine, stabilizes p53, results in higher expression of Bcl-2, and causes a dominant-negative effect on cell viability (Xiao et al., 2014).

An activation-resistant Fas complex in B-cell lymphomas, which include nucleolin, and that is absent in B-lymphocytes from healthy donors has been characterized (Wise et al., 2013). FAS (also known as APO-1) belongs to the tumor necrosis factor (TNF) receptor superfamily that contains an intracellular "death domain" (Wajant, 2002). Its physiological ligand, FASL, is a member of the corresponding TNF cytokine family. The Fas/FasL system is known, first and foremost, as a potent apoptosis activator. Growing evidences have indicated that the Fas/FasL system can elicit nondeath signals as well. These nondeath signals can promote survival, proliferation, migration, and invasion of cells (Chakrabandhu & Hueber, 2016). In B-cell lymphomas, nucleolin colocalizes and interacts with Fas at the cell surface. This binding blocks Fas/FasL interaction, thus preventing Fas-mediated apoptosis induction, and promotes surviving of lymphoma cells (Wise et al., 2013).

Altogether, these different interactions attest for a prosurvival function of nucleolin.

7. NUCLEOLIN FUNCTION IN CHEMOKINE RECEPTOR SIGNALING

The epidermal growth factor (EGF) pathway (ERK and PI3K–Akt pathways) has been reported as canonical causes in cancer development, such as glioblastoma (Xie, Ma, Wan, Zhan, & Zhou, 2014). Using glioblastoma cell lines, Lv and coworkers demonstrated that nucleolin could interact with epidermal growth factor receptor (EGFR) to regulate the activation of the ERK and PI3K–Akt pathways, which could obviously affect the growth, viability, colony formation ability, and invasiveness of cancer cells (Lv et al., 2015).

Nucleolin was implicated in the migration and invasion of esophageal squamous cell carcinoma (ESCC) by modulating the initiation and transduction of EGFR and CXCR4 signaling. Qi and coworkers shed light on the expression and roles of the cell surface nucleolin in ESCC tissues and cells

(Qi et al., 2015). Nucleolin can be used as an important indicator to distinguish cancer metastasis clinically, which greatly allows for diagnosis and treatment of ESCC and provides novel therapeutic opportunities for proliferative disease. Indeed, the distribution of nucleolin differs greatly between metastatic and nonmetastatic carcinoma tissues compared with EGFR, Ki67, and CXCR4.

Cell surface nucleolin acts as a coreceptor for many ligands and may interact with G protein-coupled receptors (GPCRs), like chemokine receptor. Increasing studies have emphasized the significance of CXCL12/CXCR4 in various cancers including inducing aggressiveness and lymphatic metastasis (Balkwill, 2003; Schall & Proudfoot, 2011; Xue, Mao, Ren, & Chu, 2017). Based on these data, Yang and coworkers have demonstrated the relationship between CXCR4 and nucleolin in papillary thyroid carcinoma (Yang et al., 2014). Using soft-agar colony formation assay and transwell assay, their study suggested that nucleolin is crucial in the activation of CXCR4 signaling, which affects cell growth, migration, and invasiveness. Cell surface nucleolin can efficaciously bind CXCR4 receptor (Niu, Yang, Xu, Du, & Wang, 2015). Once the primary chemokine ligand CXCL12 binds its CXCR4 receptor, cell surface nucleolin will be recruited to CXCR4 receptors rapidly and trigger their downstream signaling, leading to ERK phosphorylation and CXCR4 internalization.

Similarly, Qiu and coworkers have investigated the relationship between cell surface nucleolin and chemokine CC receptor CCR6 in HCC (Qiu et al., 2015). It was found that nucleolin participated in the initiation of the CCR6 pathway to modulate the adhesion, migration, and invasion of HCC cells. Cell surface nucleolin and CCR6 were stained in most of HCC tissues and differently expressed in HCC cell lines. However, both expressions were associated with advanced stage, lymph node metastasis, and a poor 5-year prognosis. Till, further studies may be needed to provide multiple therapeutic strategies to inhibit the metastasis of HCC. Collectively, these results indicated that both proteins played a crucial role in the development of HCC pathology.

8. FUNCTION OF NUCLEOLIN IN EGF–TGF-β SIGNALING PATHWAYS

In recent study it has been reported that p85α, a regulatory subunit of a class I phosphoinositide 3-kinase (PI3K), was able to regulate the EGFR expression by increasing in EGFR mRNA stability and EGF-induced cell malignant transformation (Xie et al., 2016). EGFR plays an important role in extensive cross talk among multiple signaling pathways like PI3K/AKT pathway and Ras/Raf/MAPK pathway. EGFR also plays a significant role in tumor development and progression, including cell proliferation, regulation of apoptotic cell death, angiogenesis, and metastatic spread. Nucleolin binding to EGFR mRNA increased its stability and its translation and contributes to the promotion of EGF-induced malignant cell transformation and migration (Xie et al., 2016). On the other hand, p85α upregulates nucleolin gene transcription via enhancing C-Jun activation (Xie et al., 2016).

Farin and coworkers showed a cross talk between nucleolin, EGFR (ErbB1), and Ras proteins (Farin et al., 2011). All three typical Ras isoforms (H-, K, and N-Ras) interact with nucleolin. Moreover, they showed that GTP-bound Ras strongly enhances nucleolin–ErbB1 interaction. Finally, the expression of activated H-Ras, nucleolin, and EGFR (ErbB1) synergistically enhances cell transformation, as evidenced by increased colony formation in soft agar and increased tumor volume in nude mice (Farin et al., 2011).

The study in cases of ESCC with metastasis showed that nucleolin expression was extensively located in the nucleus, cytoplasm, and cell membrane, while expression in cases of ESCC without metastasis was merely limited into the nucleus. Moreover, the extranuclear nucleolin expression was positively correlated with the expression of EGFR, CXCR4, and Ki67 and can be used as an independent prognostic factor for ESCC patients (Qi et al., 2015).

EGF receptor activation is a major regulator of gastrin gene expression. Gastrin is a classical gut peptide hormone, which is normally produced at high levels by endocrine (G) cells located in the gastric antrum and proximal duodenal mucosa and at lower levels—in the colon and pancreas (Dockray, Dimaline, & Varro, 2005). Upregulation of gastrin gene expression is reported in numerous cancers, including colorectal, pancreatic, and lung cancers (Aly, Shulkes, & Baldwin, 2004; Ferrand & Wang, 2006; Rehfeld, Bardram, & Hilsted, 1989).

EGF signal increased the gastrin mRNA stability through nucleolin (Lee, Liao, Chang, & Tseng, 2007). Indeed, an increased association of nucleolin with gastrin mRNA under EGF treatment results in a greater gastrin mRNA stability. As nucleolin is overexpressed in colorectal cancer and gastrointestinal cancer, this may explain also the higher expression level of gastrin mRNA in colorectal cancer.

Nucleolin plays an important role in promoting cell proliferation, migration, and chemotaxis of human dermal fibroblasts during the recovery of heat-denatured dermis with a mechanism probably related to TGF-β1 (Jiang et al., 2015). TGF-β1 is one of the most potent growth factors involved in wound healing throughout the body, which influences the inflammatory response, angiogenesis, extracellular matrix (ECM) deposition, and remodeling (Kim et al., 2013; Riedel et al., 2008). Transmembrane serine/threonine kinase type I receptor (TβRI) is part of the TGF-β signaling pathway. TβRI associates with importin β1 and nucleolin in both cytoplasmic and nuclear fractions of transformed MCF10A/HER2 cells (Chandra et al., 2012). The authors showed that the ligand-inducible nuclear translocation of TβRI was observed only in transformed MCF10A/HER2 cells, and the translocation required importin β1 and nucleolin (Chandra et al., 2012).

9. NUCLEOLIN ON ANGIOGENESIS AND LYMPHANGIOGENESIS

Angiogenesis and lymphangiogenesis play an essential role in cancers. In 1971, Folkman established the importance of angiogenesis in tumorigenesis (Folkman, 1971). He suggested that tumor growth and expansion are closely related to the development of a network of blood vessels within a tumor. However, tumor angiogenesis leads to the formation of a disorganized, chaotic, and poorly functioning vasculature due to the deregulation of the balance between pro- and antiangiogenic factors such as vascular endothelial growth factor (VEGF), endostatin, or angiopoietins (Ang).

Nucleolin has been shown to play critical roles in the modulation of angiogenesis and tumorigenesis through its interaction with a variety of ligands, like VEGF and endostatin (Christian et al., 2003; Koutsioumpa & Papadimitriou, 2014).

Likewise, it has been reported that VEGF is required for nucleolin cell surface localization in endothelial cells, which reinforces its contribution to the angiogenic process (Huang et al., 2006; Quiroz-Mercado et al., 2016). Further, Ding and coworkers identified that heat-shock cognate 70 (Hsc70) is essential in both the surface translocation and the angiogenic function of nucleolin (Ding et al., 2012). Hsc70 interacted with nucleolin in endothelial cells via the peptide-binding domain of Hsc70 and the RBD of nucleolin and their expression levels were intimately correlated in human lung adenocarcinomas.

Interestingly, VEGF can regulate the surface translocation of nucleolin via a regulated interaction with the nonmuscle myosin heavy chain 9 (MyH9), an actin-based motor protein (Huang et al., 2006). These studies indicate that VEGF, ECM, and intracellular motor protein MyH9 are all essential for the novel function of nucleolin in angiogenesis. Moreover, VEGF expression was correlated with nucleolin distribution in colorectal carcinoma clinical samples and it could promote the phosphorylation and relocation of nucleolin through the PI3K/Akt pathway (Wu et al., 2014). Functional blockage or downregulation of the expression of cell surface nucleolin in endothelial cells significantly inhibits their migration and prevents capillary-tubule formation.

Overexpression of nucleolin at both mRNA and protein levels increased significantly the level of the proangiogenic factor VEGF, via interaction with the G-C-rich sequences of its promoter. They demonstrated that nucleolin could also be a potent proangiogenic factor during the recovery of denatured dermis and heat-denatured human umbilical vein endothelial cells (Liang et al., 2013).

Nucleolin is also required for the expression of the hypoxia-inducible factor 1-alpha (HIF1α) protein. HIF1α is an important regulator of tumorigenesis and tumor angiogenesis under hypoxic conditions (Hirota & Semenza, 2006). Nucleolin regulates HIF1α translation through the interaction with the G-rich region in the 5′ UTR of HIF1α mRNA. In addition, the inhibition of nucleolin led to decreased HIF1α protein and mRNA levels in osteosarcoma cell lines (Cheng, Zhao, Yang, Hu, & Yang, 2014).

Moreover, cell surface nucleolin is a receptor for the antiangiogenic peptide, endostatin, and mediates its antiangiogenic effects. Endostatin inhibits tumor angiogenesis and metastasis by limiting tumor blood supply, thereby depriving tumors of nutrients, and has been considered as a potential anticancer marker in the treatment of malignant tumors. Endostatin is internalized to the nucleus, via cell surface nucleolin, where it regulates nucleolin phosphorylation (Huang et al., 2006).

Nucleolin is also selectively expressed on the cell surface of lymphangiogenic endothelial cells both in vitro and in vivo (Zhuo et al., 2010). Blockage of nucleolin either with antinucleolin antibody or by inhibiting its expression abolishes the antilymphangiogenic function of endostatin, demonstrating that endostatin directly acts on lymphangiogenic endothelial cells via cell surface nucleolin (Shi et al., 2007).

Interestingly, aptamers or antibodies that target nucleolin have an effect on angiogenesis (Birmpas, Briand, Courty, & Katsoris, 2012; Destouches et al., 2012; Fogal, Sugahara, Ruoslahti, & Christian, 2009). The multivalent synthetic pseudopeptide N6L that selectively binds to nucleolin (Destouches et al., 2011) strongly inhibits growth of several tumor types by inducing apoptosis of tumor cells and is currently in preparation for phase II clinical trials (IPP-204106). N6L impairs both in vitro and in vivo angiogenesis by targeting endothelial cells and tumor vessels (Destouches et al., 2012; Fogal et al., 2009). In pancreatic cancer, N6L targets also the tumor microenvironment by reducing angiogenesis and induces vessel normalization by increasing pericyte coverage and vessel perfusion and by reducing intratumoral hypoxia (Gilles et al., 2016). In an orthotopic mouse model of pancreatic tumor generated by injecting PDAC cells (Panc-02) into the pancreas of wild-type ($NCL^{wt/wt}$;$Cre^{+/-}$) mice or in animals in which nucleolin was deleted in one allele ($NCL^{fl/wt}$;$Cre^{+/-}$). Remarkably, it was found a reduced tumor vascularization and increased pericyte coverage in $NCL^{+/-}$ compared with $NCL^{+/+}$ mice, indicating that the normal expression of nucleolin in the microenvironment of the tumor is required for the vascularization of the tumor (Gilles et al., 2016).

Among the vascular activators, nucleolin inhibition significantly decreased angiopoietin-2 (Ang-2) secretion and expression in endothelial cells. In vivo, N6L reduces the amount of Ang-2 present in the plasma of pancreatic cancer model. Together, this work unveiled Ang-2 as a potential target and suitable response biomarker for N6L treatment in pancreatic cancer.

Tumor-induced lymphangiogenesis is mediated by lymphangiogenic growth factors that are produced and secreted by the tumors themselves, stromal cells, tumor-infiltrating macrophages, or activated platelets (Christiansen & Detmar, 2012). VEGF-C and VEGF-D are the main inducers of lymphangiogenesis. The overexpression of either VEGF-C or VEGF-D in tumors significantly increased tumor-associated lymphatic vessel growth (primarily at the tumor margin) and increased incidence of lymph node metastasis. Nucleolin specifically and directly binds to the 5′ UTR of VEGF-D and functions as an IRES-trans-acting factor (ITAF) to induce VEGF-D mRNA translation in cells under heat-shock stress conditions (Morfoisse et al., 2016). Using RNAi-mediated silencing of nucleolin, VEGF-D translation was reduced along with a subsequent constriction of lymphatic vessels in tumors. Together, this study identifies nucleolin as a key regulator of VEGF-D expression, deepening understanding of lymphangiogenesis control during tumor formation, and it can provide a novel therapeutic target for lymphatic vessel plasticity during tumor inflammation.

10. NUCLEOLIN FUNCTION IN EPITHELIAL–MESENCHYMAL TRANSITION

The epithelial–mesenchymal transition (EMT) is known as the process of conversion of an epithelial cell into an elongated cell with a mesenchymal phenotype, which can occur in physiological and pathological processes such as embryogenesis (type 1 EMT), wound healing and/or fibrosis (type 2 EMT), and malignant tumors (type 3 EMT) (Gurzu, Turdean, Kovecsi, Contac, & Jung, 2015; Stone et al., 2016). In cancers, this process was described to favor increasing motility of the tumor cells in the invasion front and increasing cell resistance to apoptosis. Moreover, it is considered, nowadays, as the critical mechanism that underlies the initiation of cancer metastasis.

Nucleolin can also regulate the EMT process by which cancer cells increase their metastatic potential before implanting in novel tissues. Cell surface nucleolin acts as a receptor for several ligands (Carpentier et al., 2005), is involved in signal transduction on the cell surface (Losfeld et al., 2009), and is constantly induced in proliferating tumor cell (Hovanessian et al., 2010). In human gastric tumor cell lines, nucleolin is located on cell surface and is a receptor of Tipα, a TNF-α-inducing protein of *Helicobacter pylori* (Watanabe et al., 2010). Tipα first binds to cell surface nucleolin, and the complex is rapidly internalized into the cells. The signal transduction induces expression of TNF-α gene through the activation of nuclear factor kappa-light chain enhancer of activated B cells (NF-kB), resulting in tumor promotion and progression (Suganuma et al., 2005; Watanabe et al., 2010). In 2014, Watanabe and coworkers have reported, for the first time, a new function of Tipα as an EMT inducer of *H. pylori* (Watanabe et al., 2014). Induction of cell migration and elongation was found to be mediated through the binding to cell surface nucleolin, which was inhibited by the nucleolin-targeted siRNAs. Tipα induces migration and morphological changes with filopodia, reduces cell stiffness, phosphorylates cancer-related proteins, and increases expression of vimentin, a significant marker of EMT, through phosphorylation MEK–ERK proteins. Nucleolin mediates bone morphogenetic protein (BMP)2-induced EMT progression via upregulation of the ERK1/2- and Akt-signaling pathways in gastric cancer (Yang, Yang, & Zhang, 2015). BMP2 is well known as a member of TGF-β family and has been found highly expressed in different cancer tissues (McCormack & O'Dea, 2013). As reported, BMP2 can bind to their receptor complexes BMPRI/II and then exerts the protumorigenic effects on cancer cells. These results revealed that both nucleolin and BMPRII were aberrantly and constitutively expressed in gastric cancer specimens and cell lines, whose expression was positively associated with metastasis, stage and differentiation, and patented poor survival outcome of gastric cancer patients.

In lung cancer, activation of the EGFR pathway induced expression of nucleolin and also the expression of MMP7. A higher nucleolin expression correlated with poor prognosis. In human and mouse lung cancer, nucleolin was mainly expressed as a 55-kDa protein (Hsu et al., 2015). This C-terminal truncated form was generated by MMP7. Cleaved nucleolin aggressively promoted proliferation and metastasis and increased the expression of several oncogenes, such as MMP9, anaplastic lymphoma kinase, HIF1α, and casitas B-lineage lymphoma oncogene and decreased the expression of tumor suppressors such as BRD4, PCM1, TFG, and KLF6 by modulating mRNA stability through binding to the 3′ UTR of their transcripts. These findings highlight the importance of MMP7-mediated nucleolin cleavage in lung cancer development. Blocking nucleolin cleavage may provide a useful new strategy for lung cancer therapy.

In bladder cancer, Jin and coworkers have identified a novel function of nucleolin by stabilizing MMP2 mRNA in highly metastatic human bladder cancer T24T cells (Jin et al., 2015). Its overexpression and nuclear translocation are essential for ultimately increasing the MMP2 protein expression and invasion abilities of these cancer cell lines.

11. NUCLEOLIN FUNCTIONS IN STEMNESS

Stemness combines the ability of a cell to perpetuate its lineage, to give rise to differentiated cells, and to interact with its environment to maintain a balance between quiescence, proliferation, and regeneration. While adult stem cells display these properties when participating in tissue homeostasis, cancer stem cells (CSCs) behave as their malignant equivalents for sustained tumor growth (Kreso & Dick, 2014; Visvader & Lindeman, 2012). Nucleolin is highly expressed in undifferentiated embryonic stem cell (ESC) and downregulated during differentiation. Using the shRNA-mediated depletion approach, Yang and coworkers investigated the function of nucleolin in controlling cell proliferation, apoptosis, cell cycle progression, and differentiation of ESCs (Yang et al., 2011). Depletion of nucleolin resulted in enhanced ESC apoptosis, reduced cell proliferation, and extensive ESC differentiation into multiple lineages. Thus, nucleolin is required for maintenance of ESC self-renewal, adding a new factor required to sustain the ESC identity, through direct repression of p53 mRNA translation (Cinghu et al., 2014; Fonseca, Cruz, Moura, Simoes, & Moreira, 2017; Yang et al., 2011).

Additionally, a correlation between mRNA levels of nucleolin and pluripotency markers NANOG and OCT4 in breast CSCs and mouse ESC has been demonstrated (Fonseca et al., 2015). An in vivo tumorigenic assay suggested that cell surface nucleolin overexpression per se could be associated with the identification of highly tumorigenic triple-negative breast cancer cells (TNBCs). This proposed link between nucleolin expression and the stem-like phenotype in TNBC enabled 100% cell death mediated by synergistic drug combination of doxorubicin and C6-ceramide (a proapoptotic sphingolipid which inhibits the PI3K/Akt pathway), suggesting the potential to abrogate the plasticity and adaptability associated with CSCs and non-stem cancer cells (Fonseca et al., 2017, 2015).

The CSCs display high tumorigenicity and therefore are thought to be responsible for relapse of drug-resistant metastatic cancer (Jordan, Guzman, & Noble, 2006). This has elicited intense searches for biomarkers and therapeutic strategies against CSCs. Recently, Holmboe and coworkers have evaluated nucleolin targeting and somatostatin receptor subtype 2 (sst2) targeting for therapeutic delivery to CSCs from lung cancer (Holmboe et al., 2017).

Somatostatin is a native inhibitory peptide hormone distributed throughout the central nervous system and peripheral tissues, acting via five GPCR subtypes (sst1–sst5) (Chalabi et al., 2014). Somatostatin receptors are expressed to varying degrees in solid organ tumors. Indeed, sst2 expression is lost progressively during the development of pancreatic cancer leading to activation of PI3K/AKT/NF-kB pathways (Chalabi et al., 2014; Chalabi-Dchar et al., 2015); but it is overexpressed in neuroendocrine tumors and peptide-based sst2-targeting, for example, by radiolabeled DOTATATE, already used for diagnostic imaging (Chalabi et al., 2014; Xu & Zhang, 2015).

In lung cancer, AS1411 nucleolin aptamer showed efficient internalization by cancer cells and, remarkably, at even higher efficiency by CSCs. In contrast, sst2 expression levels were not sufficiently high in non-small-cell lung cancer cells to confer efficient uptake by either non-CSCs or CSCs. AS1411 aptamer might be used for therapeutic delivery to non-small-cell lung CSCs.

12. CONCLUDING REMARKS

Despite numerous studies involving nucleolin, the role of this abundant protein is still not completely fully understood. In many cases, the overexpression of this protein may just be the consequences of the need for the cell to increase the production of ribosomes to support a higher translation required for the cell proliferation. By modifying the expression of nucleolin, like the overexpression in cancer cells, ribosome biogenesis, metabolic, and many signaling pathways can be coordinately affected (Kumar et al., 2017). The striking cell surface expression of nucleolin in cancer cells merits more studies to determine the functional role of this pool of nucleolin. The recent development of a conditional knockout mice model to study nucleolin functions in different cells/tissues in normal or pathological situation should reveal the contribution of nucleolin in tumor development as recently described in pancreatic cancer (Gilles et al., 2016).

ACKNOWLEDGMENTS

This work in the laboratory of I.U. is fully supported by the Hubert Curien program RILA 2016, Grant No. DNTS-France-01/13.

In P.B. laboratory, research is funded by the Ligue contre le Cancer (Allier and Saone et Loire, France), the CNRS and Ecole Normale Supérieure de Lyon, and the Agence Nationale de la Recherche, project ANR Theranuc, ANR-16-CE17-0023 and Hubert Curien program RILA 2016, project no. 38647PM.

Conflicts of interest: The authors declare no conflicts of interest.

REFERENCES

Abdelmohsen, K., & Gorospe, M. (2012). RNA-binding protein nucleolin in disease. *RNA Biology*, *9*(6), 799–808. https://doi.org/10.4161/rna.19718.

Abdelmohsen, K., Tominaga, K., Lee, E. K., Srikantan, S., Kang, M. J., Kim, M. M., et al. (2011). Enhanced translation by nucleolin via G-rich elements in coding and non-coding regions of target mRNAs. *Nucleic Acids Research*, *39*(19), 8513–8530. https://doi.org/10.1093/nar/gkr488.

Allain, F. H., Bouvet, P., Dieckmann, T., & Feigon, J. (2000). Molecular basis of sequence-specific recognition of pre-ribosomal RNA by nucleolin. *The EMBO Journal*, *19*(24), 6870–6881. https://doi.org/10.1093/emboj/19.24.6870.

Aly, A., Shulkes, A., & Baldwin, G. S. (2004). Gastrins, cholecystokinins and gastrointestinal cancer. *Biochimica et Biophysica Acta*, *1704*(1), 1–10. https://doi.org/10.1016/j.bbcan.2004.01.004.

Andersen, J. S., Wilkinson, C. J., Mayor, T., Mortensen, P., Nigg, E. A., & Mann, M. (2003). Proteomic characterization of the human centrosome by protein correlation profiling. *Nature*, *426*(6966), 570–574. https://doi.org/10.1038/nature02166.

Angelov, D., Bondarenko, V. A., Almagro, S., Menoni, H., Mongelard, F., Hans, F., et al. (2006). Nucleolin is a histone chaperone with FACT-like activity and assists remodeling

of nucleosomes. *The EMBO Journal, 25*(8), 1669–1679. https://doi.org/10.1038/sj.emboj.7601046.
Arumugam, S., Miller, M. C., Maliekal, J., Bates, P. J., Trent, J. O., & Lane, A. N. (2010). Solution structure of the RBD1,2 domains from human nucleolin. *Journal of Biomolecular NMR, 47*(1), 79–83. https://doi.org/10.1007/s10858-010-9412-1.
Bakkenist, C. J., & Kastan, M. B. (2003). DNA damage activates ATM through intermolecular autophosphorylation and dimer dissociation. *Nature, 421*(6922), 499–506. https://doi.org/10.1038/nature01368.
Balkwill, F. (2003). Chemokine biology in cancer. *Seminars in Immunology, 15*(1), 49–55. https://doi.org/10.1016/S1044-5323(02)00127-6.
Bates, P. J., Reyes-Reyes, E. M., Malik, M. T., Murphy, E. M., O'Toole, M. G., & Trent, J. O. (2017). G-quadruplex oligonucleotide AS1411 as a cancer-targeting agent: Uses and mechanisms. *Biochimica et Biophysica Acta, 1861*(5 Pt. B), 1414–1428. https://doi.org/10.1016/j.bbagen.2016.12.015.
Belenguer, P., Caizergues-Ferrer, M., Labbe, J. C., Doree, M., & Amalric, F. (1990). Mitosis-specific phosphorylation of nucleolin by p34cdc2 protein kinase. *Molecular and Cellular Biology, 10*(7), 3607–3618.
Berger, C. M., Gaume, X., & Bouvet, P. (2015). The roles of nucleolin subcellular localization in cancer. *Biochimie, 113*, 78–85. https://doi.org/10.1016/j.biochi.2015.03.023.
Bhatt, P., d'Avout, C., Kane, N. S., Borowiec, J. A., & Saxena, A. (2012). Specific domains of nucleolin interact with Hdm2 and antagonize Hdm2-mediated p53 ubiquitination. *The FEBS Journal, 279*(3), 370–383. https://doi.org/10.1111/j.1742-4658.2011.08430.x.
Birmpas, C., Briand, J. P., Courty, J., & Katsoris, P. (2012). Nucleolin mediates the antiangiogenesis effect of the pseudopeptide N6L. *BMC Cell Biology, 13*, 32. https://doi.org/10.1186/1471-2121-13-32.
Boveri, T. H. (1914). *Zur Frage der Entstehung maligner Tumoren*. Jena: Verlag von Gustav Fisher.
Bouvet, P., Allain, F. H., Finger, L. D., Dieckmann, T., & Feigon, J. (2001). Recognition of pre-formed and flexible elements of an RNA stem-loop by nucleolin. *Journal of Molecular Biology, 309*(3), 763–775. https://doi.org/10.1006/jmbi.2001.4691.
Bouvet, P., Diaz, J. J., Kindbeiter, K., Madjar, J. J., & Amalric, F. (1998). Nucleolin interacts with several ribosomal proteins through its RGG domain. *The Journal of Biological Chemistry, 273*(30), 19025–19029.
Bywater, M. J., Poortinga, G., Sanij, E., Hein, N., Peck, A., Cullinane, C., et al. (2012). Inhibition of RNA polymerase I as a therapeutic strategy to promote cancer-specific activation of p53. *Cancer Cell, 22*(1), 51–65. https://doi.org/10.1016/j.ccr.2012.05.019.
Carpentier, M., Morelle, W., Coddeville, B., Pons, A., Masson, M., Mazurier, J., et al. (2005). Nucleolin undergoes partial N- and O-glycosylations in the extranuclear cell compartment. *Biochemistry, 44*(15), 5804–5815. https://doi.org/10.1021/bi047831s.
Chakrabandhu, K., & Hueber, A. O. (2016). Fas versatile signaling and beyond: Pivotal role of tyrosine phosphorylation in context-dependent signaling and diseases. *Frontiers in Immunology*, 7, 429. https://doi.org/10.3389/fimmu.2016.00429.
Chalabi, M., Duluc, C., Caron, P., Vezzosi, D., Guillermet-Guibert, J., Pyronnet, S., et al. (2014). Somatostatin analogs: Does pharmacology impact antitumor efficacy? *Trends in Endocrinology and Metabolism, 25*(3), 115–127. https://doi.org/10.1016/j.tem.2013.11.003.
Chalabi-Dchar, M., Cassant-Sourdy, S., Duluc, C., Fanjul, M., Lulka, H., Samain, R., et al. (2015). Loss of somatostatin receptor subtype 2 promotes growth of KRAS-induced pancreatic tumors in mice by activating PI3K signaling and overexpression of CXCL16. *Gastroenterology, 148*(7), 1452–1465. https://doi.org/10.1053/j.gastro.2015.02.009.

Chandra, M., Zang, S., Li, H., Zimmerman, L. J., Champer, J., Tsuyada, A., et al. (2012). Nuclear translocation of type I transforming growth factor beta receptor confers a novel function in RNA processing. *Molecular and Cellular Biology*, *32*(12), 2183–2195. https://doi.org/10.1128/MCB.00320-12.

Chen, J., Guo, K., & Kastan, M. B. (2012). Interactions of nucleolin and ribosomal protein L26 (RPL26) in translational control of human p53 mRNA. *The Journal of Biological Chemistry*, *287*(20), 16467–16476. https://doi.org/10.1074/jbc.M112.349274.

Cheng, D. D., Zhao, H. G., Yang, Y. S., Hu, T., & Yang, Q. C. (2014). GSK3beta negatively regulates HIF1alpha mRNA stability via nucleolin in the MG63 osteosarcoma cell line. *Biochemical and Biophysical Research Communications*, *443*(2), 598–603. https://doi.org/10.1016/j.bbrc.2013.12.020.

Christian, S., Pilch, J., Akerman, M. E., Porkka, K., Laakkonen, P., & Ruoslahti, E. (2003). Nucleolin expressed at the cell surface is a marker of endothelial cells in angiogenic blood vessels. *The Journal of Cell Biology*, *163*(4), 871–878. https://doi.org/10.1083/jcb.200304132.

Christiansen, A., & Detmar, M. (2012). Lymphangiogenesis and cancer. *Genes & Cancer*, *2*(12), 1146–1158. https://doi.org/10.1177/1947601911423028.

Ciccia, A., & Elledge, S. J. (2010). The DNA damage response: Making it safe to play with knives. *Molecular Cell*, *40*(2), 179–204. https://doi.org/10.1016/j.molcel.2010.09.019.

Cinghu, S., Yellaboina, S., Freudenberg, J. M., Ghosh, S., Zheng, X., Oldfield, A. J., et al. (2014). Integrative framework for identification of key cell identity genes uncovers determinants of ES cell identity and homeostasis. *Proceedings of the National Academy of Sciences of the United States of America*, *111*(16), E1581–1590. https://doi.org/10.1073/pnas.1318598111.

Cong, R., Das, S., & Bouvet, P. (2011). The multiple properties and functions of nucleolin. In M. O. J. Olson (Ed.), *The nucleolus, protein reviews Vol. 15*. (pp. 185–212). Springer.

Cong, R., Das, S., Ugrinova, I., Kumar, S., Mongelard, F., Wong, J., et al. (2012). Interaction of nucleolin with ribosomal RNA genes and its role in RNA polymerase I transcription. *Nucleic Acids Research*, *40*(19), 9441–9454. https://doi.org/10.1093/nar/gks720.

Daniely, Y., & Borowiec, J. A. (2000). Formation of a complex between nucleolin and replication protein A after cell stress prevents initiation of DNA replication. *The Journal of Cell Biology*, *149*(4), 799–810.

Daniely, Y., Dimitrova, D. D., & Borowiec, J. A. (2002). Stress-dependent nucleolin mobilization mediated by p53-nucleolin complex formation. *Molecular and Cellular Biology*, *22*(16), 6014–6022.

Das, S., Cong, R., Shandilya, J., Senapati, P., Moindrot, B., Monier, K., et al. (2013). Characterization of nucleolin K88 acetylation defines a new pool of nucleolin colocalizing with pre-mRNA splicing factors. *FEBS Letters*, *587*(5), 417–424. https://doi.org/10.1016/j.febslet.2013.01.035.

De, A., Donahue, S. L., Tabah, A., Castro, N. E., Mraz, N., Cruise, J. L., et al. (2006). A novel interaction [corrected] of nucleolin with Rad51. *Biochemical and Biophysical Research Communications*, *344*(1), 206–213. https://doi.org/10.1016/j.bbrc.2006.03.113.

Derenzini, M., Trere, D., Pession, A., Govoni, M., Sirri, V., & Chieco, P. (2000). Nucleolar size indicates the rapidity of cell proliferation in cancer tissues. *The Journal of Pathology*, *191*(2), 181–186. https://doi.org/10.1002/(SICI)1096-9896(200006)191:2<181::AID-PATH607>3.0.CO;2-V [pii].

Destouches, D., Huet, E., Sader, M., Frechault, S., Carpentier, G., Ayoul, F., et al. (2012). Multivalent pseudopeptides targeting cell surface nucleoproteins inhibit cancer cell invasion through tissue inhibitor of metalloproteinases 3 (TIMP-3) release. *The Journal of Biological Chemistry*, *287*(52), 43685–43693. https://doi.org/10.1074/jbc.M112.380402.

Destouches, D., Page, N., Hamma-Kourbali, Y., Machi, V., Chaloin, O., Frechault, S., et al. (2011). A simple approach to cancer therapy afforded by multivalent pseudopeptides that target cell-surface nucleoproteins. *Cancer Research*, *71*(9), 3296–3305. https://doi.org/10.1158/0008-5472.CAN-10-3459.

Ding, Y., Song, N., Liu, C., He, T., Zhuo, W., He, X., et al. (2012). Heat shock cognate 70 regulates the translocation and angiogenic function of nucleolin. *Arteriosclerosis, Thrombosis, and Vascular Biology*, *32*, e126–34. https://doi.org/10.1161/ATVBAHA.112.247502.

Dockray, G., Dimaline, R., & Varro, A. (2005). Gastrin: Old hormone, new functions. *Pflügers Archiv*, *449*(4), 344–355. https://doi.org/10.1007/s00424-004-1347-5.

Erard, M. S., Belenguer, P., Caizergues-Ferrer, M., Pantaloni, A., & Amalric, F. (1988). A major nucleolar protein, nucleolin, induces chromatin decondensation by binding to histone H1. *European Journal of Biochemistry*, *175*(3), 525–530.

Farin, K., Schokoroy, S., Haklai, R., Cohen-Or, I., Elad-Sfadia, G., Reyes-Reyes, M. E., et al. (2011). Oncogenic synergism between ErbB1, nucleolin, and mutant Ras. *Cancer Research*, *71*(6), 2140–2151. https://doi.org/10.1158/0008-5472.CAN-10-2887.

Ferguson, L. R., Chen, H., Collins, A. R., Connell, M., Damia, G., Dasgupta, S., et al. (2015). Genomic instability in human cancer: Molecular insights and opportunities for therapeutic attack and prevention through diet and nutrition. *Seminars in Cancer Biology*, *35*(Suppl), S5–24. https://doi.org/10.1016/j.semcancer.2015.03.005.

Ferrand, A., & Wang, T. C. (2006). Gastrin and cancer: A review. *Cancer Letters*, *238*(1), 15–29. https://doi.org/10.1016/j.canlet.2005.06.025.

Fogal, V., Sugahara, K. N., Ruoslahti, E., & Christian, S. (2009). Cell surface nucleolin antagonist causes endothelial cell apoptosis and normalization of tumor vasculature. *Angiogenesis*, *12*(1), 91–100. https://doi.org/10.1007/s10456-009-9137-5.

Folkman, J. (1971). Tumor angiogenesis: Therapeutic implications. *The New England Journal of Medicine*, *285*(21), 1182–1186. https://doi.org/10.1056/NEJM197111182852108.

Fonseca, N. A., Cruz, A. F., Moura, V., Simoes, S., & Moreira, J. N. (2017). The cancer stem cell phenotype as a determinant factor of the heterotypic nature of breast tumors. *Critical Reviews in Oncology/Hematology*, *113*, 111–121. https://doi.org/10.1016/j.critrevonc.2017.03.016.

Fonseca, N. A., Rodrigues, A. S., Rodrigues-Santos, P., Alves, V., Gregorio, A. C., Valerio-Fernandes, A., et al. (2015). Nucleolin overexpression in breast cancer cell subpopulations with different stem-like phenotype enables targeted intracellular delivery of synergistic drug combination. *Biomaterials*, *69*, 76–88. https://doi.org/10.1016/j.biomaterials.2015.08.007.

Gadde, S., & Heald, R. (2004). Mechanisms and molecules of the mitotic spindle. *Current Biology*, *14*(18), R797–805. https://doi.org/10.1016/j.cub.2004.09.021.

Gaume, X., Place, C., Delage, H., Mongelard, F., Monier, K., & Bouvet, P. (2016). Expression of nucleolin affects microtubule dynamics. *PLoS One*, *11*(6), e0157534. https://doi.org/10.1371/journal.pone.0157534.

Gaume, X., Tassin, A. M., Ugrinova, I., Mongelard, F., Monier, K., & Bouvet, P. (2015). Centrosomal nucleolin is required for microtubule network organization. *Cell Cycle*, *14*(6), 902–919. https://doi.org/10.1080/15384101.2014.1000197.

Ghisolfi-Nieto, L., Joseph, G., Puvion-Dutilleul, F., Amalric, F., & Bouvet, P. (1996). Nucleolin is a sequence-specific RNA-binding protein: Characterization of targets on pre-ribosomal RNA. *Journal of Molecular Biology*, *260*(1), 34–53. https://doi.org/10.1006/jmbi.1996.0380.

Gilles, M. E., Maione, F., Cossutta, M., Carpentier, G., Caruana, L., Di Maria, S., et al. (2016). Nucleolin targeting impairs the progression of pancreatic cancer and promotes the normalization of tumor vasculature. *Cancer Research*, *76*(24), 7181–7193. https://doi.org/10.1158/0008-5472.CAN-16-0300.

Ginisty, H., Sicard, H., Roger, B., & Bouvet, P. (1999). Structure and functions of nucleolin. *Journal of Cell Science*, *112*(Pt. 6), 761–772.

Goldstein, M., Derheimer, F. A., Tait-Mulder, J., & Kastan, M. B. (2013). Nucleolin mediates nucleosome disruption critical for DNA double-strand break repair. *Proceedings of the National Academy of Sciences of the United States of America*, *110*(42), 16874–16879. https://doi.org/10.1073/pnas.1306160110.

Goodarzi, A. A., Jeggo, P., & Lobrich, M. (2010). The influence of heterochromatin on DNA double strand break repair: Getting the strong, silent type to relax. *DNA Repair (Amst)*, *9*(12), 1273–1282. https://doi.org/10.1016/j.dnarep.2010.09.013.

Griffin, S. V., Olivier, J. P., Pippin, J. W., Roberts, J. M., & Shankland, S. J. (2006). Cyclin I protects podocytes from apoptosis. *The Journal of Biological Chemistry*, *281*(38), 28048–28057. https://doi.org/10.1074/jbc.M513336200.

Grinstein, E., Wernet, P., Snijders, P. J., Rosl, F., Weinert, I., Jia, W., et al. (2002). Nucleolin as activator of human papillomavirus type 18 oncogene transcription in cervical cancer. *The Journal of Experimental Medicine*, *196*(8), 1067–1078.

Guo, X., Xiong, L., Yu, L., Li, R., Wang, Z., Ren, B., et al. (2014). Increased level of nucleolin confers to aggressive tumor progression and poor prognosis in patients with hepatocellular carcinoma after hepatectomy. *Diagnostic Pathology*, *9*, 175. https://doi.org/10.1186/s13000-014-0175-y.

Gurzu, S., Turdean, S., Kovecsi, A., Contac, A. O., & Jung, I. (2015). Epithelial-mesenchymal, mesenchymal-epithelial, and endothelial-mesenchymal transitions in malignant tumors: An update. *World Journal of Clinical Cases*, *3*(5), 393–404. https://doi.org/10.12998/wjcc.v3.i5.393.

Hanahan, D., & Weinberg, R. A. (2000). The hallmarks of cancer. *Cell*, *100*(1), 57–70.

Hanakahi, L. A., Bu, Z., & Maizels, N. (2000). The C-terminal domain of nucleolin accelerates nucleic acid annealing. *Biochemistry*, *39*(50), 15493–15499.

Hers, I., Vincent, E. E., & Tavare, J. M. (2011). Akt signalling in health and disease. *Cellular Signalling*, *23*(10), 1515–1527. https://doi.org/10.1016/j.cellsig.2011.05.004.

Hirota, K., & Semenza, G. L. (2006). Regulation of angiogenesis by hypoxia-inducible factor 1. *Critical Reviews in Oncology/Hematology*, *59*(1), 15–26. https://doi.org/10.1016/j.critrevonc.2005.12.003.

Holmboe, S., Hansen, P. L., Thisgaard, H., Block, I., Muller, C., Langkjaer, N., et al. (2017). Evaluation of somatostatin and nucleolin receptors for therapeutic delivery in non-small cell lung cancer stem cells applying the somatostatin-analog DOTATATE and the nucleolin-targeting aptamer AS1411. *PLoS One*, *12*(5), e0178286. https://doi.org/10.1371/journal.pone.0178286.

Hovanessian, A. G., Soundaramourty, C., El Khoury, D., Nondier, I., Svab, J., & Krust, B. (2010). Surface expressed nucleolin is constantly induced in tumor cells to mediate calcium-dependent ligand internalization. *PLoS One*, *5*(12), e15787. https://doi.org/10.1371/journal.pone.0015787.

Hsu, T. I., Lin, S. C., Lu, P. S., Chang, W. C., Hung, C. Y., Yeh, Y. M., et al. (2015). MMP7-mediated cleavage of nucleolin at Asp255 induces MMP9 expression to promote tumor malignancy. *Oncogene*, *34*(7), 826–837. https://doi.org/10.1038/onc.2014.22.

Huang, Y., Shi, H., Zhou, H., Song, X., Yuan, S., & Luo, Y. (2006). The angiogenic function of nucleolin is mediated by vascular endothelial growth factor and nonmuscle myosin. *Blood*, *107*(9), 3564–3571. https://doi.org/10.1182/blood-2005-07-2961.

Iftode, C., Daniely, Y., & Borowiec, J. A. (1999). Replication protein A (RPA): The eukaryotic SSB. *Critical Reviews in Biochemistry and Molecular Biology*, *34*(3), 141–180. https://doi.org/10.1080/10409239991209255.

Indig, F. E., Rybanska, I., Karmakar, P., Devulapalli, C., Fu, H., Carrier, F., et al. (2012). Nucleolin inhibits G4 oligonucleotide unwinding by Werner helicase. *PLoS One*, *7*(6), e35229. https://doi.org/10.1371/journal.pone.0035229.

Ishimaru, D., Ramalingam, S., Sengupta, T. K., Bandyopadhyay, S., Dellis, S., Tholanikunnel, B. G., et al. (2009). Regulation of Bcl-2 expression by HuR in HL60 leukemia cells and A431 carcinoma cells. *Molecular Cancer Research*, 7(8), 1354–1366. https://doi.org/10.1158/1541-7786.MCR-08-0476.

Ishimaru, D., Zuraw, L., Ramalingam, S., Sengupta, T. K., Bandyopadhyay, S., Reuben, A., et al. (2010). Mechanism of regulation of bcl-2 mRNA by nucleolin and A+U-rich element-binding factor 1 (AUF1). *The Journal of Biological Chemistry*, *285*(35), 27182–27191. https://doi.org/10.1074/jbc.M109.098830.

Jiang, B., Li, Y., Liang, P., Liu, Y., Huang, X., Tong, Z., et al. (2015). Nucleolin enhances the proliferation and migration of heat-denatured human dermal fibroblasts. *Wound Repair and Regeneration*, *23*(6), 807–818. https://doi.org/10.1111/wrr.12339.

Jin, H., Yu, Y., Hu, Y., Lu, C., Li, J., Gu, J., et al. (2015). Divergent behaviors and underlying mechanisms of cell migration and invasion in non-metastatic T24 and its metastatic derivative T24T bladder cancer cell lines. *Oncotarget*, *6*(1), 522–536. https://doi.org/10.18632/oncotarget.2680.

Jordan, C. T., Guzman, M. L., & Noble, M. (2006). Cancer stem cells. *The New England Journal of Medicine*, *355*(12), 1253–1261. https://doi.org/10.1056/NEJMra061808.

Kakarougkas, A., & Jeggo, P. A. (2014). DNA DSB repair pathway choice: An orchestrated handover mechanism. *The British Journal of Radiology*, *87*(1035), 20130685. https://doi.org/10.1259/bjr.20130685.

Kim, Y. J., Carvalho, F. C., Souza, J. A., Goncalves, P. C., Nogueira, A. V., Spolidorio, L. C., et al. (2013). Topical application of the lectin Artin M accelerates wound healing in rat oral mucosa by enhancing TGF-beta and VEGF production. *Wound Repair and Regeneration*, *21*(3), 456–463. https://doi.org/10.1111/wrr.12041.

Kim, K., Dimitrova, D. D., Carta, K. M., Saxena, A., Daras, M., & Borowiec, J. A. (2005). Novel checkpoint response to genotoxic stress mediated by nucleolin-replication protein a complex formation. *Molecular and Cellular Biology*, *25*(6), 2463–2474. https://doi.org/10.1128/MCB.25.6.2463-2474.2005.

Kobayashi, J., Fujimoto, H., Sato, J., Hayashi, I., Burma, S., Matsuura, S., et al. (2012). Nucleolin participates in DNA double-strand break-induced damage response through MDC1-dependent pathway. *PLoS One*, 7(11), e49245. https://doi.org/10.1371/journal.pone.0049245.

Koutsioumpa, M., & Papadimitriou, E. (2014). Cell surface nucleolin as a target for anticancer therapies. *Recent Patents on Anti-Cancer Drug Discovery*, *9*(2), 137–152. https://doi.org/10.2174/1574892808666131119095953.

Kowalczykowski, S. C. (2000). Initiation of genetic recombination and recombination-dependent replication. *Trends in Biochemical Sciences*, *25*(4), 156–165.

Kowalczykowski, S. C. (2015). An overview of the molecular mechanisms of recombinational DNA repair. *Cold Spring Harbor Perspectives in Biology*, 7(11). https://doi.org/10.1101/cshperspect.a016410.

Kreso, A., & Dick, J. E. (2014). Evolution of the cancer stem cell model. *Cell Stem Cell*, *14*(3), 275–291. https://doi.org/10.1016/j.stem.2014.02.006.

Kumar, S., Gomez, E. C., Chalabi-Dchar, M., Rong, C., Das, S., Ugrinova, I., et al. (2017). Integrated analysis of mRNA and miRNA expression in HeLa cells expressing low levels of Nucleolin. *Scientific Reports*, 7(1), 9017. https://doi.org/10.1038/s41598-017-09353-4.

Kwon, M., Godinho, S. A., Chandhok, N. S., Ganem, N. J., Azioune, A., Thery, M., et al. (2008). Mechanisms to suppress multipolar divisions in cancer cells with extra centrosomes. *Genes & Development*, *22*(16), 2189–2203. https://doi.org/10.1101/gad.1700908.

Landberg, G., Nilsson, K., Jirstrom, K., Ryden, L., Kitching, R., Burger, A. M., et al. (2005). Cyclin I is expressed in human breast cancer and closely associated with VEGF and KDR expression. *Breast Cancer Research and Treatment*, *89*(3), 313–316. https://doi.org/10.1007/s10549-004-2230-y.

Lapeyre, B., & Amalric, F. (1985). A powerful method for the preparation of cDNA libraries: Isolation of cDNA encoding a 100-kDal nucleolar protein. *Gene*, *37*(1–3), 215–220. https://doi.org/10.1016/0378-1119(85)90275-6.

Lee, P. T., Liao, P. C., Chang, W. C., & Tseng, J. T. (2007). Epidermal growth factor increases the interaction between nucleolin and heterogeneous nuclear ribonucleoprotein K/poly(C) binding protein 1 complex to regulate the gastrin mRNA turnover. *Molecular Biology of the Cell*, *18*(12), 5004–5013. https://doi.org/10.1091/mbc.E07-04-0384.

Liang, H., Chen, X., Yin, Q., Ruan, D., Zhao, X., Zhang, C., et al. (2017). PTENbeta is an alternatively translated isoform of PTEN that regulates rDNA transcription. *Nature Communications*, *8*, 14771. https://doi.org/10.1038/ncomms14771.

Liang, P., Jiang, B., Lv, C., Huang, X., Sun, L., Zhang, P., et al. (2013). The expression and proangiogenic effect of nucleolin during the recovery of heat-denatured HUVECs. *Biochimica et Biophysica Acta*, *1830*(10), 4500–4512. https://doi.org/10.1016/j.bbagen.2013.05.027.

Liao, P., Wang, W., Shen, M., Pan, W., Zhang, K., Wang, R., et al. (2014). A positive feedback loop between EBP2 and c-Myc regulates rDNA transcription, cell proliferation, and tumorigenesis. *Cell Death & Disease*, *5*, e1032. https://doi.org/10.1038/cddis.2013.536.

Lin, K., Farahani, M., Yang, Y., Johnson, G. G., Oates, M., Atherton, M., et al. (2014). Loss of MIR15A and MIR16-1 at 13q14 is associated with increased TP53 mRNA, de-repression of BCL2 and adverse outcome in chronic lymphocytic leukaemia. *British Journal of Haematology*, *167*(3), 346–355. https://doi.org/10.1111/bjh.13043.

Lingle, W. L., Lukasiewicz, K., & Salisbury, J. L. (2005). Deregulation of the centrosome cycle and the origin of chromosomal instability in cancer. *Advances in Experimental Medicine and Biology*, *570*, 393–421. https://doi.org/10.1007/1-4020-3764-3_14.

Lischwe, M. A., Roberts, K. D., Yeoman, L. C., & Busch, H. (1982). Nucleolar specific acidic phosphoprotein C23 is highly methylated. *Journal of Biological Chemistry*, *257*(24), 4600–4602.

Lischwe, M. A., Smetana, K., Olson, M. O. J., & Busch, H. (1979). Protein-C23 and protein-B23 are the major nucleolar silver staining proteins. *Life Sciences*, *25*(8), 701–708.

Losfeld, M. E., Khoury, D. E., Mariot, P., Carpentier, M., Krust, B., Briand, J. P., et al. (2009). The cell surface expressed nucleolin is a glycoprotein that triggers calcium entry into mammalian cells. *Experimental Cell Research*, *315*(2), 357–369. https://doi.org/10.1016/j.yexcr.2008.10.039.

Lv, S., Dai, C., Liu, Y., Sun, B., Shi, R., Han, M., et al. (2015). Cell surface protein C23 affects EGF-EGFR induced activation of ERK and PI3K-AKT pathways. *Journal of Molecular Neuroscience*, *55*(2), 519–524. https://doi.org/10.1007/s12031-014-0375-7.

Ma, N., Matsunaga, S., Takata, H., Ono-Maniwa, R., Uchiyama, S., & Fukui, K. (2007). Nucleolin functions in nucleolus formation and chromosome congression. *Journal of Cell Science*, *120*(Pt. 12), 2091–2105. https://doi.org/10.1242/jcs.008771.

Manning, B. D. (2004). Balancing Akt with S6K: Implications for both metabolic diseases and tumorigenesis. *The Journal of Cell Biology*, *167*(3), 399–403. https://doi.org/10.1083/jcb.200408161.

Marcel, V., Catez, F., Berger, C. M., Perrial, E., Plesa, A., Thomas, X., et al. (2017). Expression profiling of ribosome biogenesis factors reveals nucleolin as a novel potential marker to predict outcome in AML patients. *PLoS One*, *12*(1), e0170160. https://doi.org/10.1371/journal.pone.0170160.

Marcel, V., Catez, F., & Diaz, J. J. (2015). p53, a translational regulator: Contribution to its tumour-suppressor activity. *Oncogene*, *34*(44), 5513–5523. https://doi.org/10.1038/onc.2015.25.

McCormack, N., & O'Dea, S. (2013). Regulation of epithelial to mesenchymal transition by bone morphogenetic proteins. *Cellular Signalling, 25*(12), 2856–2862. https://doi.org/10.1016/j.cellsig.2013.09.012.

Mongelard, F., & Bouvet, P. (2007). Nucleolin: A multiFACeTed protein. *Trends in Cell Biology, 17*(2), 80–86. https://doi.org/10.1016/j.tcb.2006.11.010.

Morfoisse, F., Tatin, F., Hantelys, F., Adoue, A., Helfer, A. C., Cassant-Sourdy, S., et al. (2016). Nucleolin promotes heat shock-associated translation of VEGF-D to promote tumor lymphangiogenesis. *Cancer Research, 76*(15), 4394–4405. https://doi.org/10.1158/0008-5472.CAN-15-3140.

Nakamura, T., Sanokawa, R., Sasaki, Y. F., Ayusawa, D., Oishi, M., & Mori, N. (1995). Cyclin I: A new cyclin encoded by a gene isolated from human brain. *Experimental Cell Research, 221*(2), 534–542. https://doi.org/10.1006/excr.1995.1406.

Nasirudin, K. M., Ehtesham, N. Z., Tuteja, R., Sopory, S. K., & Tuteja, N. (2005). The Gly-Arg-rich C-terminal domain of pea nucleolin is a DNA helicase that catalytically translocates in the 5′- to 3′-direction. *Archives of Biochemistry and Biophysics, 434*(2), 306–315. https://doi.org/10.1016/j.abb.2004.11.016.

Negrini, S., Gorgoulis, V. G., & Halazonetis, T. D. (2010). Genomic instability—An evolving hallmark of cancer. *Nature Reviews Molecular Cell Biology, 11*(3), 220–228. https://doi.org/10.1038/nrm2858.

Nigg, E. A. (2002). Centrosome aberrations: Cause or consequence of cancer progression? *Nature Reviews Cancer, 2*(11), 815–825. https://doi.org/10.1038/nrc924.

Nigg, E. A., Cajanek, L., & Arquint, C. (2014). The centrosome duplication cycle in health and disease. *FEBS Letters, 588*(15), 2366–2372. https://doi.org/10.1016/j.febslet.2014.06.030.

Niu, H., Yang, X., Xu, Z., Du, T., & Wang, R. (2015). Cell surface nucleolin interacts with CXCR4 receptor via the 212 c-terminal portion. *Tumour Biology, 36*(2), 1099–1104. https://doi.org/10.1007/s13277-014-2734-y.

Ochi, T., Wu, Q., & Blundell, T. L. (2014). The spatial organization of non-homologous end joining: From bridging to end joining. *DNA Repair (Amst), 17*, 98–109. https://doi.org/10.1016/j.dnarep.2014.02.010.

Orrick, L. R., Olson, M. O., & Busch, H. (1973). Comparison of nucleolar proteins of normal rat liver and Novikoff hepatoma ascites cells by two-dimensional polyacrylamide gel electrophoresis. *Proceedings of the National Academy of Sciences of the United States of America, 70*(5), 1316–1320.

Otake, Y., Sengupta, T. K., Bandyopadhyay, S., Spicer, E. K., & Fernandes, D. J. (2004). Drug-induced destabilization of bcl-2 mRNA: A new approach for inducing apoptosis in tumor cells. *Current Opinion in Investigational Drugs, 5*(6), 616–622.

Otake, Y., Soundararajan, S., Sengupta, T. K., Kio, E. A., Smith, J. C., Pineda-Roman, M., et al. (2007). Overexpression of nucleolin in chronic lymphocytic leukemia cells induces stabilization of bcl2 mRNA. *Blood, 109*(7), 3069–3075. https://doi.org/10.1182/blood-2006-08-043257.

Otterlei, M., Bruheim, P., Ahn, B., Bussen, W., Karmakar, P., Baynton, K., et al. (2006). Werner syndrome protein participates in a complex with RAD51, RAD54, RAD54B and ATR in response to ICL-induced replication arrest. *Journal of Cell Science, 119*(Pt. 24), 5137–5146. https://doi.org/10.1242/jcs.03291.

Peng, L., Liang, J., Wang, H., Song, X., Rashid, A., Gomez, H. F., et al. (2010). High levels of nucleolar expression of nucleolin are associated with better prognosis in patients with stage II pancreatic ductal adenocarcinoma. *Clinical Cancer Research, 16*(14), 3734–3742. https://doi.org/10.1158/1078-0432.CCR-09-3411.

Pich, A., Chiusa, L., & Margaria, E. (2000). Prognostic relevance of AgNORs in tumor pathology. *Micron, 31*(2), 133–141. https://doi.org/10.1016/S0968-4328(99)00070-0.

Pickard, A. J., & Bierbach, U. (2013). The cell's nucleolus: An emerging target for chemotherapeutic intervention. *ChemMedChem*, *8*(9), 1441–1449. https://doi.org/10.1002/cmdc.201300262.
Pihan, G. A. (2013). Centrosome dysfunction contributes to chromosome instability, chromoanagenesis, and genome reprograming in cancer. *Frontiers in Oncology*, *3*, 277. https://doi.org/10.3389/fonc.2013.00277.
Qi, J., Li, H., Liu, N., Xing, Y., Zhou, G., Wu, Y., et al. (2015). The implications and mechanisms of the extra-nuclear nucleolin in the esophageal squamous cell carcinomas. *Medical Oncology*, *32*(3), 45. https://doi.org/10.1007/s12032-015-0484-3.
Qiu, W., Wang, G., Sun, X., Ye, J., Wei, F., Shi, X., et al. (2015). The involvement of cell surface nucleolin in the initiation of CCR6 signaling in human hepatocellular carcinoma. *Medical Oncology*, *32*(3), 75. https://doi.org/10.1007/s12032-015-0530-1.
Qiu, W., Zhou, F., Zhang, Q., Sun, X., Shi, X., Liang, Y., et al. (2013). Overexpression of nucleolin and different expression sites both related to the prognosis of gastric cancer. *APMIS*, *121*(10), 919–925. https://doi.org/10.1111/apm.12131.
Quintyne, N. J., Reing, J. E., Hoffelder, D. R., Gollin, S. M., & Saunders, W. S. (2005). Spindle multipolarity is prevented by centrosomal clustering. *Science*, *307*(5706), 127–129. https://doi.org/10.1126/science.1104905.
Quiroz-Mercado, J., Ramirez-Velazquez, N., Partido, G., Zenteno, E., Chavez, R., Agundis-Mata, C., et al. (2016). Tissue and cellular characterisation of nucleolin in a murine model of corneal angiogenesis. *Graefe's Archive for Clinical and Experimental Ophthalmology*, *254*(9), 1753–1763. https://doi.org/10.1007/s00417-016-3409-8.
Rehfeld, J. F., Bardram, L., & Hilsted, L. (1989). Gastrin in human bronchogenic carcinomas: Constant expression but variable processing of progastrin. *Cancer Research*, *49*(11), 2840–2843.
Riedel, K., Kremer, T., Ryssel, H., Riedel, F., Goessler, U. R., Koellensperger, E., et al. (2008). TGF-beta antisense oligonucleotides modulate expression of matrix metalloproteinases in isolated fibroblasts from radiated skin. *In Vivo (Athens, Greece)*, *22*(1), 1–7.
Ring, D., Hubble, R., & Kirschner, M. (1982). Mitosis in a cell with multiple centrioles. *The Journal of Cell Biology*, *94*(3), 549–556.
Ruggero, D. (2013). Translational control in cancer etiology. *Cold Spring Harbor Perspectives in Biology*, *5*(2). https://doi.org/10.1101/cshperspect.a015891.
Salvetti, A., Coute, Y., Epstein, A., Arata, L., Kraut, A., Navratil, V., et al. (2016). Nuclear functions of nucleolin through global proteomics and interactomic approaches. *Journal of Proteome Research*, *15*(5), 1659–1669. https://doi.org/10.1021/acs.jproteome.6b00126.
Sato, H., Kusumoto-Matsuo, R., Ishii, Y., Mori, S., Nakahara, T., Shinkai-Ouchi, F., et al. (2009). Identification of nucleolin as a protein that binds to human papillomavirus type 16 DNA. *Biochemical and Biophysical Research Communications*, *387*(3), 525–530. https://doi.org/10.1016/j.bbrc.2009.07.055.
Sauer, G., Korner, R., Hanisch, A., Ries, A., Nigg, E. A., & Sillje, H. H. (2005). Proteome analysis of the human mitotic spindle. *Molecular & Cellular Proteomics*, *4*(1), 35–43. https://doi.org/10.1074/mcp.M400158-MCP200.
Saxena, A., Rorie, C. J., Dimitrova, D., Daniely, Y., & Borowiec, J. A. (2006). Nucleolin inhibits Hdm2 by multiple pathways leading to p53 stabilization. *Oncogene*, *25*(55), 7274–7288. https://doi.org/10.1038/sj.onc.1209714.
Schall, T. J., & Proudfoot, A. E. (2011). Overcoming hurdles in developing successful drugs targeting chemokine receptors. *Nature Reviews. Immunology*, *11*(5), 355–363. https://doi.org/10.1038/nri2972.
Scott, D. D., & Oeffinger, M. (2016). Nucleolin and nucleophosmin: Nucleolar proteins with multiple functions in DNA repair. *Biochemistry and Cell Biology*, *94*, 419–432. https://doi.org/10.1139/bcb-2016-0068.

Sengupta, T. K., Bandyopadhyay, S., Fernandes, D. J., & Spicer, E. K. (2004). Identification of nucleolin as an AU-rich element binding protein involved in bcl-2 mRNA stabilization. *The Journal of Biological Chemistry*, *279*(12), 10855–10863. https://doi.org/10.1074/jbc.M309111200.

Shi, H., Huang, Y., Zhou, H., Song, X., Yuan, S., Fu, Y., et al. (2007). Nucleolin is a receptor that mediates antiangiogenic and antitumor activity of endostatin. *Blood*, *110*(8), 2899–2906. https://doi.org/10.1182/blood-2007-01-064428.

Stone, R. C., Pastar, I., Ojeh, N., Chen, V., Liu, S., Garzon, K. I., et al. (2016). Epithelial-mesenchymal transition in tissue repair and fibrosis. *Cell and Tissue Research*, *365*(3), 495–506. https://doi.org/10.1007/s00441-016-2464-0.

Storck, S., Thiry, M., & Bouvet, P. (2009). Conditional knockout of nucleolin in DT40 cells reveals the functional redundancy of its RNA-binding domains. *Biology of the Cell*, *101*(3), 153–167. https://doi.org/10.1042/BC20080054.

Suganuma, M., Kurusu, M., Suzuki, K., Nishizono, A., Murakami, K., Fujioka, T., et al. (2005). New tumor necrosis factor-alpha-inducing protein released from Helicobacter pylori for gastric cancer progression. *Journal of Cancer Research and Clinical Oncology*, *131*(5), 305–313. https://doi.org/10.1007/s00432-004-0652-x.

Sun, Z. L., Zhu, Y., Wang, F. Q., Chen, R., Peng, T., Fan, Z. N., et al. (2007). Serum proteomic-based analysis of pancreatic carcinoma for the identification of potential cancer biomarkers. *Biochimica et Biophysica Acta*, *1774*(6), 764–771. https://doi.org/10.1016/j.bbapap.2007.04.001.

Takagi, M., Absalon, M. J., McLure, K. G., & Kastan, M. B. (2005). Regulation of p53 translation and induction after DNA damage by ribosomal protein L26 and nucleolin. *Cell*, *123*(1), 49–63. https://doi.org/10.1016/j.cell.2005.07.034.

Ugrinova, I., Monier, K., Ivaldi, C., Thiry, M., Storck, S., Mongelard, F., et al. (2007). Inactivation of nucleolin leads to nucleolar disruption, cell cycle arrest and defects in centrosome duplication. *BMC Molecular Biology*, *8*, 66. https://doi.org/10.1186/1471-2199-8-66.

Visvader, J. E., & Lindeman, G. J. (2012). Cancer stem cells: Current status and evolving complexities. *Cell Stem Cell*, *10*(6), 717–728. https://doi.org/10.1016/j.stem.2012.05.007.

Wajant, H. (2002). The Fas signaling pathway: More than a paradigm. *Science*, *296*(5573), 1635–1636. https://doi.org/10.1126/science.1071553.

Wang, W. H., Childress, M. O., & Geahlen, R. L. (2014). Syk interacts with and phosphorylates nucleolin to stabilize Bcl-x(L) mRNA and promote cell survival. *Molecular and Cellular Biology*, *34*(20), 3788–3799. https://doi.org/10.1128/MCB.00937-14.

Wang, W., Luo, J., Xiang, F., Liu, X., Jiang, M., Liao, L., et al. (2014). Nucleolin down-regulation is involved in ADP-induced cell cycle arrest in S phase and cell apoptosis in vascular endothelial cells. *PLoS One*, *9*(10), e110101. https://doi.org/10.1371/journal.pone.0110101.

Watanabe, T., Takahashi, A., Suzuki, K., Kurusu-Kanno, M., Yamaguchi, K., Fujiki, H., et al. (2014). Epithelial-mesenchymal transition in human gastric cancer cell lines induced by TNF-alpha-inducing protein of Helicobacter pylori. *International Journal of Cancer*, *134*(10), 2373–2382. https://doi.org/10.1002/ijc.28582.

Watanabe, T., Tsuge, H., Imagawa, T., Kise, D., Hirano, K., Beppu, M., et al. (2010). Nucleolin as cell surface receptor for tumor necrosis factor-alpha inducing protein: A carcinogenic factor of Helicobacter pylori. *Journal of Cancer Research and Clinical Oncology*, *136*(6), 911–921. https://doi.org/10.1007/s00432-009-0733-y.

Willimott, S., & Wagner, S. D. (2010). Post-transcriptional and post-translational regulation of Bcl2. *Biochemical Society Transactions*, *38*(6), 1571–1575. https://doi.org/10.1042/BST0381571.

Wise, J. F., Berkova, Z., Mathur, R., Zhu, H., Braun, F. K., Tao, R. H., et al. (2013). Nucleolin inhibits Fas ligand binding and suppresses Fas-mediated apoptosis in vivo via a surface nucleolin-Fas complex. *Blood*, *121*(23), 4729–4739. https://doi.org/10.1182/blood-2012-12-471094.

Wu, D. M., Zhang, P., Liu, R. Y., Sang, Y. X., Zhou, C., Xu, G. C., et al. (2014). Phosphorylation and changes in the distribution of nucleolin promote tumor metastasis via the PI3K/Akt pathway in colorectal carcinoma. *FEBS Letters*, *588*(10), 1921–1929. https://doi.org/10.1016/j.febslet.2014.03.047.

Xiao, S., Caglar, E., Maldonado, P., Das, D., Nadeem, Z., Chi, A., et al. (2014). Induced expression of nucleolin phosphorylation-deficient mutant confers dominant-negative effect on cell proliferation. *PLoS One*, *9*(10), e109858. https://doi.org/10.1371/journal.pone.0109858.

Xie, Q., Guo, X., Gu, J., Zhang, L., Jin, H., Huang, H., et al. (2016). p85alpha promotes nucleolin transcription and subsequently enhances EGFR mRNA stability and EGF-induced malignant cellular transformation. *Oncotarget*, 7(13), 16636–16649. https://doi.org/10.18632/oncotarget.7674.

Xie, J., Ma, Y. H., Wan, M., Zhan, R. Y., & Zhou, Y. Q. (2014). Expression of dedifferentiation markers and multilineage markers in U251 glioblastoma cells with silenced EGFR and FGFR genes. *Oncology Letters*, 7(1), 131–136. https://doi.org/10.3892/ol.2013.1685.

Xu, C., & Zhang, H. (2015). Somatostatin receptor based imaging and radionuclide therapy. *BioMed Research International*, *2015*, 917968. https://doi.org/10.1155/2015/917968.

Xue, L. J., Mao, X. B., Ren, L. L., & Chu, X. Y. (2017). Inhibition of CXCL12/CXCR4 axis as a potential targeted therapy of advanced gastric carcinoma. *Cancer Medicine*, *6*(6), 1424–1436. https://doi.org/10.1002/cam4.1085.

Yang, C., Kim, M. S., Chakravarty, D., Indig, F. E., & Carrier, F. (2009). Nucleolin binds to the proliferating cell nuclear antigen and inhibits nucleotide excision repair. *Molecular and Cellular Pharmacology*, *1*(3), 130–137. https://doi.org/10.4255/mcpharmacol.09.17.

Yang, A., Shi, G., Zhou, C., Lu, R., Li, H., Sun, L., et al. (2011). Nucleolin maintains embryonic stem cell self-renewal by suppression of p53 protein-dependent pathway. *The Journal of Biological Chemistry*, *286*(50), 43370–43382. https://doi.org/10.1074/jbc.M111.225185.

Yang, X., Xu, Z., Li, D., Cheng, S., Fan, K., Li, C., et al. (2014). Cell surface nucleolin is crucial in the activation of the CXCL12/CXCR4 signaling pathway. *Tumour Biology*, *35*(1), 333–338. https://doi.org/10.1007/s13277-013-1044-0.

Yang, Y., Yang, C., & Zhang, J. (2015). C23 protein meditates bone morphogenetic protein-2-mediated EMT via up-regulation of Erk1/2 and Akt in gastric cancer. *Medical Oncology*, *32*(3), 76. https://doi.org/10.1007/s12032-015-0547-5.

Zhang, J., Tsaprailis, G., & Bowden, G. T. (2008). Nucleolin stabilizes Bcl-X L messenger RNA in response to UVA irradiation. *Cancer Research*, *68*(4), 1046–1054. https://doi.org/10.1158/0008-5472.CAN-07-1927.

Zhao, H., Huang, Y., Xue, C., Chen, Y., Hou, X., Guo, Y., et al. (2013). Prognostic significance of the combined score of endothelial expression of nucleolin and CD31 in surgically resected non-small cell lung cancer. *PLoS One*, *8*(1), e54674. https://doi.org/10.1371/journal.pone.0054674.

Zhuo, W., Luo, C., Wang, X., Song, X., Fu, Y., & Luo, Y. (2010). Endostatin inhibits tumour lymphangiogenesis and lymphatic metastasis via cell surface nucleolin on lymphangiogenic endothelial cells. *The Journal of Pathology*, *222*(3), 249–260. https://doi.org/10.1002/path.2760.

zur Hausen, H., & de Villiers, E. M. (1994). Human papillomaviruses. *Annual Review of Microbiology*, *48*, 427–447. https://doi.org/10.1146/annurev.mi.48.100194.002235.

Zyss, D., & Gergely, F. (2009). Centrosome function in cancer: Guilty or innocent? *Trends in Cell Biology*, *19*(7), 334–346. https://doi.org/10.1016/j.tcb.2009.04.001.

CHAPTER FIVE

Subcellular Targeting of Nitric Oxide Synthases Mediated by Their N-Terminal Motifs

Carlos Costas-Insua, Javier Merino-Gracia, Clara Aicart-Ramos, Ignacio Rodríguez-Crespo[1]
Universidad Complutense de Madrid, Madrid, Spain
[1]Corresponding author: e-mail address: jirodrig@quim.ucm.es

Contents

Abstract

From a catalytic point of view, the three mammalian nitric oxide synthases (NOSs) function in an almost identical way. The N-terminal oxygenase domain catalyzes the conversion of L-arginine to L-citrulline plus ·NO in two sequential oxidation steps. Once L-arginine binds to the active site positioned above the heme moiety, two consecutive monooxygenation reactions take place. In the first step, L-arginine is hydroxylated to make *N*ω-hydroxy-L-arginine in a process that requires 1 molecule of NADPH and 1 molecule of O_2 per mol of L-arginine reacted. In the second step, *N*ω-hydroxy-L-arginine, never leaving the active site, is oxidized to ·NO plus L-citrulline and 1 molecule of O_2 and 0.5 molecules of NADPH are consumed. Since nitric oxide is an important signaling molecule that participates in a number of biological processes, including neurotransmission, vasodilation, and immune response, synthesis and release of ·NO in vivo must be exquisitely regulated both in time and in space. Hence, NOSs have evolved introducing in their amino acid sequences subcellular targeting motifs, most of them located at their N-termini. Deletion studies performed on recombinant, purified NOSs have revealed that part of the N-terminus of all three NOS can be eliminated with the resulting

Advances in Protein Chemistry and Structural Biology, Volume 111
ISSN 1876-1623
https://doi.org/10.1016/bs.apcsb.2017.07.002
© 2018 Elsevier Inc.
All rights reserved.

mutant enzymes still being catalytically active. Likewise, NOS isoforms lacking part of their N-terminus when transfected in cells render mislocalized, active proteins. In this review we will comment on the current knowledge of these subcellular targeting signals present in nNOS, iNOS, and eNOS.

1. NITRIC OXIDE AND NITRIC OXIDE SYNTHASES

Nitric oxide (·NO), initially described as an *endothelium-derived relaxing factor* (Furchgott & Zawadzki, 1980) is a small and gaseous molecule that acts as an ubiquitous secondary messenger throughout the human body. It participates in a variety of biological functions, such as vasodilation of blood vessels, enhancement of learning, and memory consolidation through long-term potentiation in the brain hippocampus or pathogen removal in the innate immune response, to name but a few (Forstermann & Sessa, 2012). ·NO exerts these diverse roles through different mechanisms, being the activation of soluble guanylate cyclase (sGC) and the direct nitrosylation of proteins the canonical ones (Palacios, Knowles, Palmer, & Moncada, 1989; Stamler, Lamas, & Fang, 2001). Regarding the former, ·NO binding to sGC leads to an increment in the intracellular cGMP levels, which subsequently activates protein kinase G (PKG) and, therefore, enhances PKG-mediated phosphorylation. On the other hand, protein nitrosylation requires the posttranslational modification of a cysteine residue, typically within either a transition metal center or an acid–base or hydrophobic structural motif (Martinez-Ruiz et al., 2013). This posttranslational modification is extremely precise and in general, it ultimately leads to a conformational change in the target protein. For example, ·NO on the skeletal muscle only nitrosylates 1 out of 50 free cysteines per ryanodine receptor subunit, which is sufficient to alter the interaction of this protein with calmodulin. Extensive nitrosylation is, on the whole, the same strategy utilized to fight invasion by microbes and to damage cancer cells (Stamler et al., 2001).

Nitric oxide synthases (NOSs) (EC.1.14.13.39) are the main source of this gas in vivo, although low levels of ·NO can be generated non-enzymatically (Zhao, Vanhoutte, & Leung, 2015). To date, three different mammalian ·NO synthases have been identified: the neuronal NOS (also called nNOS or NOS-I), the inducible NOS (also called iNOS or NOS-II), and the endothelial NOS (also called eNOS or NOS-III). These proteins, usually referred as homodimeric enzymes, are, strictly speaking, tetramers, since the active state requires two NOS monomers and two molecules of Ca^{2+}-bound calmodulin (CaM) (Alderton, Cooper, & Knowles,

2001). A NOS monomer comprises two independent domains: the heme oxygenase and the reductase domain, joined by a flexible region that harbors the Ca^{2+}/CaM-binding site. The heme oxygenase domain, located toward the N-terminus of the protein, contains the binding sites for the substrate, L-arginine, two cofactors, the heme group (iron protoporphyrin IX), (6R)-5,6,7,8-tetrahydrobiopterin (H_4B), and two cysteines which coordinate a zinc ion together with the other monomer and have been proven crucial for dimer stability and enzyme activity (Hemmens, Goessler, Schmidt, & Mayer, 2000; Martasek et al., 1998). Catalytically, this domain resembles cytochromes P450, but there is neither sequence homology nor a similar fold whatsoever. In contrast, the reductase domain binds NADPH and the two cofactors, FAD and FMN, and indeed possesses sequence homology with NADPH-cytochrome P450 reductases.

The molecular mass of NOSs varies between ~130 and ~160 kDa, all three proteins present ~55% sequence identity and catalyze the same complex reaction. Briefly, using molecular oxygen and NADPH as cosubstrates, they mediate the oxidation of L-arginine to generate L-citrulline and ·NO through the intermediate *N*ω-hydroxy-L-arginine. Ca^{2+}/CaM binding elicits conformational changes that enhance flavin to flavin and flavin to heme electron transfers, hence, favoring the reaction kinetically rather than thermodynamically, since Ca^{2+}/CaM binding does not alter any redox potential (Noble et al., 1999).

The main difference between the three isoforms relies on the activation process. While eNOS and nNOS require an increase in the intracellular calcium levels to bind Ca^{2+}/CaM reversibly, iNOS is active at basal concentration of this ion, thanks to its great avidity toward calmodulin. The dimerization process also diverges between isoforms. Protein–protein interaction assays have shown that whereas the heme oxygenase by itself mediates the dimeric architecture of iNOS, for nNOS and eNOS, there are interactions between the reductase domains and between the reductase and oxygenase domains across the dimer (Venema, Ju, Zou, Ryan, & Venema, 1997). Finally, a sequence alignment of all three NOSs also shows variations in the length of the N-terminal and C-terminal ends (Fig. 1). Structure–function experiments have shown that while the N-terminal extension is dispensable for activity in all cases (Brenman, Chao, Xia, Aldape, & Bredt, 1995; Ghosh et al., 1997; Rodriguez-Crespo, Moenne-Loccoz, Loehr, & Ortiz de Montellano, 1997), the length of the C-termini is inversely proportional to ·NO synthesis (Nishida & de Montellano, 2001; Roman, Miller, de La Garza, Kim, & Siler Masters, 2000).

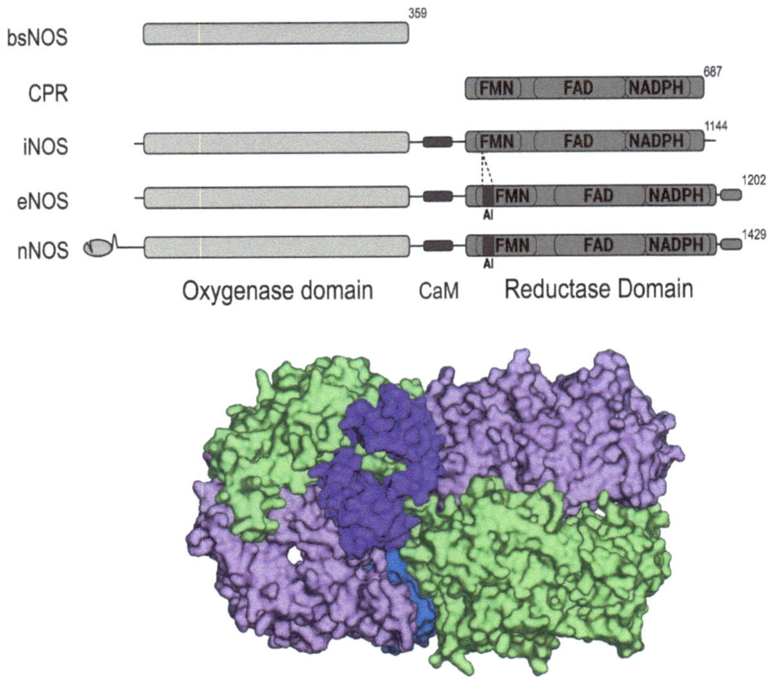

Fig. 1 Bacterial nitric oxide synthase (*Bacillus subtilis* NOS) is homologous to the oxygenase domain of mammalian NOSs. Cytochrome P450 reductase (CPR) is homologous to the reductase domain of mammalian NOSs. The PDZ domain located at the N-terminus of nNOS is also shown. The *bottom part* shows a hypothetical architecture of the NOSs heterotetramer in which the reductase domains transfer electrons in *trans* to the oxygenase domains of the other monomer. The position of the calmodulin moieties is also shown.

Since NO is a free radical, i.e., its bonding structure includes an unpaired electron within the last orbital, it is a potentially hazardous molecule whose synthesis must be tightly regulated in order to occur only *on demand*. Excess of ·NO production can rapidly lead to nitrosative stress, given that ·NO can react with superoxide ion (O_2^-) to form peroxynitrite ($ONOO^-$) which also accounts for oxidative stress, because it can decompose in hydroxyl ion (·OH) and nitrogen dioxide (·NO_2) (Mikkelsen & Wardman, 2003). Many mechanisms underlie NOSs activation to dictate controlled ·NO release and limit its toxicity. Posttranslational NOSs modifications, such as phosphorylation or acylation, as well as cellular protein–protein interactions tightly modulate ·NO production. In this regard, one of the important mechanisms that helps to control spatiotemporal release of ·NO is the proper subcellular localization of the synthesizing enzymes. This process is extensively governed by acylation and through protein–protein interactions which will be the focus of this review.

2. SUBCELLULAR TARGETING OF iNOS MEDIATED BY N-TERMINAL PALMITOYLATION

iNOS was initially identified in macrophages, but its expression can be induced mostly with lipopolysaccharide or cytokines in smooth muscle cells, hepatocytes, chondrocytes, and other cell types (Forstermann et al., 1994; Forstermann & Sessa, 2012). Although the gene is induced by endotoxin and cytokines, some cells can constitutively express the iNOS gene (Park, Park, & Krishna, 1996). iNOS-derived ·NO plays a major role in immune responses to bacteria and tumors. However, iNOS has also different roles in cell signaling, for example, in induced osteoclastic bone resorption (van't Hof et al., 2000). In this isoform Ca^{2+}/CaM is bound even at resting Ca^{2+} concentrations, and the enzyme is therefore constitutively active after expression. The irreversibility of the binding is due to the large amount of contacts between the two proteins (Xia, Misra, Iyanagi, & Kim, 2009).

At the posttranslational level, iNOS activity is mostly regulated through subcellular localization, palmitoylation, protein–protein interactions, and phosphorylation. Interestingly, both the N-terminus and C-terminus of iNOS display strong subcellular targeting motifs. Whereas the N-terminus becomes palmitoylated, its C-terminus associates with PDZ domains.

The C-terminal end of iNOS constitutes a ligand for class I PDZ domains (Hung & Sheng, 2002). Elegant studies have shown that the C-terminus of iNOS can bind to EBP50/NHERF-1, a member of the NHERF (Na^+/H^+ exchanger regulatory factors) family (Glynne, Darling, Picot, & Evans, 2002). iNOS also interacts with CAP70/NHERF-3 (CFTR-associated protein of 70 kDa), a member of NHERF family that contains four PDZ modules and is expressed in kidney, liver, small intestine, and some epithelia (Wang, Yue, Derin, Guggino, & Li, 2000). Binding of recombinant EBP50 or CAP70 augments iNOS activity and also results in an increase in the population of active iNOS dimers. Hence, both EBP50 and CAP70 participate in the correct subcellular targeting of iNOS to the cellular apical domain necessary for the vectorial release of NO in polarized cells (Aicart-Ramos & Rodriguez-Crespo, 2015; Navarro-Lerida, Martinez-Moreno, Ventoso, Alvarez-Barrientos, & Rodriguez-Crespo, 2007).

In the absence of previous myristoylation iNOS becomes palmitoylated on Cys3. This modification is necessary for full activity of the enzyme as well as for the progression of the protein along the intracellular sorting pathways and for its proper subcellular localization (Navarro-Lerida, Alvarez-Barrientos, & Rodriguez-Crespo, 2006; Navarro-Lerida et al., 2004). Acylation of iNOS is completely indispensable for intracellular sorting and ·NO synthesis. Consistently, a Cys3Ser mutant of iNOS is completely inactive and accumulates to intracellular membranes that almost totally colocalize with the Golgi marker β-cop. Somehow expectedly, using radiolabel incorporation assays it was shown that iNOS thioacylation at Cys3 was lost upon mutation of this single residue. Likewise, low concentrations of palmitoylation-blocking agents, such as 2-Br-palmitate or 8-Br-palmitate, severely affected the ·NO synthesis of both iNOS induced in muscular myotubes and transfected iNOS. However, unlike eNOS, iNOS palmitoylation is not involved in its targeting to caveolae (Navarro-Lerida et al., 2004).

3. N-TERMINAL eNOS MYRISTOYLATION AND PALMITOYLATION MEDIATE ITS SUBCELLULAR LOCALIZATION

eNOS is responsible for ·NO synthesis in the cardiovascular system, being crucial in the regulation of vascular tone, platelet aggregation, and angiogenesis. Therefore, eNOS-derived ·NO is one of the key players in vascular homeostasis, modulating blood pressure and vascular tone, but also displaying multiple antiatherogenic roles including antithrombotic, antiproliferative, antiinflammatory, and antioxidant effects (Forstermann & Sessa, 2012). Under physiological conditions the cellular sensitivity toward the ·NO produced by the endothelium is really remarkable. It has been determined that puffs of ·NO close to 1 pM lead to a rise in cGMP to ~30 nM, which is more than necessary to activate cGMP-dependent protein kinases (Batchelor et al., 2010). This responsiveness to ·NO makes cells the most sensitive ·NO detectors yet encountered, but it also implies that eNOS has to be a highly regulated enzyme whose activity must be tightly modulated in very different ways.

At the posttranslational level, the activity of eNOS is chiefly regulated through subcellular localization, myristoylation, palmitoylation, protein–protein interactions, Ca^{2+} levels, and phosphorylation. As in the case of iNOS, both the N-terminus and C-terminus of eNOS display strong subcellular targeting motifs. The eNOS N-terminus becomes irreversibly myristoylated and reversibly palmitoylated, whereas its C-terminus becomes phosphorylated in a process that triggers ·NO synthesis.

The best characterized phosphorylation site of eNOS is Ser1177, which stimulates the flux of electrons within the reductase domain and leads to enzymatic activation (Haque, Ray, & Stuehr, 2016). Comparison with the nNOS crystal structure (Garcin et al., 2004) suggests that the regulatory eNOS C-terminal tail very likely adopts a helical conformation, and the activation observed upon eNOS phosphorylation on Ser1177 is probably due to the displacement of this α-helix and the concomitant increased electron transfer within the reductase domain. Depending on the stimulus, eNOS Ser1177 phosphorylation is performed by Akt/PKB (Michell et al., 1999), AMPK (Chen et al., 1999), PKA (Zhang et al., 2006), cGK-I/PKG (Butt et al., 2000), Chk1 (Park et al., 2011), CaMKII (Fleming, Fisslthaler, Dimmeler, Kemp, & Busse, 2001), or PKD (Aicart-Ramos et al., 2014). In addition, eNOS phosphorylation at this position diminishes the Ca^{2+} sensitivity of the enzyme (Fulton et al., 1999; McCabe, Fulton, Roman, & Sessa, 2000; Schneider et al., 2003).

In cultured vascular endothelial cells, eNOS resides principally in association with the Golgi complex and with plasmalemmal caveolae. In addition, eNOS can traffic between different intracellular compartments and its subcellular targeting can affect ·NO production in response to various stimuli. Experimental results from the Michel and Sessa groups have demonstrated that the N-terminus of eNOS is acylated by N-myristoylation on Gly2 and palmitoylation on Cys15 and Cys26 (Michel & Vanhoutte, 2010; Sessa, 2004). N-myristoylation, a prerequisite for palmitoylation, occurs cotranslationally and is necessary for eNOS membrane association and targeting into the Golgi complex of transfected cells. Therefore, myristoylation-deficient eNOS mutants are distributed diffusely in the cytosol. Palmitoylation, on the other hand, mostly influences the targeting of eNOS into caveolae (Garcia-Cardena, Oh, Liu, Schnitzer, & Sessa, 1996). Hence, myristoylated, palmitoylated eNOS typically appear associated to Golgi endomembranes and cholesterol-rich microdomains of the plasma membranes, including caveolae and lipid rafts. It is generally assumed that the activity of caveolae-targeted eNOS is decreased because of its binding to caveolin-1 (Trane et al., 2014). Although several DHHC palmitoyl acyltransferases are known to use eNOS as a substrate, it has been reported that, in vivo, DHHC21 is responsible for eNOS palmitoylation (Fernandez-Hernando et al., 2006).

Myristoylation of eNOS takes place on a glycine located within a consensus sequence at the N-terminal end that is absent in the other two isoforms, and contrary to palmitoylation, that is a reversible process dynamically regulated, myristoylation is irreversible. Those modified residues are located in a region outside the catalytic core of the enzyme that is not required for full catalytic activity (Rodriguez-Crespo et al., 1997). HEK293 cells stably transfected with palmitoylation-deficient mutants of eNOS release less ·NO than wild-type enzyme does, suggesting that the appropriate subcellular localization mediated by palmitoylation plays an important role in determining the optimal release of ·NO from intact cells. After activation, eNOS is thought to be released from the inhibitory clamp of caveolin-1 and phosphorylated by cellular kinases (Church & Fulton, 2006). The diminished release of ·NO from the palmitoylation-deficient mutants was not attributable to alterations in its catalytic properties as the purified mutant and wild-type enzymes were kinetically identical. Since palmitoylation is necessary for localization of eNOS in caveolae, such localization could regulate the frequency and magnitude of ·NO release in response to stimuli in vivo (Sessa, 2004).

Interestingly, the actions of CaM are thought to be facilitated in cells not only by the dissociation of eNOS from caveolin-1 but also by the recruitment of the molecular chaperone Hsp90 to eNOS. In fact, eNOS-mediated S-nitrosylation of Hsp90 promotes the inhibition of its ATPase and eNOS regulatory activities as a feedback mechanism for limiting eNOS activation (Martinez-Ruiz et al., 2005).

4. nNOS AND ITS MULTIPLE N-TERMINAL LOCALIZATION SIGNALS

nNOS, originally found in rat cerebellum (Bredt et al., 1991; Bredt & Snyder, 1990), is constitutively expressed in specific neurons of the brain and the spinal cord, peripheral nitrergic nerves, epithelial cells of various organs, pancreatic islet cells, and vascular smooth muscle (Forstermann & Sessa, 2012). Encoded on chromosome 12 (12q24.2–12q24.3), it is the largest NOS isoform, composed of ~1434 amino acids, with a predicted molecular mass of ~160.8 kDa (Alderton et al., 2001). Although this is true for nNOSα, the most abundant isoform which predominates in brain tissue, muscle cells express nNOSμ, an isoform that presents a 34-amino acid insertion between the CaM and flavin-binding domains (Fig. 2). In addition, smaller nNOS proteins may be generated through alternative splicing

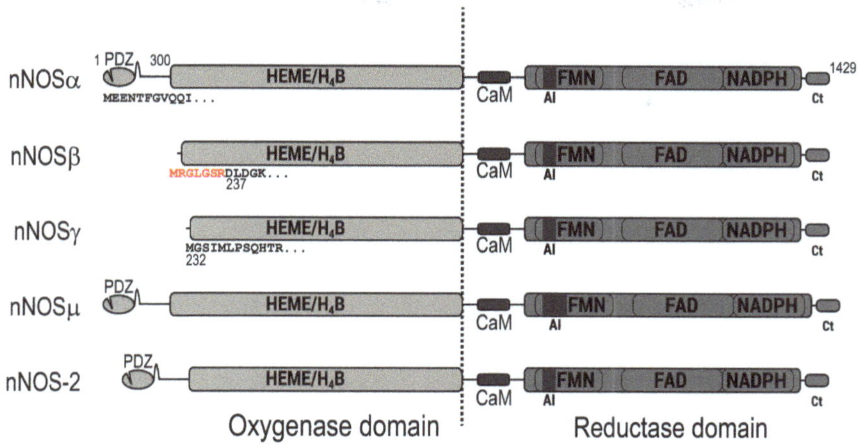

Fig. 2 Various nNOS isoforms found in neuronal and muscular tissues. A full description can be found elsewhere (Luo & Zhu, 2011).

(Bredt & Snyder, 1990; Silvagno, Xia, & Bredt, 1996). Two alternative transcripts, nNOSβ and nNOSγ, were identified in nNOS knockout mice generated by deletion of exon 2 in embryonic stem cells. Finally, nNOS-2 was identified in neuroblastoma cell lines (Brenman et al., 1996; Fujisawa et al., 1994). Both nNOSβ and nNOSγ are cytoplasmic proteins that do not associate with synaptic membranes and lack the first ~236 amino acids of nNOSα. On the other hand, nNOS-2 presents a deletion within the catalytic region (Δ504–608) that would render an inactive protein which might function as a dominant-negative regulator of nNOS activity (Alderton et al., 2001). However, it is not known if these three proteins are expressed in physiological conditions and if so, what their functions might be. Notwithstanding the potentially interesting role of these alternative transcripts, this review will focus on the common known aspects of nNOSα and nNOSμ.

At the posttranslational level nNOS is regulated through phosphorylation, degradation, subcellular localization, and dimerization. Phosphorylation of nNOS is controlled by PKA, CaMKs, PKC, PKD, and protein phosphatase 1 (PP1) (Alderton et al., 2001). Phosphorylation of Ser1412 by PKD leads to an increase in ·NO synthesis, whereas phosphorylation at Ser847 by CaMKII inhibits CaM binding and, therefore, decreases ·NO production (Rameau, Chiu, & Ziff, 2004; Sanchez-Ruiloba et al., 2014). By contrast, PP1 reduces the phosphorylation state of Ser847, hence augmenting ·NO levels. In addition, dimerization of nNOS is regulated through different protein interactions. Hsp90 is important for nNOS stabilization and regulates the dimerization by helping heme insertion in the monomer. Simultaneously, Hsp90 impedes Hsp70 binding and Hsp70-dependent ubiquitination (Peng, Morishima, Pratt, & Osawa, 2012). Because Hsp90 can be inhibited by ·NO-mediated nitrosylation, nNOS displays a feedback regulatory mechanism to control its own levels.

As stated earlier, both nNOSα and nNOSμ are the largest isoforms among NOSs. This difference in size is due to an N-terminal extension of approximately 300 amino acids absent in the other two counterparts, iNOS and eNOS. This stretch is extensively involved in the subcellular localization of the protein and does not intervene in ·NO production (Brenman et al., 1996). This specific targeting is accomplished through different modules that will be discussed later. Within the 300 N-terminal amino acids, there can be found a PDZ domain that binds to the C-termini of other proteins (Bredt et al., 1991), a β-hairpin module that associates to the PDZ domain of proteins such as PSD95 and α_1-syntrophin (Brenman et al., 1996), a DYNLL1-binding site (Fan, Zhang, Tochio, Li, & Zhang, 2001; Rodriguez-Crespo, Straub, Gavilanes, & Ortiz de Montellano, 1998), and a short stretch known to bind to the spectrin repeats R16/R17 of dystrophin whose exact position within the nNOS sequence is not yet clear (Lai, Zhao, Yue, & Duan, 2013).

One feature that is quite specific of the N-terminus of nNOS is the presence of the β-hairpin moiety (amino acids 105–128), adjacent to the PDZ domain, that is able to behave as a PDZ ligand in a noncanonical fashion, acting as a *pseudocarboxylate* end (Hillier, Christopherson, Prehoda, Bredt, & Lim, 1999; Tochio, Mok, et al., 2000). This N-terminal stretch of nNOS binds to a selected group of PDZ domains such as the second PDZ domain of PSD95, PSD93, and SAP97 (Brenman et al., 1996; Cui et al., 2007) as well as to the single PDZ domain of α1-syntrophin, β1-syntrophin, and β2-syntrophin (Brenman et al., 1996; Gee et al., 1998). Early reports suggested that both the nNOS PDZ domain and the β-hairpin modules were functionally dependent, that is, the occupancy of the hydrophobic groove of the PDZ domain would result in the inability of the and β-hairpin to associate to a subsequent PDZ domain (Jaffrey, Snowman, Eliasson, Cohen, & Snyder, 1998). More recent studies have challenged this mechanism and proposed that a ternary complex might indeed be formed (Courtney, Li, & Lai, 2014; Merino-Gracia, Costas-Insua, Canales, & Rodriguez-Crespo, 2016).

5. THE PDZ DOMAIN OF nNOS AND ITS INTERACTING PROTEINS

PDZ domains are the most common protein–protein interaction modules. To date, over 330 individual PDZ domains have been identified within the human genome (Tonikian et al., 2008). These domains mediate the assembly of large multiprotein complexes such as the one present at the postsynaptic density of neurons or at the sarcolemma of skeletal muscle. The prototypical structure of a PDZ domain consists of a β-sandwich composed of six β-strands and two α-helices, with the α-helices capping both sides of the structure. Between the β2 strand and the α2 helix there is a hydrophobic groove that can be occupied by the C-terminus of the interacting partner protein as an additional antiparallel strand, hence stabilizing the preformed β-sheet. The extensive structural data available have revealed that canonical PDZ binding implies at least the last four (in some cases, even up to the last eight) amino acids of the target protein. PDZ domains show clear binding pockets for the amino acid present in the penultimate amino acid of the incoming peptide (named P_{-2}), for the amino acid present in the last position (named P_0), and also for the terminal carboxylate group which interacts with the "GLGF" loop, an extremely conserved feature of all PDZ domains (Daniels, Cohen, Anderson, & Brunger, 1998; Morais Cabral et al., 1996; Tochio, Hung, Li, Bredt, & Zhang, 2000). Accordingly, the interactions between C-terminal peptides and PDZ domains have been divided into three different classes based on the specific amino acid sequences present in the C-terminal stretch. That is, class I PDZ domains recognize proteins ending in -[S/T]-X-ϕ-COOH (where ϕ is any hydrophobic amino acid), whereas class II prefer proteins ending in -ϕ-X-ϕ-COOH, and class III in turn bind acidic motifs (-[D/E]-X-ϕ-COOH).

Although binding to C-terminal peptides appears to be the typical mode of interaction, PDZ domains can also interact with internal peptide sequences. This is the case of the PDZ domain of α1-syntrophin and the second PDZ of PSD95 (Fig. 3) (Hillier et al., 1999). In addition, PDZ domains can associate with other PDZ domains to form homo- and heterooligomers. Examples include GRIP (Dong et al., 1999), ZO-1 (Fanning, Jameson, Jesaitis, & Anderson, 1998), or InaD (Xu, Choudhury, Li, & Montell, 1998). PDZ domains are also reported to interact with other distinct protein-binding motifs, including ankyrin repeats (Maekawa et al., 1999), spectrin repeats (Lai et al., 2013) (see below), and LIM domains (Cuppen, Gerrits, Pepers, Wieringa, & Hendriks, 1998). The structural bases for these interactions are not yet clear.

The binding properties of the PDZ domain of nNOS (amino acids 14–105) have been the object of intense study. We (Merino-Gracia, Costas-Insua, et al., 2016) and others (Schepens, Cuppen, Wieringa, & Hendriks, 1997; Stricker et al., 1997) have analyzed the sequence requirements for the C-terminus of a protein to bind to the PDZ domain of nNOS. As the founding member of class III, the PDZ domain of NOS does indeed bind to peptides that end with the sequence -G-[D/E]-X-V-COOH. However, it also binds to class II and class I sequences. Nevertheless, as it will be discussed further, the behavior of this domain does not fall entirely in any of the three PDZ classes described earlier, and, therefore, we have proposed that the nNOS PDZ domain is a hybrid PDZ domain (Merino-Gracia, Costas-Insua, et al., 2016). The list of nNOS PDZ interaction partners available in the published literature is based on yeast two-hybrid assays but also on observations from in vitro binding experiments (i.e., pull-down assays) which can render both direct and indirect associations. Hence, proposed interactions of nNOS with cellular proteins should be taken with care.

When the PDZ domain of nNOS was used as bait and confronted with a rat hippocampal cDNA library comprising 6 million clones, NOS1AP (initially called CAPON) was retrieved as the only interacting protein. NOS1AP is a 337-amino acid protein that contains a C-terminal class II PDZ-motif (-I-A-V-COOH). It was clearly shown that the PDZ domain of nNOS binds to the final 13 amino acids of NOS1AP and subtle mutations of the final four amino acids of NOS1AP abrogated this binding (Jaffrey et al., 1998). NOS1AP has recently emerged as a key regulator of multiple pathologic conditions, such as cerebral ischemia, anxiety, or schizophrenia, a role thought to be dependent on nNOS interaction (Courtney et al., 2014). A lateral association between NOS1AP and the nNOS PDZ domain that increases the overall binding has also been recently described (Li, Cisek, & Courtney, 2017; Li et al., 2015). Binding through this secondary site is important in terms of controlling dendritic development and filopodial outgrowth (Candemir et al., 2016).

Using amino acids 1–195 of nNOS in an affinity chromatography assay, the muscle isoform of phosphofructokinase (PFK-M) was retrieved as another class II interactor of the PDZ domain of nNOS (Firestein & Bredt, 1999). nNOS-positive neurons are enriched in PFK-M, which could safeguard against ·NO production since its product, fructose-1,-6-bisphosphate (FBP), has a neuroprotective profile in animal models. Although the mechanism of action of FBP is not completely understood, it attenuates LPS-induced iNOS expression in microglia, although the question of whether FBP affects nNOS expression in neurons remains unanswered (Kim, Park, Baik, & Lee, 2012). Unfortunately, we (Merino-Gracia, Costas-Insua, et al., 2016) and others (Stiffler et al., 2007) have failed to detect binding using the C-terminal end of this protein. This could be due to a low affinity interaction, the absence of a bridging protein between nNOS and PFK-M (indirect association) or to the necessity of secondary contacts between both proteins. Following this line, it has been shown that amelioration in a Duchenne Muscular Dystrophy murine model by introducing a muscle-specific nNOS transgene was related to an increase in PFK activity, effect that was partially mediated by a 36-amino acid peptide in the FAD-binding domain of nNOS (Wehling-Henricks, Oltmann, Rinaldi, Myung, & Tidball, 2009). Since this peptide is exposed on the surface of the protein, a secondary contact with this region cannot be discarded.

Vac14, the activator of the PtdIns(3)P 5-kinase PIKfyve, has also been proposed as a nNOS-interacting partner (Lemaire & McPherson, 2006). The importance of this partner relies on its singular mode of interaction. Despite presenting the archetypical class II sequence (-V-V-L-COOH), the last three amino acids of Vac14 were dispensable for binding to occur, inasmuch their deletion or substitution with Ala residues rendered a chimera that was still able to bind to amino acids 1–100 of nNOS. Such an unexpected binding mode is possible because of the presence of an internal motif (-G-D-H-L-D-), that resembles the preferred class III nNOS sequence (-G-D-A-V-COOH), with one of the aspartic residues mimicking the carboxylate end. In this context, it is interesting to mention that other proteins which have been suggested to bind the nNOS PDZ or have been related somehow with ·NO synthesis present internal binding motifs. That is the case of α_{1C}-adrenergic receptor (-G-E-N-G-E-) (Schepens et al., 1997), NOS1AP (-G-D-S-L-D-) (Jaffrey et al., 1998), and the 5-hydroxytryptamine receptor 2B (-G-D-K-T-E-) (Lemaire & McPherson, 2006).

Regardless of this plethora of class II ligands, the initial studies with the nNOS PDZ domain indicated it behaves as a bona fide class III domain. Both a screening of an unbiased library of 13 billion C-terminal peptides, or a high-throughput yeast two-hybrid library assay resulted in the predominant binding of the PDZ domain of nNOS to peptides displaying the -G-D-X-V-COOH motif, where X was predominantly a hydrophobic amino acid (Schepens et al., 1997; Stricker et al., 1997). In the absence of biochemical information regarding cellular proteins displaying this binding motif, it was suggested that the melatonin receptor (-V-D-S-V-COOH) (Tochio, Zhang, Mandal, Li, & Zhang, 1999) and the α_{1C}-adrenergic receptor (-GEEV-COOH) (Schepens et al., 1997) could represent two binding partners. A few years later, one group identified a palmitoyl transferase of the DHHC family, NIDD (nNOS interacting DHHC domain-containing protein), now renamed DHHC23, as a class III ligand (-E-D-I-V-COOH) (Saitoh et al., 2004). Interestingly, in this case, the interaction not only regulated the subcellular location of the protein but also enhanced ·NO synthesis, a fact that encases clinical relevance. *Carboxy-terminal-binding protein* (-S-D-Q-L-COOH) has been also identified as a nNOS ligand (Riefler & Firestein, 2001). This protein is a transcriptional coregulator that down-regulates the expression of many tumor suppressor genes. It is believed that nNOS interaction sequesters CtBP from the nucleus by impeding SUMOylation in Lys432, which is a nuclear localization signal (Lin et al., 2003). Nevertheless, the clinical relevance of this nNOS-mediated prevention of transcription repression remains unknown. In concordance with all these results, cytochrome *c* oxidase V (-L-D-K-V-COOH) (Persichini et al., 2005) has also been depicted as a class III ligand. Finally, we identified another class III ligand with potential physiological relevance given its tissue distribution, claudin-3 (-K-D-Y-V-COOH). Although we were able to prove binding abrogation with phosphomimetic constructs, the question about how claudin-3 functionality becomes regulated when attached to active nNOS and what role phosphorylation plays, as well as the kinase implicated, if so, is still open. Taken together, these data strengthen the conception that this PDZ domain displays a true class III specificity.

To make things even more complicated, the PDZ domain of nNOS has also been described to bind class I sequences. That would be the case of the ubiquitous calcium ATPase, PMCA-4b. This interaction was initially discovered by immunoprecipitation techniques (Schuh, Uldrijan, Telkamp, Rothlein, & Neyses, 2001) and confirmed later on by FRET assays (Duan et al., 2013) with both studies agreeing in PMCA-4b being a negative regulator of ·NO synthesis. In our laboratory, we were also able to reproduce this interaction in vitro using synthetic peptides mimicking the C-terminus of the ATPase (Merino-Gracia, Costas-Insua, et al., 2016). Albeit this interaction occurs mainly in lipid rafts of neurons of the CNS, its functionality has only been investigated in heart muscle, were some results have showed that the nNOS–PMCA-4b complex regulates contractility via cAMP and phosphorylation of both phospholamban and cardiac troponin I (Mohamed et al., 2009).

6. UNDERSTANDING THE PROMISCUITY OF BINDING SHOWN BY THE PDZ DOMAIN OF nNOS

The nNOS PDZ domain can accommodate within its binding pocket class I, class II, and class III C-terminal sequences. In order to try to rationalize this promiscuousness, we have put forward two different explanations. It is conceivable that the PDZ domain of nNOS has evolved to bind to the C-terminus of multiple sequences at the expense of interaction affinity. Alternatively, it is also feasible that despite binding multiple C-terminal sequences, some of them might have a different interaction mode with the PDZ domain, such as some C-terminal sequences binding in the main hydrophobic groove and others extended beyond toward a secondary binding surface. The scrutiny of the amino acid sequence of the nNOS PDZ domain seem to favor the former explanation, although the vast differences in binding affinities observed for certain C-terminal peptides when extended in length from 4–12 residues also seem to indicate that certain interacting areas lie outside the canonical-binding groove. In this regard, we have been able to measure accurate binding constants of proteins to small peptides using isothermal titration calorimetry and through polarization anisotropy determinations (Merino-Gracia, Costas-Insua, et al., 2016; Merino-Gracia, Zamora-Carreras, Bruix, & Rodriguez-Crespo, 2016).

A canonical interaction between PDZ domain and C-terminal ligand is defined as one in which the peptidic main chain adopts a β-strand conformation interacting in an antiparallel fashion with strand β2. This conformation places the peptide P_{-2} side chain proximal to the side chains at positions α2-1 and α2-5 of the PDZ domain, which serve to determine the ligand residue types that are favored at this position (Ernst et al., 2014; Hung & Sheng, 2002; Ye & Zhang, 2013). In addition, the residue at position α2-2 also contributes to site P_{-2} specificity by helping to orient the incoming peptide in the binding groove. In contrast, in most of PDZ domains studied to date the predominant specificity at site P_{-3} is for hydrophobic residues, albeit a few prefer Asp/Glu, and some prefer Thr/Ser (Ernst et al., 2014). In most cases, peptides interacting with an Asp or Glu residue at P_{-3} use a similar mechanism for ligand recognition. In general, positions α2-2, β3-4, or β3-5 are occupied by Lys, Arg, or Ser residues, which are capable of forming salt bridges or hydrogen bonds with the carboxylate group of the lateral chain of the acidic residue.

A sequence comparison of the nNOS PDZ domain with those of the PDZ domain of ZO1–3 and MUPP1–10, two proteins which bind peptides with acidic residues at P_{-2}, reveals that instead of a polar residue, a Leu residue occupies the nNOS α2-2 position and a Tyr residue (Tyr77) occupies position α2-1 instead of Lys or Arg. Hence, nNOS only partially fulfills the sequence requirement necessary for acidic P_{-2} recognition (class III specificity). Accordingly, a sequence comparison with the PDZ domains of ERBB2IP1, Dlg1–2, Dlg4, and SNTA1, four proteins which recognize peptides with acidic residues occupying P_{-3} position, reveals that nNOS lacks a residue capable of hydrogen bonding the incoming peptide in α2-2 position (which, as has been stated earlier, is a Leu residue in nNOS). Nevertheless, Ser44, situated in α3-4, by hydrogen bonding the carboxylate from the lateral chain in P_{-3}, could be involved in the recognition of this type of peptides. With these data in mind, we performed an NMR experiment in which we titrated a ^{15}N-labeled nNOS PDZ with two different peptides, a synthetic class III, and a peptide derived from the C-terminus of NOS1AP (Merino-Gracia, Costas-Insua, et al., 2016).

Somehow expectedly, we found differences in the chemical shifts of some amino acids of both peptides, which implies that different residues are involved in the accommodation of these peptides. Briefly, class III binding was dependent on the presence of Tyr77, which is positioned in α2-2, since its chemical shift varied notably when the peptide was added, and if substituted for an alanine, binding was abrogated. Regarding class II binding specificity, things are more complicated. On the one hand, it is feasible that the acidic residue present in P_{-3} could interact with Ser44, given that the signal from amino acids surrounding it changed noticeably when a class II peptide with this characteristic was present, and, in concordance with Tyr77, its substitution for an alanine diminished the binding (*data not published*). Nevertheless, the NOS1AP peptide shows an internal sequence similar to that of Vac14 that could be occupying the same region within the PDZ. Accordingly, we detected a chemical shift in the signal of Ala51, which is situated far ahead from the hydrophobic groove, which is consistent with a peptide extension outside the boundaries. To summarize, according to our data, the PDZ domain of nNOS displays three different docking modes. The first one would involve Tyr77 and would accommodate peptides displaying the sequence -[D/E]-X-ϕ-COOH. The second one would require Ser44 hydrogen bonding peptides with an acidic residue in P_{-3}

position (with preference for glutamate). Thus, class I or class II ligands with the sequence -E-[S/T/ϕ]-X-ϕ-COOH would fall within this binding mode. Finally, the third one would imply peptide extension outside the boundaries, with an important role for Ala51. This could be the case of Vac14, NOS1AP, and even α_{1C}-adrenergic receptor and 5-HT_{2B}. It is important to note that the second and third docking modes are not mutually exclusive (Merino-Gracia, Costas-Insua, et al., 2016), and it is feasible that NOS1AP could associate in both ways, so more experiments to clarify which determines each type of binding are necessary. Consistently with our results, a recent study proposed a similar mode of association for the PDZ domain of PICK1 (Erlendsson et al., 2014).

In neurons, the interaction with PSD95 places nNOS in the vicinity of NMDA receptors, association that couples calcium influx with ·NO production (Cao et al., 2005) and explains ·NO implication in NMDAR-evoked excitotoxicity. In contrast, α1-syntrophin mainly recruits nNOS to the sarcolemma in striated muscle, which serves to join muscle contractibility with ·NO-mediated vasodilation (Bhat, Adams, & Khanday, 2013). A recent reports suggests a role for α_1-syntrophin in the nuclear translocation of nNOS to promote mitochondria biogenesis as well (Aquilano, Baldelli, & Ciriolo, 2014). In heart muscle, it is believed that the sarcolemmal localization of nNOS enhances the late sodium current through direct nitrosylation of a sodium channel, SCNA5, mechanism that is inhibited by caveolin-3 (Cheng et al., 2013). Mutations in Cav-3 have been linked to type 9 long QT syndrome and sudden infant death syndrome, which could be due to ·NO-mediated hyperexcitability. Interestingly, α1-syntrophin also binds dystrophin, another protein that has been reported to interact with nNOS (Bhat et al., 2013).

Extensive research has been made over this stretch in order to assess its structural and functional independence from the canonical PDZ domain. Although a native-like β-finger structure of a peptide from the PDZ domain of nNOS can be observed, this peptide represented a minor population of the peptide conformers (Wang, Zhang, Tochio, Fan, & Zhang, 2000). This might indicate that the C-terminal extension peptide of the nNOS PDZ domain may represent a relatively independent structural unit; this result could also be read as an increase in the instability of the structure when the PDZ is absent. In concordance, the same group elegantly proved the importance of the salt bridge between Asp62 (within the PDZ domain) and Arg121 (within the β-hairpin) to maintain the β-hairpin conformation (Tochio, Mok, et al., 2000) which further strengthens the conception that the folding is dependent on the PDZ. In agreement, our laboratory has not detected interaction when confronting α1-syntrophin with amino acids 85–130 of nNOS in a yeast two-hybrid assay (*data not published*), which highlights the key role of this salt bridge. In contrast, the possibility of this module binding PSD95 or α1-syntrophin when the hydrophobic groove of the PDZ is occupied has remained controversial. Jaffrey et al. (1998) suggested that NOS1AP binding would displace PSD95 from the β-hairpin which would imply a negative role for NOS1AP in ·NO synthesis, given that NMDAR-mediated Ca^{2+} influx is necessary. Nevertheless, this same group demonstrated ·NO-mediated activation of Dexras1 via NOS1AP after NMDAR stimulation, which challenges this mutually excluding model (Fang et al., 2000). Moreover, some studies have detected a ternary complex, composed of PSD95, nNOS, and a peptide, by immunoprecipitation techniques (Christopherson, Hillier, Lim, & Bredt, 1999; Courtney et al., 2014; Li et al., 2013). We also tested the possibility of formation of a tripartite complex and rationalized that if the occupancy of the nNOS PDZ domain affects the binding properties of the nNOS β-hairpin module (and vice versa), the K_d value in a fluorescence polarization assay toward any C-terminal peptide should differ if using a nNOS/α1-syntrophin complex or nNOS alone. In contrast, we did not detect significant changes in the K_d values under assay conditions in which the fluorescence polarization values were higher for a peptide binding to a nNOS/α1-syntrophin that to nNOS. Detection of higher polarization values indicates that the nNOS/α1–syntrophin complex has been formed and the absence in noticeable changes in the K_d values indicate that binding of the nNOS PDZ domain to C-terminal peptides and binding of the nNOS β-hairpin module to other PDZ are fundamentally independent events (Merino-Gracia, Costas-Insua, et al., 2016). Likewise, using nNOS PDZ domain protein chimeras self-saturated with a class III peptide connected through a linker, we were able to detect binding to PSD95 and α_1-syntrophin in a yeast two-hybrid assay (*data not published*) consistent with the formation of a ternary complex, whereas binding to classical PDZ ligands (e.g., NOS1AP) was disrupted.

7. nNOS AS A TARGET IN EXCITOTOXICITY

Compared to other tissues and organs, the brain is particularly prone to ischemic damage so that in the past decades, hypoxia–ischemia stroke research has scrutinized in detail glutamate neurotoxicity. One explanation for this susceptibility might be the high levels of glutamate known to trigger the so-called excitotoxic cascade, and to which many neurons present receptors (Lai, Zhang, & Wang, 2014). Briefly, an interruption in the glucose and oxygen input to the brain causes a primary energetic malfunction and ionic gradient imbalance because of failure of ionic pumps, which originates the massive release of glutamate due to cell depolarization. This excess glutamate, mainly through activation of NMDA receptors, causes a large calcium influx within neurons and the activation of many calcium-dependent enzymes, such as nNOS, PLA2, or certain caspases. In addition, nNOS activation leads to nitrosative and oxidative stress. Excess glutamate also causes microglia activation, which secretes cytokines such as TNF-α that in turn increase blood–brain barrier permeability and enhance neuroinflammation. Hence, excitotoxicity, inflammation, and oxidative stress are considered the "deadly triad" leading to hypoxia–ischemia-induced brain damage (Lafuente et al., 2016). When oxygen and glucose supply is restored, many oxygen-dependent enzymes (such as nNOS) remain hyperactive, which causes a secondary injury that accounts for neuronal cell death outside the core of the lesion. This biphasic pattern of damage allows for a time-window for clinical intervention during which appropriate therapeutic compounds could be used. To date, alleviation of vascular occlusion on thrombotic cases is the only therapy approved, but new therapeutic approaches are being investigated. Among them, the NMDAR–PSD95–nNOS axis has been proposed as a possible target and, specifically, chemical compounds that might be able to interfere with the NMDAR–PSD95 interaction.

Early studies demonstrated that nNOS inhibition is neuroprotective in animal models of ischemic stroke (Dawson, Dawson, Bartley, Uhl, & Snyder, 1993; Dawson, Dawson, London, Bredt, & Snyder, 1991; Dawson, Kizushi, Huang, Snyder, & Dawson, 1996). Consistent with the hypothesis of NMDAR-mediated ·NO production as a main event in hypoxia–ischemia induced damage, it was shown that knockdown of PSD95 also attenuated excitotoxicity (Sattler et al., 1999). Altogether, these results led to a clinical trial with a peptide designed to disrupt this axis, given that this strategy has several advantages over pharmacological inhibition of either NMDAR or nNOS. Ideally, impairment of NMDAR-mediated nNOS activation without affecting normal NMDAR function would be the ultimate goal, since NMDAR has important functions in the brain. TAT-NR2B9c, a small interference peptide resembling the last nine amino acids of the NMDAR GluN2B subunit, was designed to selectively block NMDAR–PSD95 interaction (Aarts et al., 2002). Neuron loading with this peptide had no effect on NMDAR current but protected from neurotoxicity in murine and nonprimate animal models (Aarts et al., 2002; Cook, Teves, & Tymianski, 2012; Sun et al., 2008), which led to its approval in a clinical trial in humans. The results of the phase II trial suggested that a NMDAR-base treatment is indeed clinically feasible for the management of stroke (Hill et al., 2012). Nevertheless, despite the neuroprotective efficacy of the Tat-NR2B9c peptide in preclinical and clinical trials, its mechanism of action might be different from what was originally thought since this permeable peptide disrupted the PSD95–nNOS interaction with higher potency than that of PSD95 with NMDAR (Cui et al., 2007). Given this finding and, in order to overcome TAT-NR2B9c limitations (poor pharmacokinetics, excessive production costs, and immunological concerns) (Lai & Wang, 2010) a high-throughput screening to identify small drugs that could effectively disrupt de PSD95–nNOS interaction was conducted (Florio et al., 2009). This study rendered the precursor of the compound IC87201, which has an IC_{50} of 31 μM. An independent study designed and synthesized the compound ZL006, which, despite the structural similarity, has higher specificity and potency without noticeable effects on the interaction of PSD95 with other proteins (Zhou et al., 2010). A third promising compound SCR-4026, with much lower IC_{50} value and clear ability to disrupt the nNOS–PSD95 interaction, has also been recently described (Mo et al., 2016). Interestingly, a recent report proposed that compounds ZL006 and IC87201, although quite promising as drugs, do not interact with the PDZ domains of nNOS or PSD-95, nor inhibit the nNOS–PDZ/PSD95–PDZ interface by interacting with the β-finger of nNOS (Bach et al., 2015). This behavior might reflect the necessity of using quite large concentrations of synthetic compounds in order to obtain a clear interference in the association between both proteins (Mo et al., 2016).

8. nNOS AS A TARGET IN DUCHENNE MUSCULAR DYSTROPHY

Duchenne muscular dystrophy is the most common of heritable muscular dystrophies and results from mutations in the dystrophin gene (Tidball & Wehling-Henricks, 2014). Dystrophin deficiency causes muscle fatigability, misregulated blood flow, fibrosis, inflammation, impaired muscle regeneration, abnormalities in synaptic structure and function, and even defects in specific features of learning and memory, a plethora of symptoms that cannot imply just one protein. nNOS is part of the varied number of proteins that associate to dystrophin and form the dystrophin complex, an association that is disrupted in the pathology (Brenman et al., 1995). Subsequent studies have confirmed that nNOS is indeed absent from the sarcolemma in both DMD patients and animal models (Chang et al., 1996; Judge, Haraguchiln, & Chamberlain, 2006; Thomas et al., 1998; Wells et al., 2003). Although these results suggest that dystrophin deficiency causes nNOS absence from muscle, rather than mislocalization, recovery of the dystrophic muscle by strategies aiming to relocalize nNOS to the sarcolemma has proven successful (Wells et al., 2003) and sarcolemmal targeting of nNOSμ clearly improved contractile function of mdx muscle (Rebolledo, Kim, Whitehead, Adams, & Froehner, 2016).

Until a few years ago, the prevailing model was that nNOS was recruited to the sarcolemma by syntrophin, a process altered in dystrophic patients. Nevertheless, in some cases, despite the sarcolemmal localization of syntrophin, nNOS was still absent from the membrane. Finally, spectrin repeats R16/R17 of dystrophin were identified as the docking site for the N-terminus of nNOS (Fig. 3) (Lai et al., 2009). This docking site has been subsequently delimited to a 10-amino acid microdomain, present in helix α1 of R17 and framed by helices α2 and α3 (Lai et al., 2013), a region lost in most dystrophin mutations causing the disease. Very likely, the concerted binding of the β-hairpin of nNOS to syntrophin together with the lateral association of the PDZ domain of nNOS to dystrophin would result in the correct nNOS localization to the sarcolemma (Zhao et al., 2016). Apparently, the nNOS secondary elements involved in dystrophin binding include the β-hairpin, together with β-strand 1, 4, and 5., which constitute a flat and hydrophobic front surface that strongly associates with a hydrophobic patch in the dystrophin R16–17 surface (Molza et al., 2015).

The treatment of the disease is based on the knowledge acquired from a milder form of the pathology, the Becker Muscular Dystrophy, which has shown that large in-frame deletions do not result in a severe phenotype. Hence, smaller versions of the dystrophin gene (mini- and micro-dystrophins), genome have been developed to facilitate gene therapy. Given the relatively recent discovery of nNOS association to repeats 16/17 of dystrophin, few studies have compared the effect of smaller with and without the nNOS-binding region (Reza, Laval, Roos, Carr, & Lochmuller, 2016). Although treatment with ·NO donors does not seem to ameliorate symptoms but instead worsen them (Timpani et al., 2017), the gene therapy aimed to restore nNOS at the sarcolemma could have a completely opposite effect. In this line, nNOS restoration using autologous skeletal muscle-derived cells treated with lentiviral vectors proved effective (Meng et al., 2016).

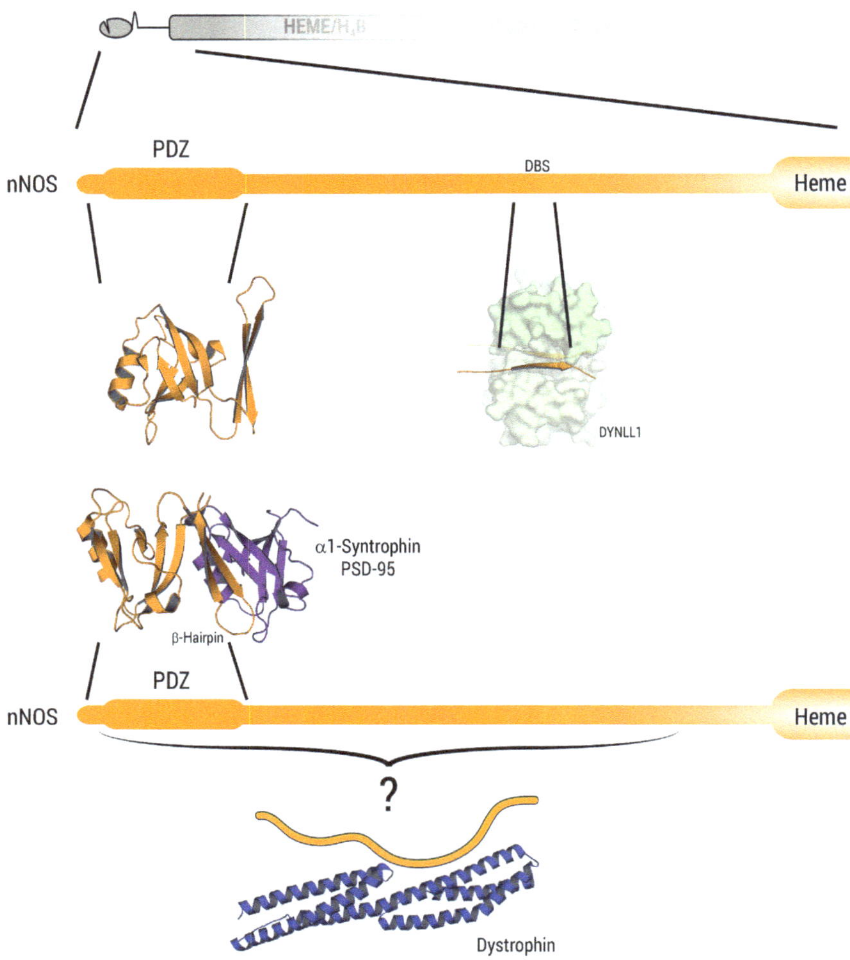

Fig. 3 Representative *cartoon* of the protein binding properties of the N-terminus of nNOS. High-resolution structures are available for the PDZ domain of nNOS (Tochio et al., 1999) and for the short stretch (dynein binding sequence, DBS) known to bind to DLC8/DYNLL1 (Fan et al., 2001). In addition, the nNOS β-hairpin associates to other PDZ domains (Hillier et al., 1999), and two spectrin repeats of dystrophin (R16/R17) bind to the N-terminus of nNOS as well (Lai et al., 2009).

9. FINAL REMARKS

The N-terminus region of all three mammalian NOSs significantly determine their subcellular localization, activity, interaction with other proteins and correct coupling with effectors. Although the amount of available data regarding the targeting signals present within these N-terminal extensions, it would not be a surprise if novel protein binders were identified in the future (Luo & Zhu, 2011).

REFERENCES

Aarts, M., Liu, Y., Liu, L., Besshoh, S., Arundine, M., Gurd, J. W., et al. (2002). Treatment of ischemic brain damage by perturbing NMDA receptor-PSD-95 protein interactions. *Science*, *298*, 846–850.

Aicart-Ramos, C., & Rodriguez-Crespo, I. (2015). Binding of PDZ domains to the carboxy terminus of inducible nitric oxide synthase boosts electron transfer and NO synthesis. *FEBS Letters*, *589*, 2207–2212.

Aicart-Ramos, C., Sanchez-Ruiloba, L., Gomez-Parrizas, M., Zaragoza, C., Iglesias, T., & Rodriguez-Crespo, I. (2014). Protein kinase D activity controls endothelial nitric oxide synthesis. *Journal of Cell Science*, *127*, 3360–3372.

Alderton, W. K., Cooper, C. E., & Knowles, R. G. (2001). Nitric oxide synthases: Structure, function and inhibition. *The Biochemical Journal*, *357*, 593–615.

Aquilano, K., Baldelli, S., & Ciriolo, M. R. (2014). Nuclear recruitment of neuronal nitric-oxide synthase by alpha-syntrophin is crucial for the induction of mitochondrial biogenesis. *The Journal of Biological Chemistry*, *289*, 365–378.

Bach, A., Pedersen, S. W., Dorr, L. A., Vallon, G., Ripoche, I., Ducki, S., et al. (2015). Biochemical investigations of the mechanism of action of small molecules ZL006 and IC87201 as potential inhibitors of the nNOS-PDZ/PSD-95-PDZ interactions. *Scientific Reports*, *5*, 12157.

Batchelor, A. M., Bartus, K., Reynell, C., Constantinou, S., Halvey, E. J., Held, K. F., et al. (2010). Exquisite sensitivity to subsecond, picomolar nitric oxide transients conferred on cells by guanylyl cyclase-coupled receptors. *Proceedings of the National Academy of Sciences of the United States of America*, *107*, 22060–22065.

Bhat, H. F., Adams, M. E., & Khanday, F. A. (2013). Syntrophin proteins as Santa Claus: Role(s) in cell signal transduction. *Cellular and Molecular Life Sciences*, *70*, 2533–2554.

Bredt, D. S., Hwang, P. M., Glatt, C. E., Lowenstein, C., Reed, R. R., & Snyder, S. H. (1991). Cloned and expressed nitric oxide synthase structurally resembles cytochrome P-450 reductase. *Nature*, *351*, 714–718.

Bredt, D. S., & Snyder, S. H. (1990). Isolation of nitric oxide synthetase, a calmodulin-requiring enzyme. *Proceedings of the National Academy of Sciences of the United States of America*, *87*, 682–685.

Brenman, J. E., Chao, D. S., Gee, S. H., McGee, A. W., Craven, S. E., Santillano, D. R., et al. (1996). Interaction of nitric oxide synthase with the postsynaptic density protein PSD-95 and alpha1-syntrophin mediated by PDZ domains. *Cell*, *84*, 757–767.

Brenman, J. E., Chao, D. S., Xia, H., Aldape, K., & Bredt, D. S. (1995). Nitric oxide synthase complexed with dystrophin and absent from skeletal muscle sarcolemma in Duchenne muscular dystrophy. *Cell*, *82*, 743–752.

Butt, E., Bernhardt, M., Smolenski, A., Kotsonis, P., Frohlich, L. G., Sickmann, A., et al. (2000). Endothelial nitric-oxide synthase (type III) is activated and becomes calcium independent upon phosphorylation by cyclic nucleotide-dependent protein kinases. *The Journal of Biological Chemistry*, *275*, 5179–5187.

Candemir, E., Kollert, L., Weissflog, L., Geis, M., Muller, A., Post, A. M., et al. (2016). Interaction of NOS1AP with the NOS-I PDZ domain: Implications for schizophrenia-related alterations in dendritic morphology. *European Neuropsychopharmacology*, *26*, 741–755.

Cao, J., Viholainen, J. I., Dart, C., Warwick, H. K., Leyland, M. L., & Courtney, M. J. (2005). The PSD95-nNOS interface: A target for inhibition of excitotoxic p38 stress-activated protein kinase activation and cell death. *The Journal of Cell Biology*, *168*, 117–126.

Chang, W. J., Iannaccone, S. T., Lau, K. S., Masters, B. S., McCabe, T. J., McMillan, K., et al. (1996). Neuronal nitric oxide synthase and dystrophin-deficient muscular dystrophy. *Proceedings of the National Academy of Sciences of the United States of America*, *93*, 9142–9147.

Chen, Z. P., Mitchelhill, K. I., Michell, B. J., Stapleton, D., Rodriguez-Crespo, I., Witters, L. A., et al. (1999). AMP-activated protein kinase phosphorylation of endothelial NO synthase. *FEBS Letters*, *443*, 285–289.

Cheng, J., Valdivia, C. R., Vaidyanathan, R., Balijepalli, R. C., Ackerman, M. J., & Makielski, J. C. (2013). Caveolin-3 suppresses late sodium current by inhibiting nNOS-dependent S-nitrosylation of SCN5A. *Journal of Molecular and Cellular Cardiology*, *61*, 102–110.

Christopherson, K. S., Hillier, B. J., Lim, W. A., & Bredt, D. S. (1999). PSD-95 assembles a ternary complex with the N-methyl-D-aspartic acid receptor and a bivalent neuronal NO synthase PDZ domain. *The Journal of Biological Chemistry*, *274*, 27467–27473.

Church, J. E., & Fulton, D. (2006). Differences in eNOS activity because of subcellular localization are dictated by phosphorylation state rather than the local calcium environment. *The Journal of Biological Chemistry*, *281*, 1477–1488.

Cook, D. J., Teves, L., & Tymianski, M. (2012). Treatment of stroke with a PSD-95 inhibitor in the gyrencephalic primate brain. *Nature*, *483*, 213–217.

Courtney, M. J., Li, L. L., & Lai, Y. Y. (2014). Mechanisms of NOS1AP action on NMDA receptor-nNOS signaling. *Frontiers in Cellular Neuroscience*, *8*, 252.

Cui, H., Hayashi, A., Sun, H. S., Belmares, M. P., Cobey, C., Phan, T., et al. (2007). PDZ protein interactions underlying NMDA receptor-mediated excitotoxicity and neuroprotection by PSD-95 inhibitors. *The Journal of Neuroscience*, *27*, 9901–9915.

Cuppen, E., Gerrits, H., Pepers, B., Wieringa, B., & Hendriks, W. (1998). PDZ motifs in PTP-BL and RIL bind to internal protein segments in the LIM domain protein RIL. *Molecular Biology of the Cell*, *9*, 671–683.

Daniels, D. L., Cohen, A. R., Anderson, J. M., & Brunger, A. T. (1998). Crystal structure of the hCASK PDZ domain reveals the structural basis of class II PDZ domain target recognition. *Nature Structural Biology*, *5*, 317–325.

Dawson, V. L., Dawson, T. M., Bartley, D. A., Uhl, G. R., & Snyder, S. H. (1993). Mechanisms of nitric oxide-mediated neurotoxicity in primary brain cultures. *The Journal of Neuroscience*, *13*, 2651–2661.

Dawson, V. L., Dawson, T. M., London, E. D., Bredt, D. S., & Snyder, S. H. (1991). Nitric oxide mediates glutamate neurotoxicity in primary cortical cultures. *Proceedings of the National Academy of Sciences of the United States of America*, *88*, 6368–6371.

Dawson, V. L., Kizushi, V. M., Huang, P. L., Snyder, S. H., & Dawson, T. M. (1996). Resistance to neurotoxicity in cortical cultures from neuronal nitric oxide synthase-deficient mice. *The Journal of Neuroscience*, *16*, 2479–2487.

Dong, H., Zhang, P., Song, I., Petralia, R. S., Liao, D., & Huganir, R. L. (1999). Characterization of the glutamate receptor-interacting proteins GRIP1 and GRIP2. *The Journal of Neuroscience*, *19*, 6930–6941.

Duan, W., Zhou, J., Li, W., Zhou, T., Chen, Q., Yang, F., et al. (2013). Plasma membrane calcium ATPase 4b inhibits nitric oxide generation through calcium-induced dynamic interaction with neuronal nitric oxide synthase. *Protein & Cell*, *4*, 286–298.

Erlendsson, S., Rathje, M., Heidarsson, P. O., Poulsen, F. M., Madsen, K. L., Teilum, K., et al. (2014). Protein interacting with C-kinase 1 (PICK1) binding promiscuity relies on unconventional PSD-95/discs-large/ZO-1 homology (PDZ) binding modes for non-class II PDZ ligands. *The Journal of Biological Chemistry*, *289*, 25327–25340.

Ernst, A., Appleton, B. A., Ivarsson, Y., Zhang, Y., Gfeller, D., Wiesmann, C., et al. (2014). A structural portrait of the PDZ domain family. *Journal of Molecular Biology, 426*, 3509–3519.

Fan, J., Zhang, Q., Tochio, H., Li, M., & Zhang, M. (2001). Structural basis of diverse sequence-dependent target recognition by the 8 kDa dynein light chain. *Journal of Molecular Biology, 306*, 97–108.

Fang, M., Jaffrey, S. R., Sawa, A., Ye, K., Luo, X., & Snyder, S. H. (2000). Dexras1: A G protein specifically coupled to neuronal nitric oxide synthase via CAPON. *Neuron, 28*, 183–193.

Fanning, A. S., Jameson, B. J., Jesaitis, L. A., & Anderson, J. M. (1998). The tight junction protein ZO-1 establishes a link between the transmembrane protein occludin and the actin cytoskeleton. *The Journal of Biological Chemistry, 273*, 29745–29753.

Fernandez-Hernando, C., Fukata, M., Bernatchez, P. N., Fukata, Y., Lin, M. I., Bredt, D. S., et al. (2006). Identification of Golgi-localized acyl transferases that palmitoylate and regulate endothelial nitric oxide synthase. *The Journal of Cell Biology, 174*, 369–377.

Firestein, B. L., & Bredt, D. S. (1999). Interaction of neuronal nitric-oxide synthase and phosphofructokinase-M. *The Journal of Biological Chemistry, 274*, 10545–10550.

Fleming, I., Fisslthaler, B., Dimmeler, S., Kemp, B. E., & Busse, R. (2001). Phosphorylation of Thr(495) regulates Ca2+/calmodulin-dependent endothelial nitric oxide synthase activity. *Circulation Research, 88*, E68–E75.

Florio, S. K., Loh, C., Huang, S. M., Iwamaye, A. E., Kitto, K. F., Fowler, K. W., et al. (2009). Disruption of nNOS-PSD95 protein-protein interaction inhibits acute thermal hyperalgesia and chronic mechanical allodynia in rodents. *British Journal of Pharmacology, 158*, 494–506.

Forstermann, U., Closs, E. I., Pollock, J. S., Nakane, M., Schwarz, P., Gath, I., et al. (1994). Nitric oxide synthase isozymes. Characterization, purification, molecular cloning, and functions. *Hypertension, 23*, 1121–1131.

Forstermann, U., & Sessa, W. C. (2012). Nitric oxide synthases: Regulation and function. *European Heart Journal, 33*, 829–837. 837a-837d.

Fujisawa, H., Ogura, T., Kurashima, Y., Yokoyama, T., Yamashita, J., & Esumi, H. (1994). Expression of two types of nitric oxide synthase mRNA in human neuroblastoma cell lines. *Journal of Neurochemistry, 63*, 140–145.

Fulton, D., Gratton, J. P., McCabe, T. J., Fontana, J., Fujio, Y., Walsh, K., et al. (1999). Regulation of endothelium-derived nitric oxide production by the protein kinase Akt. *Nature, 399*, 597–601.

Furchgott, R. F., & Zawadzki, J. V. (1980). The obligatory role of endothelial cells in the relaxation of arterial smooth muscle by acetylcholine. *Nature, 288*, 373–376.

Garcia-Cardena, G., Oh, P., Liu, J., Schnitzer, J. E., & Sessa, W. C. (1996). Targeting of nitric oxide synthase to endothelial cell caveolae via palmitoylation: Implications for nitric oxide signaling. *Proceedings of the National Academy of Sciences of the United States of America, 93*, 6448–6453.

Garcin, E. D., Bruns, C. M., Lloyd, S. J., Hosfield, D. J., Tiso, M., Gachhui, R., et al. (2004). Structural basis for isozyme-specific regulation of electron transfer in nitric-oxide synthase. *The Journal of Biological Chemistry, 279*, 37918–37927.

Gee, S. H., Madhavan, R., Levinson, S. R., Caldwell, J. H., Sealock, R., & Froehner, S. C. (1998). Interaction of muscle and brain sodium channels with multiple members of the syntrophin family of dystrophin-associated proteins. *The Journal of Neuroscience, 18*, 128–137.

Ghosh, D. K., Wu, C., Pitters, E., Moloney, M., Werner, E. R., Mayer, B., et al. (1997). Characterization of the inducible nitric oxide synthase oxygenase domain identifies a 49 amino acid segment required for subunit dimerization and tetrahydrobiopterin interaction. *Biochemistry, 36*, 10609–10619.

Glynne, P. A., Darling, K. E., Picot, J., & Evans, T. J. (2002). Epithelial inducible nitric-oxide synthase is an apical EBP50-binding protein that directs vectorial nitric oxide output. *The Journal of Biological Chemistry*, *277*, 33132–33138.

Haque, M. M., Ray, S. S., & Stuehr, D. J. (2016). Phosphorylation controls endothelial nitric-oxide synthase by regulating its conformational dynamics. *The Journal of Biological Chemistry*, *291*, 23047–23057.

Hemmens, B., Goessler, W., Schmidt, K., & Mayer, B. (2000). Role of bound zinc in dimer stabilization but not enzyme activity of neuronal nitric-oxide synthase. *The Journal of Biological Chemistry*, *275*, 35786–35791.

Hill, M. D., Martin, R. H., Mikulis, D., Wong, J. H., Silver, F. L., Terbrugge, K. G., et al. (2012). Safety and efficacy of NA-1 in patients with iatrogenic stroke after endovascular aneurysm repair (ENACT): A phase 2, randomised, double-blind, placebo-controlled trial. *Lancet Neurology*, *11*, 942–950.

Hillier, B. J., Christopherson, K. S., Prehoda, K. E., Bredt, D. S., & Lim, W. A. (1999). Unexpected modes of PDZ domain scaffolding revealed by structure of nNOS-syntrophin complex. *Science*, *284*, 812–815.

Hung, A. Y., & Sheng, M. (2002). PDZ domains: Structural modules for protein complex assembly. *The Journal of Biological Chemistry*, *277*, 5699–5702.

Jaffrey, S. R., Snowman, A. M., Eliasson, M. J., Cohen, N. A., & Snyder, S. H. (1998). CAPON: A protein associated with neuronal nitric oxide synthase that regulates its interactions with PSD95. *Neuron*, *20*, 115–124.

Judge, L. M., Haraguchiln, M., & Chamberlain, J. S. (2006). Dissecting the signaling and mechanical functions of the dystrophin-glycoprotein complex. *Journal of Cell Science*, *119*, 1537–1546.

Kim, Y. C., Park, T. Y., Baik, E., & Lee, S. H. (2012). Fructose-1,6-bisphosphate attenuates induction of nitric oxide synthase in microglia stimulated with lipopolysaccharide. *Life Sciences*, *90*, 365–372.

Lafuente, H., Pazos, M. R., Alvarez, A., Mohammed, N., Santos, M., Arizti, M., et al. (2016). Effects of cannabidiol and hypothermia on short-term brain damage in new-born piglets after acute hypoxia-ischemia. *Frontiers in Neuroscience*, *10*, 323.

Lai, Y., Thomas, G. D., Yue, Y., Yang, H. T., Li, D., Long, C., et al. (2009). Dystrophins carrying spectrin-like repeats 16 and 17 anchor nNOS to the sarcolemma and enhance exercise performance in a mouse model of muscular dystrophy. *The Journal of Clinical Investigation*, *119*, 624–635.

Lai, T. W., & Wang, Y. T. (2010). Fashioning drugs for stroke. *Nature Medicine*, *16*, 1376–1378.

Lai, T. W., Zhang, S., & Wang, Y. T. (2014). Excitotoxicity and stroke: Identifying novel targets for neuroprotection. *Progress in Neurobiology*, *115*, 157–188.

Lai, Y., Zhao, J., Yue, Y., & Duan, D. (2013). Alpha2 and alpha3 helices of dystrophin R16 and R17 frame a microdomain in the alpha1 helix of dystrophin R17 for neuronal NOS binding. *Proceedings of the National Academy of Sciences of the United States of America*, *110*, 525–530.

Lemaire, J. F., & McPherson, P. S. (2006). Binding of Vac14 to neuronal nitric oxide synthase: Characterisation of a new internal PDZ-recognition motif. *FEBS Letters*, *580*, 6948–6954.

Li, L. L., Cisek, K., & Courtney, M. J. (2017). Efficient binding of the NOS1AP C-terminus to the nNOS PDZ pocket requires the concerted action of the PDZ ligand motif, the internal ExF site and structural integrity of an independent element. *Frontiers in Molecular Neuroscience*, *10*, 58.

Li, L. L., Ginet, V., Liu, X., Vergun, O., Tuittila, M., Mathieu, M., et al. (2013). The nNOS-p38MAPK pathway is mediated by NOS1AP during neuronal death. *The Journal of Neuroscience*, *33*, 8185–8201.

Li, L. L., Melero-Fernandez de Mera, R. M., Chen, J., Ba, W., Kasri, N. N., Zhang, M., et al. (2015). Unexpected heterodivalent recruitment of NOS1AP to nNOS reveals multiple sites for pharmacological intervention in neuronal disease models. *The Journal of Neuroscience*, *35*, 7349–7364.

Lin, X., Sun, B., Liang, M., Liang, Y.-Y., Gast, A., Hildebrand, J., et al. (2003). Opposed regulation of corepressor CtBP by SUMOylation and PDZ binding. *Molecular Cell*, *11*, 1389–1396.

Luo, C. X., & Zhu, D. Y. (2011). Research progress on neurobiology of neuronal nitric oxide synthase. *Neuroscience Bulletin*, *27*, 23–35.

Maekawa, K., Imagawa, N., Naito, A., Harada, S., Yoshie, O., & Takagi, S. (1999). Association of protein-tyrosine phosphatase PTP-BAS with the transcription-factor-inhibitory protein IkappaBalpha through interaction between the PDZ1 domain and ankyrin repeats. *The Biochemical Journal*, *337*(Pt. 2), 179–184.

Martasek, P., Miller, R. T., Liu, Q., Roman, L. J., Salerno, J. C., Migita, C. T., et al. (1998). The C331A mutant of neuronal nitric-oxide synthase is defective in arginine binding. *The Journal of Biological Chemistry*, *273*, 34799–34805.

Martinez-Ruiz, A., Araujo, I. M., Izquierdo-Alvarez, A., Hernansanz-Agustin, P., Lamas, S., & Serrador, J. M. (2013). Specificity in S-nitrosylation: A short-range mechanism for NO signaling? *Antioxidants & Redox Signaling*, *19*, 1220–1235.

Martinez-Ruiz, A., Villanueva, L., Gonzalez de Orduna, C., Lopez-Ferrer, D., Higueras, M. A., Tarin, C., et al. (2005). S-nitrosylation of Hsp90 promotes the inhibition of its ATPase and endothelial nitric oxide synthase regulatory activities. *Proceedings of the National Academy of Sciences of the United States of America*, *102*, 8525–8530.

McCabe, T. J., Fulton, D., Roman, L. J., & Sessa, W. C. (2000). Enhanced electron flux and reduced calmodulin dissociation may explain "calcium-independent" eNOS activation by phosphorylation. *The Journal of Biological Chemistry*, *275*, 6123–6128.

Meng, J., Counsell, J. R., Reza, M., Laval, S. H., Danos, O., Thrasher, A., et al. (2016). Autologous skeletal muscle derived cells expressing a novel functional dystrophin provide a potential therapy for duchenne muscular dystrophy. *Scientific Reports*, *6*, 19750.

Merino-Gracia, J., Costas-Insua, C., Canales, M. A., & Rodriguez-Crespo, I. (2016). Insights into the C-terminal peptide binding specificity of the PDZ domain of neuronal nitric-oxide synthase: Characterization of the interaction with the tight junction protein claudin-3. *The Journal of Biological Chemistry*, *291*, 11581–11595.

Merino-Gracia, J., Zamora-Carreras, H., Bruix, M., & Rodriguez-Crespo, I. (2016). Molecular basis for the protein recognition specificity of the dynein light chain DYNLT1/Tctex1: Characterization of the interaction with activin receptor IIB. *The Journal of Biological Chemistry*, *291*, 20962–20975.

Michel, T., & Vanhoutte, P. M. (2010). Cellular signaling and NO production. *Pflügers Archiv*, *459*, 807–816.

Michell, B. J., Griffiths, J. E., Mitchelhill, K. I., Rodriguez-Crespo, I., Tiganis, T., Bozinovski, S., et al. (1999). The Akt kinase signals directly to endothelial nitric oxide synthase. *Current Biology*, *9*, 845–848.

Mikkelsen, R. B., & Wardman, P. (2003). Biological chemistry of reactive oxygen and nitrogen and radiation-induced signal transduction mechanisms. *Oncogene*, *22*, 5734–5754.

Mo, S. F., Liao, G. Y., Yang, J., Wang, M. Y., Hu, Y., Lian, G. N., et al. (2016). Protection of neuronal cells from excitotoxicity by disrupting nNOS-PSD95 interaction with a small molecule SCR-4026. *Brain Research*, *1648*, 250–256.

Mohamed, T. M., Oceandy, D., Prehar, S., Alatwi, N., Hegab, Z., Baudoin, F. M., et al. (2009). Specific role of neuronal nitric-oxide synthase when tethered to the plasma membrane calcium pump in regulating the beta-adrenergic signal in the myocardium. *The Journal of Biological Chemistry*, *284*, 12091–12098.

Molza, A. E., Mangat, K., Le Rumeur, E., Hubert, J. F., Menhart, N., & Delalande, O. (2015). Structural basis of neuronal nitric-oxide synthase interaction with dystrophin repeats 16 and 17. *The Journal of Biological Chemistry, 290*, 29531–29541.

Morais Cabral, J. H., Petosa, C., Sutcliffe, M. J., Raza, S., Byron, O., Poy, F., et al. (1996). Crystal structure of a PDZ domain. *Nature, 382*, 649–652.

Navarro-Lerida, I., Alvarez-Barrientos, A., & Rodriguez-Crespo, I. (2006). N-terminal palmitoylation within the appropriate amino acid environment conveys on NOS2 the ability to progress along the intracellular sorting pathways. *Journal of Cell Science, 119*, 1558–1569.

Navarro-Lerida, I., Corvi, M. M., Barrientos, A. A., Gavilanes, F., Berthiaume, L. G., & Rodriguez-Crespo, I. (2004). Palmitoylation of inducible nitric-oxide synthase at Cys-3 is required for proper intracellular traffic and nitric oxide synthesis. *The Journal of Biological Chemistry, 279*, 55682–55689.

Navarro-Lerida, I., Martinez-Moreno, M., Ventoso, I., Alvarez-Barrientos, A., & Rodriguez-Crespo, I. (2007). Binding of CAP70 to inducible nitric oxide synthase and implications for the vectorial release of nitric oxide in polarized cells. *Molecular Biology of the Cell, 18*, 2768–2777.

Nishida, C. R., & de Montellano, P. R. (2001). Control of electron transfer in nitric-oxide synthases. Swapping of autoinhibitory elements among nitric-oxide synthase isoforms. *The Journal of Biological Chemistry, 276*, 20116–20124.

Noble, M. A., Munro, A. W., Rivers, S. L., Robledo, L., Daff, S. N., Yellowlees, L. J., et al. (1999). Potentiometric analysis of the flavin cofactors of neuronal nitric oxide synthase. *Biochemistry, 38*, 16413–16418.

Palacios, M., Knowles, R. G., Palmer, R. M., & Moncada, S. (1989). Nitric oxide from L-arginine stimulates the soluble guanylate cyclase in adrenal glands. *Biochemical and Biophysical Research Communications, 165*, 802–809.

Park, J. H., Kim, W. S., Kim, J. Y., Park, M. H., Nam, J. H., Yun, C. W., et al. (2011). Chk1 and Hsp90 cooperatively regulate phosphorylation of endothelial nitric oxide synthase at serine 1179. *Free Radical Biology & Medicine, 51*, 2217–2226.

Park, C. S., Park, R., & Krishna, G. (1996). Constitutive expression and structural diversity of inducible isoform of nitric oxide synthase in human tissues. *Life Sciences, 59*, 219–225.

Peng, H. M., Morishima, Y., Pratt, W. B., & Osawa, Y. (2012). Modulation of heme/substrate binding cleft of neuronal nitric-oxide synthase (nNOS) regulates binding of Hsp90 and Hsp70 proteins and nNOS ubiquitination. *The Journal of Biological Chemistry, 287*, 1556–1565.

Persichini, T., Mazzone, V., Polticelli, F., Moreno, S., Venturini, G., Clementi, E., et al. (2005). Mitochondrial type I nitric oxide synthase physically interacts with cytochrome c oxidase. *Neuroscience Letters, 384*, 254–259.

Rameau, G. A., Chiu, L. Y., & Ziff, E. B. (2004). Bidirectional regulation of neuronal nitric-oxide synthase phosphorylation at serine 847 by the N-methyl-D-aspartate receptor. *The Journal of Biological Chemistry, 279*, 14307–14314.

Rebolledo, D. L., Kim, M. J., Whitehead, N. P., Adams, M. E., & Froehner, S. C. (2016). Sarcolemmal targeting of nNOSmu improves contractile function of mdx muscle. *Human Molecular Genetics, 25*, 158–166.

Reza, M., Laval, S. H., Roos, A., Carr, S., & Lochmuller, H. (2016). Optimization of internally deleted dystrophin constructs. *Human Gene Therapy Methods, 27*, 174–186.

Riefler, G. M., & Firestein, B. L. (2001). Binding of neuronal nitric-oxide synthase (nNOS) to carboxyl-terminal-binding protein (CtBP) changes the localization of CtBP from the nucleus to the cytosol: A novel function for targeting by the PDZ domain of nNOS. *The Journal of Biological Chemistry, 276*, 48262–48268.

Rodriguez-Crespo, I., Moenne-Loccoz, P., Loehr, T. M., & Ortiz de Montellano, P. R. (1997). Endothelial nitric oxide synthase: Modulations of the distal heme site produced by progressive N-terminal deletions. *Biochemistry, 36*, 8530–8538.

Rodriguez-Crespo, I., Straub, W., Gavilanes, F., & Ortiz de Montellano, P. R. (1998). Binding of dynein light chain (PIN) to neuronal nitric oxide synthase in the absence of inhibition. *Archives of Biochemistry and Biophysics, 359*, 297–304.

Roman, L. J., Miller, R. T., de La Garza, M. A., Kim, J. J., & Siler Masters, B. S. (2000). The C terminus of mouse macrophage inducible nitric-oxide synthase attenuates electron flow through the flavin domain. *The Journal of Biological Chemistry, 275*, 21914–21919.

Saitoh, F., Tian, Q. B., Okano, A., Sakagami, H., Kondo, H., & Suzuki, T. (2004). NIDD, a novel DHHC-containing protein, targets neuronal nitric-oxide synthase (nNOS) to the synaptic membrane through a PDZ-dependent interaction and regulates nNOS activity. *The Journal of Biological Chemistry, 279*, 29461–29468.

Sanchez-Ruiloba, L., Aicart-Ramos, C., Garcia-Guerra, L., Pose-Utrilla, J., Rodriguez-Crespo, I., & Iglesias, T. (2014). Protein kinase D interacts with neuronal nitric oxide synthase and phosphorylates the activatory residue serine 1412. *PloS One, 9*, e95191.

Sattler, R., Xiong, Z., Lu, W. Y., Hafner, M., MacDonald, J. F., & Tymianski, M. (1999). Specific coupling of NMDA receptor activation to nitric oxide neurotoxicity by PSD-95 protein. *Science, 284*, 1845–1848.

Schepens, J., Cuppen, E., Wieringa, B., & Hendriks, W. (1997). The neuronal nitric oxide synthase PDZ motif binds to -G(D,E)XV* carboxyterminal sequences. *FEBS Letters, 409*, 53–56.

Schneider, J. C., El Kebir, D., Chereau, C., Lanone, S., Huang, X. L., Roessingh, A. S. D., et al. (2003). Involvement of Ca2+/calmodulin-dependent protein kinase II in endothelial NO production and endothelium-dependent relaxation. *American Journal of Physiology. Heart and Circulatory Physiology, 284*, H2311–H2319.

Schuh, K., Uldrijan, S., Telkamp, M., Rothlein, N., & Neyses, L. (2001). The plasmamembrane calmodulin-dependent calcium pump: A major regulator of nitric oxide synthase I. *The Journal of Cell Biology, 155*, 201–205.

Sessa, W. C. (2004). eNOS at a glance. *Journal of Cell Science, 117*, 2427–2429.

Silvagno, F., Xia, H., & Bredt, D. S. (1996). Neuronal nitric-oxide synthase-mu, an alternatively spliced isoform expressed in differentiated skeletal muscle. *The Journal of Biological Chemistry, 271*, 11204–11208.

Stamler, J. S., Lamas, S., & Fang, F. C. (2001). Nitrosylation. The prototypic redox-based signaling mechanism. *Cell, 106*, 675–683.

Stiffler, M. A., Chen, J. R., Grantcharova, V. P., Lei, Y., Fuchs, D., Allen, J. E., et al. (2007). PDZ domain binding selectivity is optimized across the mouse proteome. *Science, 317*, 364–369.

Stricker, N. L., Christopherson, K. S., Yi, B. A., Schatz, P. J., Raab, R. W., Dawes, G., et al. (1997). PDZ domain of neuronal nitric oxide synthase recognizes novel C-terminal peptide sequences. *Nature Biotechnology, 15*, 336–342.

Sun, H. S., Doucette, T. A., Liu, Y., Fang, Y., Teves, L., Aarts, M., et al. (2008). Effectiveness of PSD95 inhibitors in permanent and transient focal ischemia in the rat. *Stroke, 39*, 2544–2553.

Thomas, G. D., Sander, M., Lau, K. S., Huang, P. L., Stull, J. T., & Victor, R. G. (1998). Impaired metabolic modulation of alpha-adrenergic vasoconstriction in dystrophin-deficient skeletal muscle. *Proceedings of the National Academy of Sciences of the United States of America, 95*, 15090–15095.

Tidball, J. G., & Wehling-Henricks, M. (2014). Nitric oxide synthase deficiency and the pathophysiology of muscular dystrophy. *The Journal of Physiology, 592*, 4627–4638.

Timpani, C. A., Trewin, A. J., Stojanovska, V., Robinson, A., Goodman, C. A., Nurgali, K., et al. (2017). Attempting to compensate for reduced neuronal nitric oxide synthase protein with nitrate supplementation cannot overcome metabolic dysfunction but rather has detrimental effects in dystrophin-deficient mdx muscle. *Neurotherapeutics, 14*, 429–446.

Tochio, H., Hung, F., Li, M., Bredt, D. S., & Zhang, M. (2000). Solution structure and backbone dynamics of the second PDZ domain of postsynaptic density-95. *Journal of Molecular Biology, 295*, 225–237.

Tochio, H., Mok, Y. K., Zhang, Q., Kan, H. M., Bredt, D. S., & Zhang, M. (2000). Formation of nNOS/PSD-95 PDZ dimer requires a preformed beta-finger structure from the nNOS PDZ domain. *Journal of Molecular Biology, 303*, 359–370.

Tochio, H., Zhang, Q., Mandal, P., Li, M., & Zhang, M. (1999). Solution structure of the extended neuronal nitric oxide synthase PDZ domain complexed with an associated peptide. *Nature Structural Biology, 6*, 417–421.

Tonikian, R., Zhang, Y., Sazinsky, S. L., Currell, B., Yeh, J. H., Reva, B., et al. (2008). A specificity map for the PDZ domain family. *PLoS Biology, 6*, e239.

Trane, A. E., Pavlov, D., Sharma, A., Saqib, U., Lau, K., van Petegem, F., et al. (2014). Deciphering the binding of caveolin-1 to client protein endothelial nitric-oxide synthase (eNOS): Scaffolding subdomain identification, interaction modeling, and biological significance. *The Journal of Biological Chemistry, 289*, 13273–13283.

van't Hof, R. J., Armour, K. J., Smith, L. M., Armour, K. E., Wei, X. Q., Liew, F. Y., et al. (2000). Requirement of the inducible nitric oxide synthase pathway for IL-1-induced osteoclastic bone resorption. *Proceedings of the National Academy of Sciences of the United States of America, 97*, 7993–7998.

Venema, R. C., Ju, H., Zou, R., Ryan, J. W., & Venema, V. J. (1997). Subunit interactions of endothelial nitric-oxide synthase. Comparisons to the neuronal and inducible nitric-oxide synthase isoforms. *The Journal of Biological Chemistry, 272*, 1276–1282.

Wang, S., Yue, H., Derin, R. B., Guggino, W. B., & Li, M. (2000). Accessory protein facilitated CFTR-CFTR interaction, a molecular mechanism to potentiate the chloride channel activity. *Cell, 103*, 169–179.

Wang, P., Zhang, Q., Tochio, H., Fan, J. S., & Zhang, M. (2000). Formation of a native-like beta-hairpin finger structure of a peptide from the extended PDZ domain of neuronal nitric oxide synthase in aqueous solution. *European Journal of Biochemistry, 267*, 3116–3122.

Wehling-Henricks, M., Oltmann, M., Rinaldi, C., Myung, K. H., & Tidball, J. G. (2009). Loss of positive allosteric interactions between neuronal nitric oxide synthase and phosphofructokinase contributes to defects in glycolysis and increased fatigability in muscular dystrophy. *Human Molecular Genetics, 18*, 3439–3451.

Wells, K. E., Torelli, S., Lu, Q., Brown, S. C., Partridge, T., Muntoni, F., et al. (2003). Relocalization of neuronal nitric oxide synthase (nNOS) as a marker for complete restoration of the dystrophin associated protein complex in skeletal muscle. *Neuromuscular Disorders, 13*, 21–31.

Xia, C., Misra, I., Iyanagi, T., & Kim, J. J. (2009). Regulation of interdomain interactions by calmodulin in inducible nitric-oxide synthase. *The Journal of Biological Chemistry, 284*, 30708–30717.

Xu, X. Z., Choudhury, A., Li, X., & Montell, C. (1998). Coordination of an array of signaling proteins through homo- and heteromeric interactions between PDZ domains and target proteins. *The Journal of Cell Biology, 142*, 545–555.

Ye, F., & Zhang, M. (2013). Structures and target recognition modes of PDZ domains: Recurring themes and emerging pictures. *The Biochemical Journal, 455*, 1–14.

Zhang, Y., Lee, T. S., Kolb, E. M., Sun, K., Lu, X., Sladek, F. M., et al. (2006). AMP-activated protein kinase is involved in endothelial NO synthase activation in response to shear stress. *Arteriosclerosis, Thrombosis, and Vascular Biology, 26*, 1281–1287.

Zhao, J., Kodippili, K., Yue, Y., Hakim, C. H., Wasala, L., Pan, X., et al. (2016). Dystrophin contains multiple independent membrane-binding domains. *Human Molecular Genetics*, *25*, 3647–3653.
Zhao, Y., Vanhoutte, P. M., & Leung, S. W. (2015). Vascular nitric oxide: Beyond eNOS. *Journal of Pharmacological Sciences*, *129*, 83–94.
Zhou, L., Li, F., Xu, H. B., Luo, C. X., Wu, H. Y., Zhu, M. M., et al. (2010). Treatment of cerebral ischemia by disrupting ischemia-induced interaction of nNOS with PSD-95. *Nature Medicine*, *16*, 1439–1443.

CHAPTER SIX

Development of Protein–Protein Interaction Inhibitors for the Treatment of Infectious Diseases

Andrew F. Voter, James L. Keck[1]
University of Wisconsin School of Medicine and Public Health, Madison, WI, United States
[1]Corresponding author: e-mail address: jlkeck@wisc.edu

Contents

Abstract

Protein–protein interaction (PPI) inhibitors are a rapidly expanding class of therapeutics. Recent advances in our understanding of PPIs and success of early examples of PPI inhibitors demonstrate the feasibility of targeting PPIs. This review summarizes the techniques used for the discovery and optimization of a diverse set PPI inhibitors, focusing on the development of PPI inhibitors as new antibacterial and antiviral agents. We close with a summary of the advances responsible for making PPI inhibitors realistic targets for therapeutic intervention and brief outlook of the field.

The increasing prevalence of treatment-resistant pathogens is one of the biggest challenges facing the biomedical community today. Antiinfective agents have revolutionized medicine, allowing for curative therapies that vastly reduced the morbidity and mortality of diseases caused by pathogenic

Advances in Protein Chemistry and Structural Biology, Volume 111
ISSN 1876-1623
https://doi.org/10.1016/bs.apcsb.2017.07.005
© 2018 Elsevier Inc.
All rights reserved.

bacteria, viruses, and parasites. Unfortunately, the widespread overuse and misuse of these agents has led to increasing levels of resistance that threatens our ability to effectively treat not only infections but also the use of therapies that requires prophylactic antibiotic treatment, such as surgeries. While better stewardship is critical for extending the usable life of the existing repertoire of antibacterial and antivirals agents (Goff et al., 2017), classes of agents with novel mechanisms of action are urgently needed ("The 10 x '20 Initiative," 2010).

Traditionally, antibiotic development has focused primarily on inhibitors of essential bacterial enzymes for understandable reasons (Walsh, 2016). The mechanism of action was easily understood, biochemical assays existed to test for activity, and substrate analog inhibitors were relatively easy to produce. Unfortunately, the supply of easily inhibited targets appears to be limited as the output of new antiinfective agents has dwindled (Ventola, 2015). Deepening our understanding of the underlying biology of pathogens will undoubtedly uncover new potential enzymatic targets, but the rate of discovery has not kept pace with the development of resistance (Ventola, 2015).

In the face of a dry antiinfective pipeline, researchers and pharmaceutic companies have begun to turn their attention to a different class of targets, protein–protein interactions (PPI) (Arkin & Wells, 2004; Wells & McClendon, 2007). Many essential cellular functions rely on the precise and timely recruitment of proteins, often accomplished through a PPI. Disruption of protein interfaces, either through genetic proof-of-principle studies or small-molecule inhibitors, can kill pathogens or render them nonvirulent, making PPI inhibitors an exciting new research area in the antiinfective world. There are also plenty of PPIs with which to work. The number of cataloged PPIs varies by database, but at a minimum there are tens of thousands human PPIs and the *Escherichia coli* interactions number in the thousands (Lehne & Schlitt, 2009). These counts do not include a sizeable number of host–pathogen interactions, with over 50,000 human–pathogen interactions cataloged in HPIDB (Kumar & Nanduri, 2010). While not all of these interactions are feasible targets for inhibition, a sizeable number are. We will explore examples of inhibitors that target several classes of PPIs: pathogen–pathogen, host–pathogen, and host–host interactions and how they might alter the treatment of infectious diseases.

Historically, PPIs were considered undruggable targets. This reputation likely stemmed from the lack of high-throughput ready screening assays as well as the thought that most PPIs are held together by large, chemically noncomplex surfaces with a lack of easily druggable pockets (Spencer, 1998). While such difficult PPI targets undoubtedly exist, it is now appreciated that many PPIs use much smaller interfaces for their interaction, frequently consisting of an unstructured peptide bound to a well-defined groove (Arkin, Tang, & Wells, 2014). Furthermore, mutagenesis studies of several PPIs have revealed that surfaces contributing to the affinity of a given PPI are not evenly distributed across the entire interface. Rather, there tends to be a "hot spot" or a small number of critical residues that anchor two proteins together (Cukuroglu, Engin, Gursoy, & Keskin, 2014). This means that a putative inhibitor would not need to displace the entirety of a given PPI, but rather only occupy the hot spot, a more tractable problem. Recent review articles have highlighted small molecules disrupting PPIs for the treatment of oncologic targets that have reached early clinical trials, demonstrating the feasibility of the approach. Because many of these inhibitors have already been reviewed in depth (Arkin et al., 2014; Sheng, Dong, Miao, Zhang, & Wang, 2015), this review will focus on PPI inhibitors for the treatment of infectious diseases.

1. ANTIBACTERIAL AGENTS

1.1 ZipA–FtsZ

During bacterial cytokinesis, the cell contents must be properly partitioned between the two daughter cells and the cell wall sealed to prevent loss of cytoplasmic material or cell lysis. To accomplish this task, a ring, called the *Z*-ring, is formed at the site of division from the head-to-tail polymerization of the GTPase FtsZ (Adams & Errington, 2009). While the contribution that FtsZ and the *Z*-ring play in generating the force required to pinch the cell membrane is debated, it is clear that FtsZ plays an essential role in cytokinesis (Xiao & Goley, 2016).

To maintain contact with the cell wall throughout cytokinesis, FtsZ uses the 17 C-terminal most residues to bind to the membrane-associated protein ZipA (Mosyak et al., 2000). Loss of this interaction is lethal in the gammaproteobacteria (although it is absent in other bacteria; Hale & de Boer, 1997) likely due to the ability of ZipA to stabilize FtsZ polymers and localize them to the membrane (Kuchibhatla, Bhattacharya, & Panda, 2011). Additionally, alanine-scanning mutations of the FtsZ interaction site demonstrated that the majority of the affinity between the two protein is derived from only three hydrophobic residues, I374, F377, and L378 (Mosyak et al., 2000). Together these data suggest that a small molecule could block the FtsZ–ZipA interaction and that an inhibitor of this PPI would have antibacterial properties.

Researchers at Wyeth Research developed a high-throughput fluorescence polarization (FP) assay to screen for inhibitors of the FtsZ–ZipA interaction. During assay development, they realized that the relatively poor affinity of the PPI (7 μM K_D as determined by surface plasmon resonance) meant that a prohibitively large amount of ZipA would be required to screen an acceptable number of compounds. To circumvent this limitation, a phage display screen was conducted to identify a probe with a higher affinity to the ZipA. The resulting peptide, FtsZ-PD1, was found to have a K_D of 150 nM, a 45-fold improvement and a FP high-throughput screen (HTS) of 250,000 compounds was conducted using a labeled version of the FtsZ-PD1 as a probe. This screening identified a pyridylpyrimidine inhibitor with a modest 12 μM K_i in the FP assay (Fig. 1; Kenny et al., 2003), and several additional inhibitor scaffolds with weak activities were identified in the same screen. Crystallographic studies confirmed that the inhibitor occupied the FtsZ binding pocket on ZipA.

Besides reducing the protein production burden, one can imagine two possible results of using a tighter binding probe for screening. First, the higher affinity peptide may serve to exclude low potency, but still active, inhibitor scaffolds that could be improved through medical chemistry efforts. The remaining hits are more likely to be active and potent against the native PPI, although reduced in number. The trade-offs associated with this approach likely depend on the size and quality of the chemical library to be screened. With a large enough library, the reduced hit rate is unlikely to be problematic and may even reduce the burden of secondary screening and

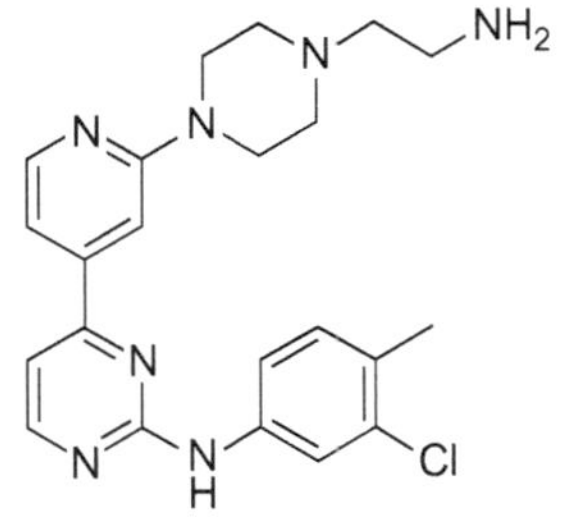

Fig. 1 Structure of the pyridylpyrimidine HTS hit.

hit validation. Presumably, the opposite approach could also be used; where a targeted interaction may be weakened to gain a foothold for later optimization. Second, the use of a higher affinity probe will introduce interaction sites not used in the wild-type interface. Ideally, a screening hit will target only the residues comprising the original interface and not the artificially introduced interactions.

Unfortunately, the pyridylpyrimidine scaffold was found to have nonspecific toxicity against both bacterial and yeast cells lines which precluded further development (Rush, Grant, Mosyak, & Nicholls, 2005). A variety of techniques were tried in attempts to develop more promising inhibitor scaffolds. First, a computational scaffold hopping method was used to identify a triazolopyridazine ring structures that mimicked the pyridylpyrimidine binding pose. While a crystal structure of a triazolopyridazine inhibitor bound to ZipA was obtained, these inhibitors were quite weak ($IC_{50} \sim 80\,\mu M$) and do not appear to have been pursued (Rush et al., 2005). Second, an additional set of weak scaffolds ($IC_{50} > 1$ mM) were discovered in the initial FP screen. Remarkably in the face of this low affinity, crystal structures of these inhibitors bound to the FtsZ pocket of ZipA were obtained (Jennings et al., 2004). This led to additional medicinal chemistry optimization, as well as an attempt to merge two weak inhibitors to fill the entirety of the binding pocket (Sutherland et al., 2003). While a 10-fold reduction in IC_{50} was obtained, none of these compounds displayed antibacterial activity against *E. coli*. These compounds did inhibit the growth of *E. coli* strains with compromised cell membranes, leading the authors to blame poor cell penetrance for the lack of bactericidal effect. However, a subset of these compounds was found to be active against several gram-positive organisms, which lack a ZipA homolog. This suggests that observed the antibacterial activity may have been a result of off-target effects (Sutherland et al., 2003). Finally, an NMR fragment screen was conducted to identify a more cell-penetrant inhibitor scaffold. While several new inhibitor scaffolds were identified, none possessed adequate activity in the FP assay to be pursued farther and the project appears to have been terminated (Tsao et al., 2006).

Despite the difficulties encountered in cell penetrance and nonspecific toxicity, this early effort was a successful demonstration of the ability to identify PPI disrupting inhibitor scaffolds. We are not aware of additional work against this target, although the combination of advances in screening libraries, techniques, and the ease of obtaining inhibitor-bound structures suggests that potent inhibitors could yet be discovered.

1.2 Pilicides

Urinary tract infections (UTI) are one of the most prevalent infections and are responsible for nearly 10 million clinic visits per year (Schappert, 2011). Treatment of UTIs is estimated to exceed $3,500,000,000 annually (Flores-Mireles, Walker, Caparon, & Hultgren, 2015). UTIs can be caused by a wide range of pathogens, but most UTIs (65%–75%) are caused by uropathogenic *E. coli* (UPEC) (Flores-Mireles et al., 2015). Most uncomplicated UTIs will resolve spontaneously within a week; however, symptoms associated with UTIs are unpleasant (dysuria, frequency, urgency, suprapubic pain, and hematuria; Bent, Nallamothu, Simel, Fihn, & Saint, 2002). Patients are typically treated with nitrofurantoin, trimethoprim–sulfamethoxazole, or fosfomycin (Gupta et al., 2011). While resistance to these commonly used antibiotics is relatively rare, resistance rates are increasing (Zowawi et al., 2015). Furthermore, treatment with broad spectrum antibiotics can lead to profound disruptions in the native microbiota with poorly understood consequences (Dethlefsen, Huse, Sogin, & Relman, 2008). Both of these problems could be circumvented with novel therapies targeting the virulence factors of UPEC specifically, sparing the nonpathogenic members of the microbiota.

To maintain an infection in the urinary tract despite the repeated outward flow of urine, UPEC anchor to the urothelium with the help of a variety of excreted pilus fibers (Wu, Sun, & Medina, 1996). One pilus, the type 1 pili, is capped with a FimH subunit that binds to mannose presented on urothelium glycolipids (Krogfelt, Bergmans, & Klemm, 1990). Type 1 pili also contributes toward biofilm formation and the accompanying resistance to therapy (Martinez, Mulvey, Schilling, Pinkner, & Hultgren, 2000; Wright, Seed, & Hultgren, 2007). To ascend into the kidney and cause pyelonephritis, UPEC must express P pili, terminating with PapG, which binds to kidney specific glycolipids (Lane & Mobley, 2007).

Because bacteria remain physically anchored in the urinary tract during an infection, inhibitors of FimH binding are predicted to prophylactically block UPEC binding or allow for the washout of an existing infection. Indeed, recently reported small-molecule inhibitors of FimH binding, called mannocides, have activity as both prophylaxis or for treatment of established infections of murine models of UTI, demonstrating the utility of targeting UPEC pili (Cusumano et al., 2011; Han et al., 2010). While an exciting development for the treatment of UTIs, mannocide therapies disrupt only a specific pilus–ligand interaction, limiting the number of susceptible organisms.

Pilus formation and export to the bacteria cellular membrane occurs through the chaperone–usher pathway, which is widely conserved among gram-negative pathogens (Busch & Waksman, 2012). Given the need for novel antibiotics targeting gram-negative pathogens, general inhibitors of the chaperone–usher pathway could find use as therapies for a range of pathogens such as *Salmonella*, *Yersinia*, and *Pseudomonas* species (Nuccio & Bäumler, 2007). The chaperone–usher pilus synthesis pathway begins with the transport of pilus subunits to the periplasmic space where they are bound by a chaperone protein, PapD, that is required for proper folding and pilin assembly. The pilin subunits contain a nearly complete β-barrel, lacking only a single strand so that donation of the missing strand by the chaperone protein allows for proper folding of the subunits. Additionally, the chaperone uses a hydrophobic groove to trap and display an unstructured N-terminal extension of the pilin subunit. To add a new subunit to the growing strand, the displayed N-terminus extension replaces the chaperone strand to complete the β-barrel and allow for the release of the chaperone (Waksman & Hultgren, 2009). Loss of binding to the chaperone blocks pilus assembly and leads to the accumulation of aggregated pilin subunits (Slonim, Pinkner, Branden, & Hultgren, 1992), suggesting that small-molecule inhibitors of the usher–chaperone pathway could have the same effect.

To test this hypothesis, a series of small molecules based on the core of the chaperone ligand, PapG, were synthesized. These so-called pilicides were observed to bind to PapD by SPR and NMR (Hedenstrom et al., 2005; Svensson et al., 2001). Initial biological results were promising, as the lead pilicide compound (Fig. 2A) was found to block bacterial binding to bladder epithelial cells, reduce the number of cells with developed pili, and disrupt biofilm formation, albeit at relatively high inhibitor

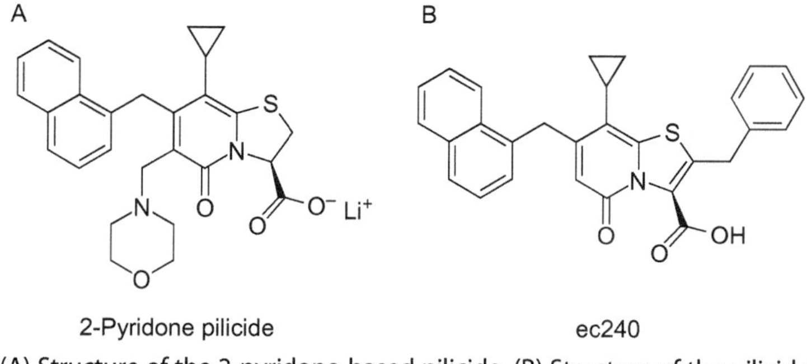

Fig. 2 (A) Structure of the 2-pyridone-based pilicide. (B) Structure of the pilicide ec240.

concentrations. To confirm the pilicide mechanism of action, the authors solved the structure of a pilicide bound PapD. The pilicide was found to obstruct the binding site for FimH, leading to the observed failure of pilus assembly (Pinkner et al., 2006). Further medicinal chemistry optimization of the inhibitors yielded a derivative carrying an additional benzyl functional group, with over a 16-fold improvement in potency in a biofilm formation assay (Chorell et al., 2010). In spite of these gains, even the most active pilicides to date are relatively low potency, (7 μM IC_{50}, Fig. 2B) and have not been tested in animal models of urinary tract infections. Furthermore, recent studies have indicated that the advanced pilicides may have additional, nonpilicidal, effects on the transcriptional regulation of UPEC virulence and motility, although these experiments were conducted at high inhibitor concentrations, where off-targets effects are more likely (Greene et al., 2014).

Despite the remaining challenges, these studies demonstrate exciting progress toward novels UTI therapies that target UPEC adhesion, both by mannosides and the pilicides. Since both classes of inhibitors target bacterial virulence rather than survival, it is possible that resistance would be slower to develop and nonvirulent microbiota would be spared, both welcome attributes in a therapeutic agent.

1.3 DnaN–DnaE1

The development of effective therapies against *Mycobacterium tuberculosis* was a major global achievement and has made it possible to imagine that this scourge may eventually be eradicated. While the number of tuberculosis-related deaths has fallen by more than 20% since 2000, it remains one of the top 10 global causes of death (Panel on Antiretroviral Guidelines for Adults and Adolescents, 2016). Unfortunately, there has been a remarkable increase in the extent of drug resistance in *M. tuberculosis* with the emergence of extensively (Jassal & Bishai, 2008) and totally drug-resistant strains (Udwadia, Amale, Ajbani, & Rodrigues, 2012). In 2015, over 30% of tuberculosis cases were found to be resistant to the first-line therapy rifampicin (Panel on Antiretroviral Guidelines for Adults and Adolescents, 2016). To continue to make strides against this pathogen, antibacterial agents acting against novel targets are needed.

To identify potential inhibitors of *M. tuberculosis*, Kling and coworkers identified a promising lead compound, griselimycin (GM), that was first identified at Rhône-Poulenc in the 1960s (Kling et al., 2015). GM is a

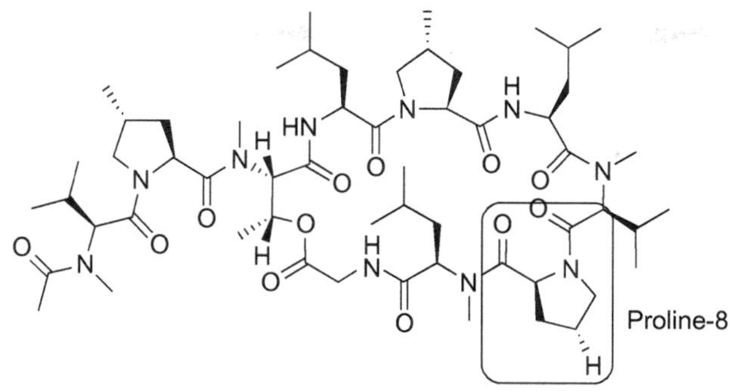

Fig. 3 Structure of the DnaN–DnaE1 PPI inhibitor griselimycin. Methylation or cyclohexylation of the labeled hydrogen of the *boxed* residue, proline-8, leads to improved potency and pharmacokinetic properties.

cyclic peptide natural product produced from *Streptomyces* species (Fig. 3), but the compound suffered from poor pharmacokinetics and the introduction of other active antituberculosis agents. Hoping to overcome these limitations, the authors undertook a preliminary structure activity relationship (SAR) study that demonstrate that minor derivatization of the Pro^8 residue leads to both an increased stability and potency. High doses of a cyclohexyl-derivatized GM (CGM) were found to be as active as clinical therapy isoniazid against *M. tuberculosis* in vitro and in an in vivo mouse lung TB model.

Because the mechanism of GM was not established during the initial discovery process, resistance studies were conducted to identify the cellular target of GM. A variety of genetic approaches lead to the determination that DnaN, the DNA polymerase sliding clamp, was the target, and GM binding to DnaN was confirmed by surface plasmon resonance with femtomolar dissociation constants. Both GM and CGM bind to mycoplasma DnaN with over a ~1000 × higher affinity than to the *E. coli* homolog or the resistance-associated protein GriR. The binding site of a methylated GM was determined by X-ray crystallographic studies of DnaN, which revealed the compounds block binding to DnaE1, the catalytic subunit of Pol III. The authors propose that disruption of this interaction leads to lethal failures in DNA replication and likely DNA breaks.

Intriguingly, previous screening campaigns have targeted the DnaN–DnaE interface with only modest results. An FP-based chemical HTS identified a 10-μM inhibitor of the interaction, and while a crystal structure of the inhibitor bound to DnaN was obtained, no measurements of antibacterial efficacy were reported. This inhibitor contains a rhodanine, a common pan-assay interference compounds motif (Baell & Holloway, 2010), suggesting off-target effects would be problematic (Georgescu et al., 2008). A later fragment screening effort identified two possible inhibitor scaffolds (Yin et al., 2014), and while further SAR efforts modestly improved the potency, only a 20-μM IC_{50} for the in vitro interaction was described (Yin et al., 2015). It seems likely that both efforts were hampered by a limited library size available for the initial screening (~30,000 and 352 compounds, respectively), whereas the natural product approach taken to discover GM was agnostic to the mechanism of action and only incidentally targeted a PPI.

2. ANTIVIRAL PPI INHIBITORS

2.1 Human Papillomavirus Ori–E1–E2 Interaction

Human papillomaviruses (HPV) are double-stranded DNA viruses that infect the human epithelium and cause wart formation from the rapid growth of the epithelial tissue. One distinguishing factor of HPVs is that they are among the few known human cancer viruses, where infection of the anogenital region by high-risk strains (HPV-16 and HPV-18) can lead to the development of cervical cancer (Bosch & De Sanjosé, 2003). The increased risk of cancer results from the incorporation of the virus into the host genome, forming a provirus. To continue propagating, the host cell is then compelled to divide, simultaneously copying the provirus. This continued dysregulated replication can eventually lead to cervical cancer (Muñoz, Castellsagué, de González, & Gissmann, 2006).

The low-risk viruses (HPV-6 and HPV-11) are rarely carcinogenic, but can lead to the development of anogenital warts. Wart removal is a painful process and, because the underlying HPV infection is not cleared, warts recur frequently. An antiviral therapy for the low-risk HPV strains could allow for the ultimate clearing of the infection and a permanent wart treatment.

Rather than reproduce as a provirus, low-risk HPV strains maintain their genomes as plasmids in infected cells. HPV relies on a host polymerase to replicate this plasmid, as the HPV genome does not encode a polymerase. Initiation of viral replication depends on the formation of a ternary DNA–protein complex to recruit the polymerase to the origin of replication (Berg & Stenlund, 1997). First, the viral protein E1 binds to the origin through a DNA-binding domain (DBD), which also serves as a dimerization site. A C-terminal helicase domain of E1 mediates an interaction to the transactivation domain of E2 (Chen & Stenlund, 1998). Once located at the origin, an E2 DBD domain tethers the E1–E2–DNA complex and the recruitment of additional E1 dimers leads to the formation of a hexameric structure that encircles and melts the origin DNA (Sedman & Stenlund, 1998). During this process, E2 is displaced allowing for the recruitment of a host DNA polymerase which replicates the plasmid (Conger, Liu, Kuo, Chow, & Wang, 1999). Disruption of any of these interactions is sufficient to block HPV replication.

To identify inhibitors of E1–E2–Ori complex formation, White and coworkers developed a bead-based scintillation proximity assay (SPA). A DNA substrate containing the HPV origin was radiolabeled and then incubated with E1 and E2. An SPA bead was tether to the complex by an anti-E1 antibody and this allowed for the recruitment of E1 to the DNA to be detected, which requires E1–E2 binding. Functionally, this allowed for the identification of inhibitors of the E1–E2, E1–DNA, or E2–DNA interactions simultaneously, since disruption of any is sufficient to block E1 recruitment. This assay was used to screen a 140,000 compound library and identified a single lead compound with an indanedione scaffold (11 μM IC_{50}) (Fig. 4A; White et al., 2003). Additional chlorination of the phenyl ring through SAR studies led to the discovery of compounds with approximately 20-fold greater potency (Yoakim et al., 2003). Activity against the interaction was confirmed by ELISA, and binding to E2 was confirmed by isothermal titration calorimetry (Wang et al., 2004). Unfortunately, when these compounds were tested in cellular assays, EC_{50} values were markedly higher than the activity in biochemical assays had suggested. Together with the fact that the indanedione class of compounds possessed relatively poor pharmacokinetic properties, this series was ultimately abandoned (White, Faucher, & Goudreau, 2010).

However, this well-characterized set of in vitro active compounds allowed the researchers to improve their HTS assay. A tritiated version of an indanedione compound was prepared, and a larger compound library was screened for compounds with the ability displace the tritiated probe. A novel inhibitor scaffold was found to both displace the tritiated probe and block the E1–E2 interaction. The scaffold showed a marked similarity to repaglinide, a diabetes drug with blood glucose-lowering properties (Fig. 4B and C). Accordingly, early members of repaglinide series shared this undesirable (for an antiviral agent) property. While no crystal structure for this class was obtained, preliminary optimization SAR suggested that

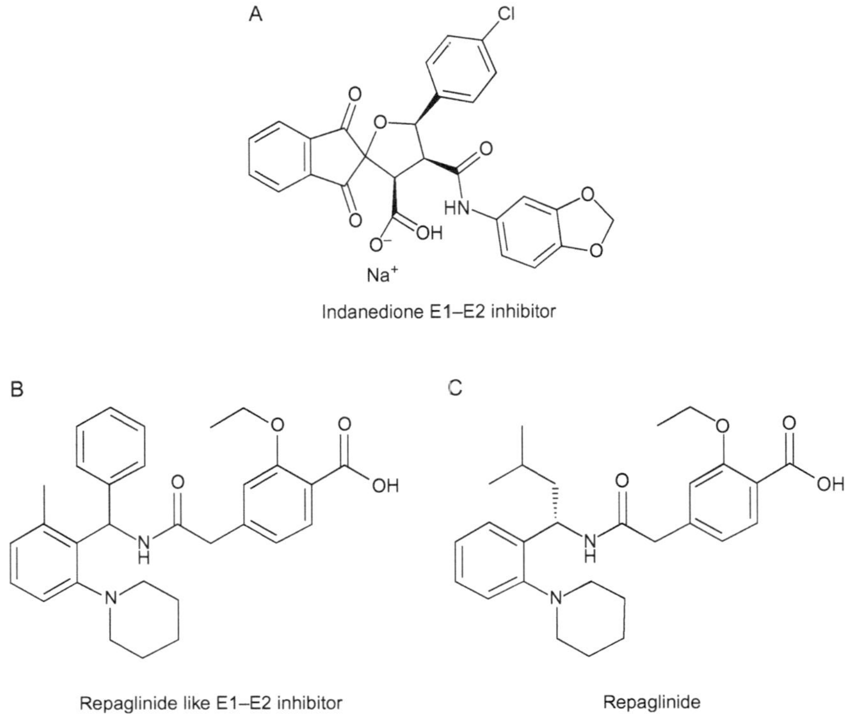

Fig. 4 Structures of the inhibitors of the E1–E2 interface. The repaglinide like class of inhibitors was found to have undesirable hypoglycemic effects, resulting from the chemical similarity to repaglinide.

antiviral activity could be maintained while minimizing the glucose-lowering properties. Despite these promising initial results, the introduction of an effective HPV vaccine was expected to vastly reduce the future demand for HPV therapies and the project was terminated. In this particular case, the failure to generate a clinical agent was not the due to the difficultly of targeting a PPI, but rather external market forces (White et al., 2010).

2.2 Human Immunodeficiency Virus

Prior to the development of antiretrovirus therapy, infection with human immunodeficiency virus (HIV), the causative agent of the acquired immunodeficiency syndrome, was a death sentence. Starting in the 1990s and continuing through today, the introduction of a broad range of HIV treatments has greatly improved patient outcomes. Unfortunately, viruses resistant to every class of HIV therapy have been observed and these can lead to treatment failure (Magambo et al., 2014; Wensing et al., 2015). Using combinations of different classes of antiretrovirals greatly diminishes the risk of developing resistance, but resistance can occur because of monotherapy or if subtherapeutic levels of antiretrovirals are maintained as a result of improper dosing. Furthermore, resistance to one member of an antiretroviral class frequently leads to cross-resistance to all members of that class, which can both reduce the number of available agents and undermine the effectiveness of preexposure prophylaxis (PrEP). In the face of these challenges, the development of novel classes of HIV treatment for both PrEP and therapy is needed (Waheed & Tachedjian, 2016).

Given the limited repertoire of viral proteins, viruses rely on hijacking host proteins to carry out functions in the viral life cycle. Viruses typically rely on protein interactions with the host cellular machinery to enter or otherwise alter the normal cellular function. As such, small-molecule blockades of these interactions can be used to disrupt the viral life cycle. Two key HIV PPIs that have been targeted for disruption as antiviral therapies play vital roles in viral–host fusion and integration into the host genome.

2.3 HIV Entry Inhibitors

To access the cellular machinery required for replication, viruses must first bypass the cellular membrane. Most viruses accomplish this by binding to a cellular receptor on the outer cellular membrane, which is used as an initiation point for membrane fusion. The structure and identity of the receptor targeted varies between viruses and determines which cell types are targeted (i.e., the tropism). To target CD4+ T cells, the gp120 subunit of the envelope protein first engages the CD4 receptor. CD4 binding causes a conformational shift that allows for binding to an additional T-cell receptor, either CXCR4 or CCR5. Once bound to both receptors, the second subunit of ENV, gp41, penetrates into the cellular membrane and begins the fusion process (Wilen, Tilton, & Doms, 2012).

There is biological support for targeting HIV fusion as a therapeutic route. In the mid-1990s, it was noted that a subset of the population lack cell surface expression of CCR5 and those lacking this receptor appear to be resistant to infection with the HIV-1 strain (Samson, Libert, Doranz, & Rucker, 1996). This is important for two principle reasons. First, this implies that blockade of the CCR5 receptor is sufficient to impede HIV infection. Second, inhibition of the normal function of CCR5 is likely to be tolerated with minimal adverse effects since the $CCR5^{-/-}$ genotype is maintained in the population.

In the early 2000s, there was much interest in developing inhibitors of CD4, CXCR4, or CCR5 receptors to block HIV entry into cells and peptide or small-molecule inhibitors of gp120 binding were developed against each of these targets (Lin et al., 2003). Despite the early success of these entry inhibitors, only two have reached the clinic, primarily due to the poor bioavailability of the CD4 antagonists (Yang et al., 2005) and limited effectiveness of the CXCR4 antagonists (Doranz et al., 2001). The first clinical agent, enfuvirtide, is a peptide that binds to gp120 and blocks entry pore formation. As a peptide inhibitor, it must be dosed intravenously, is difficult to self-administer and is not recommended as first-line therapy (Panel on Antiretroviral Guidelines for Adults and Adolescents, 2016). But because enfuvirtide targets the viral protein gp120 rather than the cellular receptor, it is active against both CCR5 and CXCR4 trophic viruses (Matthews et al., 2004).

The second entry inhibitor approved to date is maraviroc. The lead that was eventually developed into maraviroc was originally identified from a Pfizer HTS. The compound (UK-107,543) was found to block binding of a radiolabeled substrate to CCR5. A heroic medicinal chemistry optimization campaign yielded maraviroc (Fig. 5), with low nanomolar IC_{50} values against three CCR5 peptide substrates. Additionally, maraviroc was found

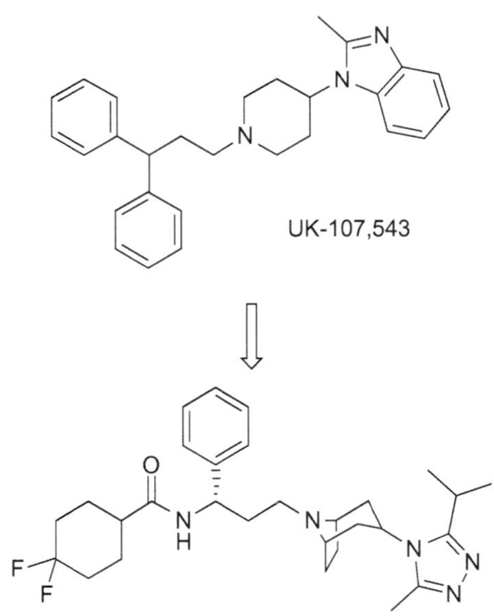

Fig. 5 Structure of the initial screening hit for CCR5 antagonist, UK-107,543, and the optimized inhibitor, Maraviroc.

to have nanomolar antiviral activity against a variety of both lab-adapted and clinical CCR5 tropic HIV-1 strains. As expected, no activity was observed against CXCR4 tropic strains. Phase I clinical trials demonstrated that therapeutic doses could be reached safely, and after a series of phase III trials demonstrated efficacy, maraviroc was approved for clinical use. Currently, maraviroc is rarely used in treatment-naive patients but is reserved for patients with treatment resistant, CCR5-specific HIV strains (Panel on Antiretroviral Guidelines for Adults and Adolescents, 2016). Despite these limitations, the development of maraviroc represented a major accomplishment in the field of PPI inhibitors.

2.4 HIV Integrase Inhibitors

Integration of the viral genome into the host chromosome represents another step of the HIV life cycle that has been targeted for therapy. The validity of this target has been established by the clinical success of the integrase active site inhibitor raltegravir (Eron et al., 2013). Relatively high levels of resistance to integrase inhibitors and cross-resistance within the class have limited the utility of these therapies (Malet et al., 2014). Fortunately, integrase relies on an essential PPI where lens epithelium-derived growth factor (LEDGF/p75) tethers integrase to the DNA while stimulating the integrase activity (Poeschla, 2008). This interface is essential to carry out its function and represents a potential therapeutic target.

Several groups have conducted screens to identify small-molecule inhibitors of the LEDGF/p75 interface, including both chemical and in silico screens. The first inhibitor, named D77, was found to be active in a yeast two-hybrid assay, although it proved cytotoxic to noninfected cells. This cytotoxicity likely results from the presence of a rhodanine functional group in D77 (Fig. 6A; Du et al., 2008). Small molecules containing these functional groups are common hits in HTSs as a result of their ability to covalently modify proteins in a nonspecific fashion (Baell & Holloway, 2010). Further in silico screening of clinically approved drugs identified eight inhibitors with IC_{50} values in the low- to mid-micromolar range, although further improvements have not been reported (Hu et al., 2012). A more promising inhibitor scaffold was identified from an effort that used in silico pharmacophore screening to identify commercially available compounds. SAR studies allowed for the development of more potent inhibitors possessing a 2-(*tert*-butoxy) functionality, resulting in CX014442 (Fig. 6B), with 69 nM EC_{50} and low cytotoxicity.

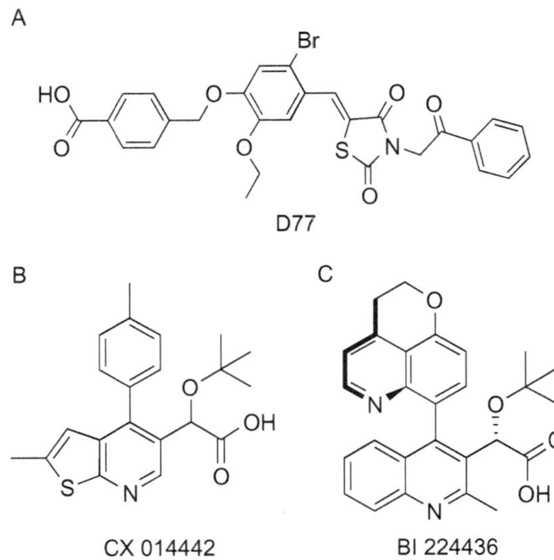

Fig. 6 Structure of representative inhibitors of integrase/LEDGF interaction.

Another series of inhibitors of this interface was developed at Boehringer Ingelheim. Rather than specifically targeting the LEDGF/p75 interface, this group screened for inhibitors of the 3′-processing activity of HIV integrase. After successfully screening their compound collection, they noticed that the lead compound bound integrase away from the active site, leading them to name this class of inhibitors the noncatalytic site integrase inhibitors (NCINI). Optimization of this series of compounds yielded a potent inhibitor, BI 224436 (Fig. 6C).

Despite the structural similarity between the NCINI and LEDGIN inhibitors, it was not immediately clear that shared a common mechanism of action since they were discovered by different methodologies. Work by Kvaratskhelia and coworkers concluded that both inhibitors blocked LEDGF binding, albeit more potently by BI 224436, and there was substantial overlap of the inhibitor binding sites (Engelman, Kessl, & Kvaratskhelia, 2013).

During in vitro analysis, BI 224436 demonstrated antiviral activity against a range of viruses resistant to other therapeutics and had a promising animal pharmacokinetic profile, which prompted the initiation of a phase I clinical trial in 2010 (Fenwick et al., 2014). While the results from this trial have not been disclosed and no phase II trials have been started, Gilead licensed the development of the NCINI class of inhibitors in 2011 ("Boehringer license novel HIV non-catalytic integrase inhibitors to

Gilead," 2011). A paper describing the medicinal chemistry optimization of another NCINI derivative for activity against a integrase-resistant mutant and improved pharmacokinetics was published in 2016, suggesting that an integrase inhibitor may yet show clinical efficacy (Fader et al., 2016).

3. TARGETING HOST–HOST PPIs TO IMPROVE INFECTION SURVIVAL

One of the most lethal aspects of bacterial infections can be the immune response to the insult. While important for the eventual clearance of the pathogens, widespread overactivation of the innate immune system and the resulting systemic inflammation can lead to multiorgan failure or even death (Marshall, 2005). Neutrophil exocytosis is a principle driver of this systemic inflammation, where neutrophils release a range of potent toxins such as reactive oxygen species and proteases (Marshall, 2005). Excessive release of these species does not improve control of the infection, but rather damages surrounding tissues (Narasaraju et al., 2011). In rats, blockade of neutrophil exocytosis was shown to reduce the extent of tissue injury (Uriarte et al., 2013) and this blockade has been proposed as a mechanism to allow for patient survival, while the infection is treated by the antibiotic therapy and the immune system. Related inhibitors would also be expected to be effective for the treatment of some autoimmune conditions.

3.1 Neutrophil Exocytosis Inhibitors (Nexinhibs)

One of the critical regulators of neutrophil exocytosis is the interaction between JFC1 and a GTPase, Rab27a. Downregulation of either of these proteins blocks the exocytosis of neutrophil granules, but neither is known to play a role in phagosome maturation (Brzezinska et al., 2008). This raises the possibility that small-molecule inhibition of the Rab27a–JFC1 interaction would lead to a targeted inhibition of the negative effects of the neutrophil granule release, while allowing for neutrophil survival and the continued clearance of the pathogens. Because this PPI is a host-specific interaction, resistance is unlikely to be a problem.

A high-throughput time-resolved fluorescence resonance energy transfer was used to screen 32,000 compounds, and lead compounds were confirmed by the reduction in neutrophil myeloperoxidase secretion in a cell-based secondary screen. In this assay, neutrophils were pretreated with the putative nexinhibs (*n*eutrophil *ex*ocytosis *inhib*itor) and then granule release was stimulated. A decrease in the generation of H_2O_2 indicated that the compounds

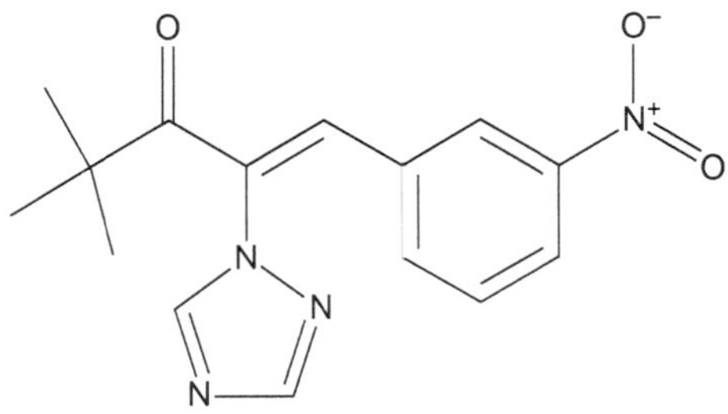

Nexinhib 20

Fig. 7 Structure of Nexinhib 20, an inhibitor of the neutrophil exocytosis.

successfully inhibited the release of the myeloperoxidase-containing granules. After excluding peroxide-scavenging compounds, the authors concluded that compounds active in this assay must penetrate the membrane, allowing the authors to eliminate nonpermeable compounds early in the hit reduction process (Johnson et al., 2016).

The most promising compound, Nexinhib 20 (Fig. 7), was found to block neutrophil degranulation in response to agonists and disrupted the Rab27a–JFC1 interaction in pulldown and ELISA assays with an IC_{50} of 2.6 μM in the ELISA. Nexinhib 20 treatment did not result in cell death or impair the ability of neutrophils to phagocytize, but did reduce exocytosis after neutrophil stimulation. To examine how nexinhibs function in a physiologic setting, mice were treated with Nexinhib 20 prior to the induction of systemic inflammation by an intraperitoneal injection of lipopolysaccharide. The total number of neutrophils and white blood cells remained unchanged relative to an untreated control, but there was a modest, albeit significant, reduction in the plasma levels of myeloperoxidase as well as neutrophil infiltration of the kidney and liver (Johnson et al., 2016). While further in vivo testing and a mouse sepsis survival study would bolster the case for nexinhibs, this preliminary study is an exciting step.

4. CONSIDERATIONS FOR TARGETING PPIs

Historically, PPIs were considered undruggable therapeutic targets. One early estimate from Pfizer suggested that while inhibitors for enzymes or GPCR targets could be discovered by high-throughput screening methods at a rate 1 per 10^5 compounds screened, a discovery rate of 1 per 10^9–10^{10} compounds was expected for PPI targets (Spencer, 1998). Given this discouraging prediction, it is obvious why pharmaceutical companies limited their exposure to PPIs. The prevailing view at the time was that protein interfaces consisted of interactions between two broad, mostly hydrophobic and featureless surfaces. The lack of pockets for a hit to gain an initial foothold was also thought to complicate screening.

Two major advances have made targeting PPIs more feasible. First was the realization that while many PPI match the above description, a sizeable number are much simpler. A useful system for classifying the nature and difficulty of targeting a PPI has been developed. Briefly, PPIs can be fit into one of the three categories based on the complexity of the simplest member: (1) primary interfaces consist of an unstructured peptide that binds to a channel. (2) A single fold of secondary structure (α-helix or β-sheet) or (3) a lengthy stretch of protein that uses a tertiary structure (Fig. 8; Blundell et al., 2006). While there are examples of successfully targeting each class of PPI, there is a significant bias toward the simpler interfaces (Arkin et al., 2014). Careful selection of a target, ideally guided by structural characterization of the PPI, can greatly increase the odds of success. To help calibrate expectations about a potential project, an online tool has been developed to aid in assessing the "drugability" of a given PPI (Basse et al., 2013; Basse, Betzi, Morelli, & Roche, 2016).

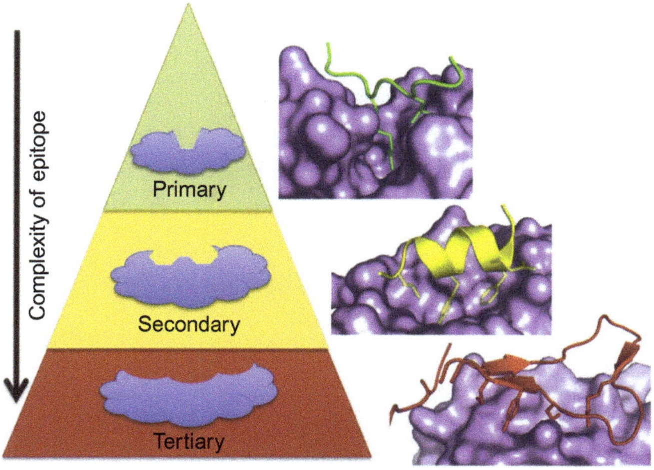

Fig. 8 Scheme for classifying PPIs by the nature of the interaction site. Most of the successful PPI inhibitors target the simpler interactions, such as the primary or secondary ones, although there are a examples of inhibitors of tertiary interfaces. *Reprinted from Arkin, M. R., Tang, Y., & Wells, J. A. (2014). Small-molecule inhibitors of protein-protein interactions: Progressing toward the reality.* Chemistry & Biology, 21, *1102–1114, with permission from Elsevier.*

The second key development in the targeting of PPI has been improvements in screening methodologies. The introduction and miniaturization of PPI HTS assays and the advent of academic screening facilities have reduced the cost, and therefore the risk, of performing a chemical HTS. There has been debate regarding the suitability of existing screening libraries for targeting PPIs, as PPI inhibitors tend to be more aromatic, hydrophobic, three-dimensional and have higher molecular weights than non-PPI inhibitors (Kuenemann, Labbe, Cerdan, & Sperandio, 2016). Armed with this information, PPI specific screening libraries have been introduced ("iPPI Focused Libraries," 2017; "Protein-Protein Interaction Libraries—Enamine," 2017; Reynès et al., 2010), although more time is needed to tell if these libraries show improved hit rates. Besides chemical HTS methods, additional hit identification strategies have been successfully implemented. Some have been highlighted in this review and include in silico screening, natural product screening, rational design, scaffold hopping, and fragment screening and elaboration (Sheng et al., 2015). There is unlikely to be a single best method for lead generation, and different approaches may succeed where prior attempts have failed. Because hit identification methods rely on varied input libraries and individual libraries may not contain a bona fide inhibitor, trying multiple screening methods allows for the sampling of a broader range of chemical space than could be achieved by solely by increasing the library size. This is also demonstrated by several of the examples presented here, where different approaches against the same target yielded unique inhibitor scaffolds. Furthermore, the use of phenotypic screens or other assay designs allowing for the discovery of unpredicted mechanisms of action may prove fruitful (Fischer, Rossmann, & Hyvonen, 2015).

While the case studies presented here represent the successful inhibition of a range of protein interactions, most or all of these compounds will still fail to reach the clinic. It is difficult to assess the true success rate of PPI-targeted drug discovery efforts as publication and survivorship bias obscures the denominator of this calculation. However, from the examples presented here and elsewhere, it appears that small-molecule disruption of many PPIs is challenging yet feasible. The development of existing inhibitors appears to be hindered by nonspecific effects, poor pharmacokinetics, or cell penetrance, which are challenges common to all therapeutics. As seen with the introduction of oncologic therapies targeting PPIs, once the pipeline of antiinfective PPI inhibitors expands, we have no reason to doubt that they will find increasing clinical success.

REFERENCES

Adams, D. W., & Errington, J. (2009). Bacterial cell division: Assembly, maintenance and disassembly of the Z ring. *Nature Reviews Microbiology*, 7(9), 642–653.

Arkin, M. R., Tang, Y., & Wells, J. A. (2014). Small-molecule inhibitors of protein-protein interactions: Progressing toward the reality. *Chemistry & Biology*, *21*(9), 1102–1114. https://doi.org/10.1016/j.chembiol.2014.09.001.

Arkin, M. R., & Wells, J. A. (2004). Small-molecule inhibitors of protein-protein interactions: Progressing towards the dream. *Nature Reviews. Drug Discovery*, *3*(4), 301–317.

Baell, J. B., & Holloway, G. A. (2010). New substructure filters for removal of pan assay interference compounds (PAINS) from screening libraries and for their exclusion in bioassays. *Journal of Medicinal Chemistry*, *53*(7), 2719–2740. https://doi.org/10.1021/jm901137j.

Basse, M. J., Betzi, S., Bourgeas, R., Bouzidi, S., Chetrit, B., Hamon, V., et al. (2013). 2P2Idb: A structural database dedicated to orthosteric modulation of protein-protein interactions. *Nucleic Acids Research*, *41*(Database issue), D824–827. https://doi.org/10.1093/nar/gks1002.

Basse, M. J., Betzi, S., Morelli, X., & Roche, P. (2016). 2P2Idb v2: Update of a structural database dedicated to orthosteric modulation of protein-protein interactions. *Database (Oxford)*, *2016*, 1–6. https://doi.org/10.1093/database/baw007.

Bent, S., Nallamothu, B. K., Simel, D. L., Fihn, S. D., & Saint, S. (2002). Does this woman have an acute uncomplicated urinary tract infection? *JAMA*, *287*(20), 2701–2710.

Berg, M., & Stenlund, A. (1997). Functional interactions between papillomavirus E1 and E2 proteins. *Journal of Virology*, *71*(5), 3853–3863.

Blundell, T. L., Sibanda, B. L., Montalvão, R. W., Brewerton, S., Chelliah, V., Worth, C. L., et al. (2006). Structural biology and bioinformatics in drug design: Opportunities and challenges for target identification and lead discovery. *Philosophical Transactions of the Royal Society of London. Series B, Biological Sciences*, *361*(1467), 413–423.

Boehringer license novel HIV non-catalytic integrase inhibitors to Gilead. (2011). Retrieved from https://www.thepharmaletter.com/article/boehringer-license-novel-hiv-non-catalytic-integrase-inhibitors-to-gilead. (accessed on May 21, 2017).

Bosch, F. X., & De Sanjosé, S. (2003). Human papillomavirus and cervical cancer—Burden and assessment of causality. *Journal of the National Cancer Institute. Monographs*, *2003*(31), 3–13.

Brzezinska, A. A., Johnson, J. L., Munafo, D. B., Crozat, K., Beutler, B., Kiosses, W. B., et al. (2008). The Rab27a effectors JFC1/Slp1 and Munc13-4 regulate exocytosis of neutrophil granules. *Traffic*, *9*(12), 2151–2164. https://doi.org/10.1111/j.1600-0854.2008.00838.x.

Busch, A., & Waksman, G. (2012). Chaperone–usher pathways: Diversity and pilus assembly mechanism. *Philosophical Transactions of the Royal Society of London. Series B, Biological Sciences*, *367*(1592), 1112–1122. https://doi.org/10.1098/rstb.2011.0206.

Chen, G., & Stenlund, A. (1998). Characterization of the DNA-binding domain of the bovine papillomavirus replication initiator E1. *Journal of Virology*, *72*(4), 2567–2576.

Chorell, E., Pinkner, J. S., Phan, G., Edvinsson, S., Buelens, F., Remaut, H., et al. (2010). Design and synthesis of C-2 substituted Thiazolo and Dihydrothiazolo ring-fused 2-pyridones: Pilicides with increased antivirulence activity. *Journal of Medicinal Chemistry*, *53*(15), 5690–5695. https://doi.org/10.1021/jm100470t.

Conger, K. L., Liu, J.-S., Kuo, S.-R., Chow, L. T., & Wang, T. S.-F. (1999). Human papillomavirus DNA replication: Interactions between the viral E1 protein and two subunits of human DNA polymerase α/primase. *Journal of Biological Chemistry*, *274*(5), 2696–2705. https://doi.org/10.1074/jbc.274.5.2696.

Cukuroglu, E., Engin, H. B., Gursoy, A., & Keskin, O. (2014). Hot spots in protein–protein interfaces: Towards drug discovery. *Progress in Biophysics and Molecular Biology, 116*(2–3), 165–173. https://doi.org/10.1016/j.pbiomolbio.2014.06.003.

Cusumano, C. K., Pinkner, J. S., Han, Z., Greene, S. E., Ford, B. A., Crowley, J. R., et al. (2011). Treatment and prevention of urinary tract infection with orally active FimH inhibitors. *Science Translational Medicine, 3*(109), 109ra115. https://doi.org/10.1126/scitranslmed.3003021.

Dethlefsen, L., Huse, S., Sogin, M. L., & Relman, D. A. (2008). The pervasive effects of an antibiotic on the human gut microbiota, as revealed by deep 16S rRNA sequencing. *PLoS Biology, 6*(11), e280 https://doi.org/10.1371/journal.pbio.0060280.

Doranz, B. J., Filion, L. G., Diaz-Mitoma, F., Sitar, D. S., Sahai, J., Baribaud, F., et al. (2001). Safe use of the CXCR4 inhibitor ALX40-4C in humans. *AIDS Research and Human Retroviruses, 17*(6), 475–486. https://doi.org/10.1089/08892220151126508.

Du, L., Zhao, Y., Chen, J., Yang, L., Zheng, Y., Tang, Y., et al. (2008). D77, one benzoic acid derivative, functions as a novel anti-HIV-1 inhibitor targeting the interaction between integrase and cellular LEDGF/p75. *Biochemical and Biophysical Research Communications, 375*(1), 139–144. https://doi.org/10.1016/j.bbrc.2008.07.139.

Engelman, A., Kessl, J. J., & Kvaratskhelia, M. (2013). Allosteric inhibition of HIV-1 integrase activity. *Current Opinion in Chemical Biology, 17*(3), 339–345.

Eron, J. J., Cooper, D. A., Steigbigel, R. T., Clotet, B., Gatell, J. M., Kumar, P. N., et al. (2013). Efficacy and safety of raltegravir for treatment of HIV for 5 years in the BENCHMRK studies: Final results of two randomised, placebo-controlled trials. *The Lancet Infectious Diseases, 13*(7), 587–596. https://doi.org/10.1016/S1473-3099(13)70093-8.

Fader, L. D., Bailey, M., Beaulieu, E., Bilodeau, F., Bonneau, P., Bousquet, Y., et al. (2016). Aligning potency and pharmacokinetic properties for pyridine-based NCINIs. *ACS Medicinal Chemistry Letters*, 7(8), 797–801.

Fenwick, C., Bailey, M. D., Bethell, R., Bös, M., Bonneau, P., Cordingley, M., et al. (2014). Preclinical profile of BI 224436, a novel HIV-1 non-catalytic-site integrase inhibitor. *Antimicrobial Agents and Chemotherapy, 58*(6), 3233–3244.

Fischer, G., Rossmann, M., & Hyvonen, M. (2015). Alternative modulation of protein-protein interactions by small molecules. *Current Opinion in Biotechnology, 35*, 78–85. https://doi.org/10.1016/j.copbio.2015.04.006.

Flores-Mireles, A. L., Walker, J. N., Caparon, M., & Hultgren, S. J. (2015). Urinary tract infections: Epidemiology, mechanisms of infection and treatment options. *Nature Reviews. Microbiology, 13*(5), 269–284. https://doi.org/10.1038/nrmicro3432. http://www.nature.com/nrmicro/journal/v13/n5/abs/nrmicro3432.html#supplementary-information.

Georgescu, R. E., Yurieva, O., Kim, S. S., Kuriyan, J., Kong, X. P., & O'Donnell, M. (2008). Structure of a small-molecule inhibitor of a DNA polymerase sliding clamp. *Proceedings of the National Academy of Sciences of the United States of America, 105*(32), 11116–11121. https://doi.org/10.1073/pnas.0804754105.

Goff, D. A., Kullar, R., Goldstein, E. J., Gilchrist, M., Nathwani, D., Cheng, A. C., et al. (2017). A global call from five countries to collaborate in antibiotic stewardship: United we succeed, divided we might fail. *The Lancet Infectious Diseases, 17*(2), e56–e63.

Greene, S. E., Pinkner, J. S., Chorell, E., Dodson, K. W., Shaffer, C. L., Conover, M. S., et al. (2014). Pilicide ec240 disrupts virulence circuits in uropathogenic Escherichia coli. *mBio, 5*(6), e02038-14. https://doi.org/10.1128/mBio.02038-14.

Gupta, K., Hooton, T. M., Naber, K. G., Wullt, B., Colgan, R., Miller, L. G., et al. (2011). International clinical practice guidelines for the treatment of acute uncomplicated cystitis and pyelonephritis in women: A 2010 update by the Infectious Diseases Society of

America and the European Society for Microbiology and Infectious Diseases. *Clinical Infectious Diseases*, *52*(5), e103–e120. https://doi.org/10.1093/cid/ciq257.

Hale, C. A., & de Boer, P. A. (1997). Direct binding of FtsZ to ZipA, an essential component of the septal ring structure that mediates cell division in E. coli. *Cell*, *88*(2), 175–185.

Han, Z., Pinkner, J. S., Ford, B., Obermann, R., Nolan, W., Wildman, S. A., et al. (2010). Structure-based drug design and optimization of mannoside bacterial FimH antagonists. *Journal of Medicinal Chemistry*, *53*(12), 4779–4792. https://doi.org/10.1021/jm100438s.

Hedenstrom, M., Emtenas, H., Pemberton, N., Aberg, V., Hultgren, S. J., Pinkner, J. S., et al. (2005). NMR studies of interactions between periplasmic chaperones from uropathogenic E. coli and pilicides that interfere with chaperone function and pilus assembly. *Organic & Biomolecular Chemistry*, *3*(23), 4193–4200. https://doi.org/10.1039/B511857C.

Hu, G., Li, X., Sun, X., Lu, W., Liu, G., Huang, J., et al. (2012). Identification of old drugs as potential inhibitors of HIV-1 integrase—Human LEDGF/p75 interaction via molecular docking. *Journal of Molecular Modeling*, *18*(12), 4995–5003. https://doi.org/10.1007/s00894-012-1494-0.

iPPI Focused Libraries. (2017). Retrieved from http://www.otavachemicals.com/products/targeted-libraries-and-focused-libraries/protein-protein-interaction. (accessed on May 21, 2017).

Jassal, M., & Bishai, W. R. (2008). Extensively drug-resistant tuberculosis. *The Lancet Infectious Diseases*, *9*(1), 19–30. https://doi.org/10.1016/S1473-3099(08)70260-3.

Jennings, L. D., Foreman, K. W., Rush, T. S., Tsao, D. H., Mosyak, L., Li, Y., et al. (2004). Design and synthesis of indolo [2, 3-a] quinolizin-7-one inhibitors of the ZipA–FtsZ interaction. *Bioorganic & Medicinal Chemistry Letters*, *14*(6), 1427–1431.

Johnson, J. L., Ramadass, M., He, J., Brown, S. J., Zhang, J., Abgaryan, L., et al. (2016). Identification of neutrophil exocytosis inhibitors (Nexinhibs), small molecule inhibitors of neutrophil exocytosis and inflammation: Druggability of the small GTPase Rab27a. *The Journal of Biological Chemistry*, *291*(50), 25965–25982. https://doi.org/10.1074/jbc.M116.741884.

Kenny, C. H., Ding, W., Kelleher, K., Benard, S., Dushin, E. G., Sutherland, A. G., et al. (2003). Development of a fluorescence polarization assay to screen for inhibitors of the FtsZ/ZipA interaction. *Analytical Biochemistry*, *323*(2), 224–233.

Kling, A., Lukat, P., Almeida, D. V., Bauer, A., Fontaine, E., Sordello, S., et al. (2015). Antibiotics. Targeting DnaN for tuberculosis therapy using novel griselimycins. *Science*, *348*(6239), 1106–1112. https://doi.org/10.1126/science.aaa4690.

Krogfelt, K. A., Bergmans, H., & Klemm, P. (1990). Direct evidence that the FimH protein is the mannose-specific adhesin of Escherichia coli type 1 fimbriae. *Infection and Immunity*, *58*(6), 1995–1998.

Kuchibhatla, A., Bhattacharya, A., & Panda, D. (2011). ZipA binds to FtsZ with high affinity and enhances the stability of FtsZ protofilaments. *PLoS One*, *6*(12), e28262.

Kuenemann, M. A., Labbe, C. M., Cerdan, A. H., & Sperandio, O. (2016). Imbalance in chemical space: How to facilitate the identification of protein-protein interaction inhibitors. *Scientific Reports*, *6*, 23815. https://doi.org/10.1038/srep23815.

Kumar, R., & Nanduri, B. (2010). HPIDB—A unified resource for host-pathogen interactions. *BMC Bioinformatics*, *11*(6), S16. https://doi.org/10.1186/1471-2105-11-s6-s16.

Lane, M., & Mobley, H. (2007). Role of P-fimbrial-mediated adherence in pyelonephritis and persistence of uropathogenic Escherichia coli (UPEC) in the mammalian kidney. *Kidney International*, *72*(1), 19–25.

Lehne, B., & Schlitt, T. (2009). Protein-protein interaction databases: Keeping up with growing interactomes. *Human Genomics*, *3*(3), 291. https://doi.org/10.1186/1479-7364-3-3-291.

Lin, P.-F., Blair, W., Wang, T., Spicer, T., Guo, Q., Zhou, N., et al. (2003). A small molecule HIV-1 inhibitor that targets the HIV-1 envelope and inhibits CD4 receptor binding. *Proceedings of the National Academy of Sciences of the United States of America, 100*(19), 11013–11018. https://doi.org/10.1073/pnas.1832214100.

Magambo, B., Nazziwa, J., Bbosa, N., Gupta, R. K., Kaleebu, P., & Parry, C. M. (2014). The arrival of untreatable multidrug-resistant HIV-1 in sub-Saharan Africa. *AIDS, 28*(9), 1373–1374.

Malet, I., Arriaga, L. G., Artese, A., Costa, G., Parrotta, L., Alcaro, S., et al. (2014). New raltegravir resistance pathways induce broad cross-resistance to all currently used integrase inhibitors. *Journal of Antimicrobial Chemotherapy, 69*(8), 2118–2122.

Marshall, J. C. (2005). Neutrophils in the pathogenesis of sepsis. *Critical Care Medicine, 33*(12), S502–S505.

Martinez, J. J., Mulvey, M. A., Schilling, J. D., Pinkner, J. S., & Hultgren, S. J. (2000). Type 1 pilus-mediated bacterial invasion of bladder epithelial cells. *The EMBO Journal, 19*(12), 2803–2812.

Matthews, T., Salgo, M., Greenberg, M., Chung, J., DeMasi, R., & Bolognesi, D. (2004). Enfuvirtide: The first therapy to inhibit the entry of HIV-1 into host CD4 lymphocytes. *Nature Reviews. Drug Discovery, 3*(3), 215–225.

Mosyak, L., Zhang, Y., Glasfeld, E., Haney, S., Stahl, M., Seehra, J., et al. (2000). The bacterial cell-division protein ZipA and its interaction with an FtsZ fragment revealed by X-ray crystallography. *The EMBO Journal, 19*(13), 3179–3191. https://doi.org/10.1093/emboj/19.13.3179.

Muñoz, N., Castellsagué, X., de González, A. B., & Gissmann, L. (2006). Chapter 1: HPV in the etiology of human cancer. *Vaccine, 24*(Suppl. 3), S1–S10. https://doi.org/10.1016/j.vaccine.2006.05.115.

Narasaraju, T., Yang, E., Samy, R. P., Ng, H. H., Poh, W. P., Liew, A.-A., et al. (2011). Excessive neutrophils and neutrophil extracellular traps contribute to acute lung injury of influenza pneumonitis. *The American Journal of Pathology, 179*(1), 199–210. https://doi.org/10.1016/j.ajpath.2011.03.013.

Nuccio, S.-P., & Bäumler, A. J. (2007). Evolution of the chaperone/usher assembly pathway: Fimbrial classification goes Greek. *Microbiology and Molecular Biology Reviews, 71*(4), 551–575. https://doi.org/10.1128/mmbr.00014-07.

Panel on Antiretroviral Guidelines for Adults and Adolescents. (2016). Guidelines for the use of antiretroviral agents in HIV-1-infected adults and adolescents. In *Department of Health and Human Services*. Available from http://aidsinfo.nih.gov/contentfiles./lvguidelines/AdultandAdolescentGL.pdf.

Pinkner, J. S., Remaut, H., Buelens, F., Miller, E., Aberg, V., Pemberton, N., et al. (2006). Rationally designed small compounds inhibit pilus biogenesis in uropathogenic bacteria. *Proceedings of the National Academy of Sciences of the United States of America, 103*(47), 17897–17902. https://doi.org/10.1073/pnas.0606795103.

Poeschla, E. M. (2008). Integrase, LEDGF/p75 and HIV replication. *Cellular and Molecular Life Sciences, 65*(9), 1403–1424. https://doi.org/10.1007/s00018-008-7540-5.

Protein-Protein Interaction Libraries—Enamine. (2017). Retrieved from http://www.enamine.net/index.php?option=com_content&task=view&id=227. (accessed on May 21, 2017).

Reynès, C., Host, H., Camproux, A.-C., Laconde, G., Leroux, F., Mazars, A., et al. (2010). Designing focused chemical libraries enriched in protein-protein interaction inhibitors using machine-learning methods. *PLoS Computational Biology, 6*(3), e1000695. https://doi.org/10.1371/journal.pcbi.1000695.

Rush, T. S., Grant, J. A., Mosyak, L., & Nicholls, A. (2005). A shape-based 3-D scaffold hopping method and its application to a bacterial protein – protein interaction. *Journal of Medicinal Chemistry, 48*(5), 1489–1495.

Samson, M., Libert, F., Doranz, B. J., & Rucker, J. (1996). Resistance to HIV-1 infection in caucasian individuals bearing mutant alleles of the CCR-5 chemokine receptor gene. *Nature*, *382*(6593), 722.

Schappert, S. M. (2011). Ambulatory medical care utilization estimates for 2007. *Vital and Health Statistics*, *13*(169), 1–38.

Sedman, J., & Stenlund, A. (1998). The papillomavirus E1 protein forms a DNA-dependent hexameric complex with ATPase and DNA helicase activities. *Journal of Virology*, *72*(8), 6893–6897.

Sheng, C., Dong, G., Miao, Z., Zhang, W., & Wang, W. (2015). State-of-the-art strategies for targeting protein-protein interactions by small-molecule inhibitors. *Chemical Society Reviews*, *44*(22), 8238–8259. https://doi.org/10.1039/C5CS00252D.

Slonim, L. N., Pinkner, J. S., Branden, C. I., & Hultgren, S. J. (1992). Interactive surface in the PapD chaperone cleft is conserved in pilus chaperone superfamily and essential in subunit recognition and assembly. *The EMBO Journal*, *11*(13), 4747–4756.

Spencer, R. W. (1998). High-throughput screening of historic collections: Observations on file size, biological targets, and file diversity. *Biotechnology and Bioengineering*, *61*(1), 61–67.

Sutherland, A. G., Alvarez, J., Ding, W., Foreman, K., Kenny, C. H., Labthavikul, P., et al. (2003). Structure-based design of carboxybiphenylindole inhibitors of the ZipA–FtsZ interaction. *Organic & Biomolecular Chemistry*, *1*(23), 4138–4140.

Svensson, A., Larsson, A., Emtenäs, H., Hedenström, M., Fex, T., Hultgren, S. J., et al. (2001). Design and evaluation of pilicides: Potential novel antibacterial agents directed against uropathogenic Escherichia coli. *Chembiochem*, *2*(12), 915–918.

The 10 x '20 Initiative. (2010). The 10 x '20 Initiative: pursuing a global commitment to develop 10 new antibacterial drugs by 2020. *Clinical Infectious Diseases*, *50*(8), 1081–1083. https://doi.org/10.1086/652237.

Tsao, D. H., Sutherland, A. G., Jennings, L. D., Li, Y., Rush, T. S., Alvarez, J. C., et al. (2006). Discovery of novel inhibitors of the ZipA/FtsZ complex by NMR fragment screening coupled with structure-based design. *Bioorganic & Medicinal Chemistry*, *14*(23), 7953–7961.

Udwadia, Z. F., Amale, R. A., Ajbani, K. K., & Rodrigues, C. (2012). Totally drug-resistant tuberculosis in India. *Clinical Infectious Diseases*, *54*(4), 579–581. https://doi.org/10.1093/cid/cir889.

Uriarte, S. M., Rane, M. J., Merchant, M. L., Jin, S., Lentsch, A. B., Ward, R. A., et al. (2013). Inhibition of neutrophil exocytosis ameliorates acute lung injury in rats. *Shock (Augusta, Ga.)*, *39*(3), 286–292. https://doi.org/10.1097/SHK.0b013e318282c9a1.

Ventola, C. L. (2015). The antibiotic resistance crisis: Part 1: Causes and threats. *Pharmacy and Therapeutics*, *40*(4), 277–283.

Waheed, A. A., & Tachedjian, G. (2016). Why do we need new drug classes for HIV treatment and prevention? *Current Topics in Medicinal Chemistry*, *16*(12), 1343–1349.

Waksman, G., & Hultgren, S. J. (2009). Structural biology of the chaperone-usher pathway of pilus biogenesis. *Nature Reviews Microbiology*, 7(11), 765–774.

Walsh, C. A. (2016). *Antibiotics: Challenges, mechanisms, opportunities*. Washington, DC: ASM Press.

Wang, Y., Coulombe, R., Cameron, D. R., Thauvette, L., Massariol, M.-J., Amon, L. M., et al. (2004). Crystal structure of the E2 transactivation domain of human papillomavirus type 11 bound to a protein interaction inhibitor. *Journal of Biological Chemistry*, *279*(8), 6976–6985.

Wells, J. A., & McClendon, C. L. (2007). Reaching for high-hanging fruit in drug discovery at protein-protein interfaces. *Nature*, *450*(7172), 1001–1009. http://www.nature.com/nature/journal/v450/n7172/suppinfo/nature06526_S1.html.

Wensing, A. M., Calvez, V., Günthard, H. F., Johnson, V. A., Paredes, R., Pillay, D., et al. (2015). 2015 Update of the drug resistance mutations in HIV-1. *Topics in Antiviral Medicine*, *23*(4), 132–141.

White, P. W., Faucher, A.-M., & Goudreau, N. (2010). Small molecule inhibitors of the human papillomavirus E1-E2 interaction. In *Small-molecule inhibitors of protein-protein interactions* (pp. 61–88): Heidelberg: Springer.

White, P. W., Titolo, S., Brault, K., Thauvette, L., Pelletier, A., Welchner, E., et al. (2003). Inhibition of human papillomavirus DNA replication by small molecule antagonists of the E1-E2 protein interaction. *Journal of Biological Chemistry*, *278*(29), 26765–26772.

Wilen, C. B., Tilton, J. C., & Doms, R. W. (2012). HIV: Cell binding and entry. *Cold Spring Harbor Perspectives in Medicine*, *2*(8), 1–13. https://doi.org/10.1101/cshperspect.a006866.

Wright, K. J., Seed, P. C., & Hultgren, S. J. (2007). Development of intracellular bacterial communities of uropathogenic Escherichia coli depends on type 1 pili. *Cellular Microbiology*, *9*(9), 2230–2241.

Wu, X.-R., Sun, T.-T., & Medina, J. J. (1996). In vitro binding of type 1-fimbriated Escherichia coli to uroplakins Ia and Ib: Relation to urinary tract infections. *Proceedings of the National Academy of Sciences of the United States of America*, *93*(18), 9630–9635.

Xiao, J., & Goley, E. D. (2016). Redefining the roles of the FtsZ-ring in bacterial cytokinesis. *Current Opinion in Microbiology*, *34*, 90–96.

Yang, Z., Zadjura, L., D'arienzo, C., Marino, A., Santone, K., Klunk, L., et al. (2005). Preclinical pharmacokinetics of a novel HIV-1 attachment inhibitor BMS-378806 and prediction of its human pharmacokinetics. *Biopharmaceutics & Drug Disposition*, *26*(9), 387–402.

Yin, Z., Whittell, L. R., Wang, Y., Jergic, S., Liu, M., Harry, E. J., et al. (2014). Discovery of lead compounds targeting the bacterial sliding clamp using a fragment-based approach. *Journal of Medicinal Chemistry*, *57*(6), 2799–2806. https://doi.org/10.1021/jm500122r.

Yin, Z., Whittell, L. R., Wang, Y., Jergic, S., Ma, C., Lewis, P. J., et al. (2015). Bacterial sliding clamp inhibitors that mimic the sequential binding mechanism of endogenous linear motifs. *Journal of Medicinal Chemistry*, *58*(11), 4693–4702. https://doi.org/10.1021/acs.jmedchem.5b00232.

Yoakim, C., Ogilvie, W. W., Goudreau, N., Naud, J., Hache, B., O'Meara, J. A., et al. (2003). Discovery of the first series of inhibitors of human papillomavirus type 11: Inhibition of the assembly of the E1–E2–origin DNA complex. *Bioorganic & Medicinal Chemistry Letters*, *13*(15), 2539–2541.

Zowawi, H. M., Harris, P. N. A., Roberts, M. J., Tambyah, P. A., Schembri, M. A., Pezzani, M. D., et al. (2015). The emerging threat of multidrug-resistant Gram-negative bacteria in urology. *Nature Reviews Urology*, *12*(10), 570–584. https://doi.org/10.1038/nrurol.2015.199. http://www.nature.com/nrurol/journal/v12/n10/abs/nrurol.2015.199.html#supplementary-information.

CHAPTER SEVEN

Defining Pharmacological Targets by Analysis of Virus–Host Protein Interactions

Manuel Llano*,[1], Mario A. Peña-Hernandez*,†
*University of Texas at El Paso, El Paso, TX, United States
†Laboratorio de Inmunología y Virología, Facultad de Ciencias Biológicas, Universidad Autónoma de Nuevo León, San Nicolas de los Garza, Mexico
[1]Correspondence author: e-mail address: mllano@utep.edu

Contents

Abstract

Viruses are obligate parasites that depend on cellular factors for replication. Pharmacological inhibition of essential viral proteins, mostly enzymes, is an effective therapeutic alternative in the absence of effective vaccines. However, this strategy commonly encounters drug resistance mechanisms that allow these pathogens to evade control. Due to the dependency on host factors for viral replication, pharmacological disruption of the host-pathogen protein–protein interactions (PPIs) is an important therapeutic alternative to block viral replication. In this review we discuss salient aspects of PPIs implicated in viral replication and advances in the development of small molecules that inhibit viral replication through antagonism of these interactions.

Advances in Protein Chemistry and Structural Biology, Volume 111
ISSN 1876-1623
https://doi.org/10.1016/bs.apcsb.2017.11.001
© 2018 Elsevier Inc.
All rights reserved.

1. CELLULAR PROTEINS EXIST IN COMPLEXES

Chromatographic resolution of human cell extracts subsequently analyzed by quantitative tandem mass spectrometry resulted in the identification of 13,993 physical interactions established by 3006 individual proteins. Computational analysis of the physical interactions led to the mapping of 622 complexes, suggesting an average of 4 proteins per complex (Havugimana et al., 2012). These estimates are also supported by other analysis indicating that most of the mammalian complexes are formed by the association of three or four different proteins. Importantly, some host proteins are found in more than one complex at a time (Wong et al., 2008). Similarly, a study in human cells combining large-scale immunoprecipitation of 331 bait proteins followed by high-throughput mass spectrometry analysis led to the discovery of 6463 interactions between 2235 unique proteins, suggesting an average of 3 proteins per complex (Ewing et al., 2007).

Protein complexes are stable assemblies that carry out most of the biochemical activities in the cell. CORUM, a public database of mammalian protein complexes, compiles complexes reported in nonhigh-throughput, physical interaction experiments. Of the proteins annotated in CORUM, 78% belong to one protein complex. Fig. 1 represents the frequency of annotated complexes in the CORUM database. Interestingly, protein complexes implicated in subcellular localization processes are notoriously overrepresented in this database, whereas others

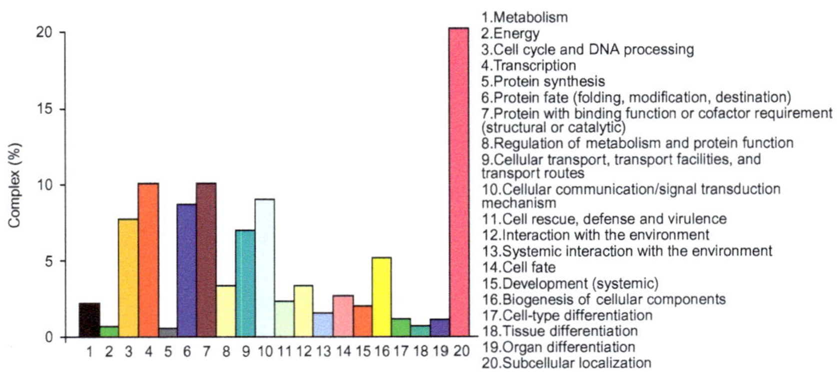

Fig. 1 Functional annotation of protein complexes from mammalian organisms reported in individual experiments complied in CORUM (10/16/2017; http://mips.helmholtz-muenchen.de/corum/#).

implicated in energy production, protein synthesis, and tissue differentiation are underrepresented.

In these collection of experimentally characterized protein complexes, approximately 16% of all predicted human open reading frames are organized in 1815 protein complexes (Ruepp et al., 2010). However, this number is expected to be only a minor fraction of the human complexome considering that in yeast 45% of the encoded proteins reside in complexes (Ruepp et al., 2010).

2. DISCOVERING VIRUS–HOST PROTEIN COMPLEXES

Hypothesis-driven experiments conducted at small scale have notably contributed to our understanding of protein–protein interactions (PPIs) implicated in viral replication. In addition, high-throughput technologies analyzing PPIs, gene loss of function, or transcriptomic profiles have been utilized to define cellular complexes implicated in different steps of the viral life cycle. Large-scale physical analysis of PPI is conducted by yeast two-hybrid (Y2H) screens (Calderwood et al., 2007; de Chassey et al., 2008; Fossum et al., 2009; Lee et al., 2011; Rajagopala, Casjens, & Uetz, 2011; Shapira et al., 2009; Trigg et al., 2017; Uetz et al., 2006) and tandem affinity purification followed by mass spectrometry (TAP–MS) (Germain et al., 2014; Jager, Cimermancic, et al., 2011; Jager, Gulbahce, et al., 2011; Pichlmair et al., 2012; Ramage et al., 2015; Rozenblatt-Rosen et al., 2012). Y2H screens can analyze only binary interactions, whereas TAP–MS allows for the identification of more complex interactors. Then, the physical interactions identified with these methods are subjected to organization in functional networks by computational approaches. Despite of their rate of success in PPIs identification, both methodologies interrogating physical interactions of proteins fail to detect multiple types of PPIs. For example, interactions involving transmembrane proteins, or transient/weak interactions escape detection.

Similar to the physical methods, computational systems biology approaches successfully predict exploitation of biological processes by viruses (Kitano, 2002a, 2002b; Navratil, de Chassey, Combe, & Lotteau, 2011; Zak, Tam, & Aderem, 2014). These methods use gene expression and gene loss-of-function systemic analysis to identify potential host factor networks implicated in viral replication (Hirsch, 2010).

The main problem that the high-throughput functional approaches discussed earlier has is the poor data overlap resulting from these analyses.

This is even though these methods have a high degree of similarity or complementary. Therefore, validation of the discovered PPIs is required (Rozenblatt-Rosen et al., 2012; Shapira et al., 2009). This validation involves the demonstration of the interaction by coimmunoprecipitation analysis, particularly during infection. Prioritization analysis of PPIs can be implemented by combining the findings from the direct interaction approaches with comprehensive gene loss-of-function screenings.

An example of the effectiveness of this multifactorial approach for the identification of PPIs is a study aimed to identify host-influenza PPIs (Shapira et al., 2009). In this study, 35% of the 1745 genes initially found to be involved in viral protein interactions were demonstrated to influence viral replication by subsequent RNAi screening. This high rate of positive validation may have been the result of the selection of the final candidate genes by a combined analysis of the results obtained in Y2H screens using viral and host proteins (direct interactors), in transcriptional profiling experiments to define genes regulated by viral infection, and in in silico predictions of genes implicated in the protein networks found in the first two experimental approaches.

3. DIFFERENT PATHOGENS EXPLOIT SIMILAR CELLULAR PROCESSES

Comparison of the cellular complexes exploited by different pathogens indicates an significant overlap, leading to the definition of cellular processes implicated in infection (Brander & Walker, 2000; Davis et al., 2015; Dyer, Murali, & Sobral, 2008; Hirsch, 2010; Hiscott, Nguyen, Arguello, Nakhaei, & Paz, 2006; Pichlmair et al., 2012; Rozenblatt-Rosen et al., 2012; Shapira et al., 2009). Therefore, shared cellular complexes could constitute broad-spectrum therapeutic targets potentially impairing more than one pathogen. An example of this strategy is discussed later when we consider the implication of protein translation in the mechanism of replication of multiple viruses.

The overlap in the utilization of cellular complexes between different pathogens also brings support to the hypothesis that the innate immune system has evolved the ability of detecting patterns of pathogenicity. These patterns are generated by the activation of similar biological processes by different pathogens, indicating the existence of pathogen-induced processes (Dyer et al., 2008; Pichlmair et al., 2012; Vance, Isberg, & Portnoy, 2009).

The viral manipulation of the activity of type I interferon (IFN)-stimulated genes (ISG) illustrates this concept. Protein Kinase, RNA-activated (PKR) is an ISG that upon activation phosphorylates the α-subunit of the translation initiation factor eIF-2, inhibiting protein synthesis. Viral dsRNA is the main PKR activator in virus-infected cells, whereas the cellular protein PACT (Protein activator of PKR), that heterodimerizes with PKR, is an important activator of the kinase in stressed, noninfected cells in the absence of dsRNA (Li et al., 2006; Patel & Sen, 1998). HIV-1 evades the robust type I IFN response mounted by plasmacytoid dendritic cells in infected individuals by rewiring the PACT/PKR complex. In infected cells, HIV-1 TAR nucleates a complex with the viral protein Tat and the cellular proteins PACT and ADAR1 (Adenosine Deaminase Acting on RNA 1). In this complex, ADAR1 inhibits PACT PKR activatory function by direct interaction with PACT (Clerzius et al., 2013). Another example of pathogen-induced host complex rewiring will be discussed later in relation to the role of HIV-1 accessory proteins in the targeting of the ubiquitin/proteasome system to degrade different restriction factors.

Importantly, multiple viruses including Middle East Respiratory Syndrome Coronavirus, Herpes Simplex Virus, Ebola Virus, Influenza Virus, and Orf Virus also inactivate PACT, inhibiting PKR activation. Similar to HIV-1, ADAR1 influences replication of Measles Virus, Vesicular Stomatitis Virus, Polyoma DNA Virus, West Nile Virus, Yellow Fever Virus, Chikungunya Virus, Venezuelan Equine Encephalitis Virus, and viruses of the Paramyxoviridae and the Rhabdoviridae families (Chukwurah, Handy, & Patel, 2017). Therefore, the PACT-ADAR1-PKR axis is a cellular pathway commonly manipulated by different viruses that could constitute a therapeutic target with broad antiviral activity or pathogen-induced processes detected by the innate immune system. The hijacking of similar cellular pathways by different viruses also allows for the identification of broader biomarkers of viral infection.

Generally viral pathogens target cell cycle regulation, nuclear transport, and immune response processes (Dyer et al., 2008; Shapira et al., 2009). ssRNA(−) viral proteins tend to interact with processes implicated in the protection of RNA from degradation and RNA processing, whereas dsRNA viruses are more connected to protein degradation processes. DNA viruses, however, target proteins connecting the cell cycle with other processes such as chromosomal and transcriptional homeostasis (Pichlmair et al., 2012; Shapira et al., 2009).

4. VIRAL PROTEINS INTERACT WITH HOST PROTEINS IN COMPLEXES

RNAi-based screens have also been useful in identifying host factors implicated in viral replication. Interestingly, the number of host factors identified for the different viruses analyzed in these screens outnumber by a factor of 10 the proteins encoded by these viruses (Hirsch, 2010). This disproportion could illustrate the multifunctional character of viral proteins. In this case viral proteins are expected to establish multiple independent binary interactions with independent host proteins. Alternatively, viral proteins could interact with host proteins organized in complexes. Therefore, RNAi-mediated downregulation of the subunits of these protein complexes will produce similar phenotypes.

Correspondingly, Y2H screenings have found that viral proteins establish a disproportionate number of binary interactions with the human proteome. These numbers of interactions registered for viral proteins exceed the predicted number of interactions estimated from the analysis of the human interaction network (Shapira et al., 2009). For example, assessment of the interaction of each of the 10 viral proteins encoded by influenza with 12,000 human ORF through Y2H screening found 135 interactions with 87 human proteins (Shapira et al., 2009). These evidences support a model indicating that viral proteins evolve multiple direct interactions. Furthermore, computational analysis of host factors implicated in these pairwise interactions indicated that the host proteins occupy central positions, hub proteins, within the cellular interactome (Shapira et al., 2009). This higher than expected connectivity suggests that the direct interactions of viral proteins with host factors allow the access of the virus to cellular complexes.

Computational analysis of the interaction of human proteins binding to different viral proteins encoded by 35 different viruses showed that approximately 97% ±9.1% of the 1396 unique targeted human proteins interact at least with one other human protein (Dyer et al., 2008). These data indicate that the majority of the host factors implicated in viral replication are in complexes. These conclusions are further supported by TAP–MS and functional genomic studies. TAP–MS studies of the virus–host interactome indicate that viral proteins tend to interact with proteins that have multiple interacting partners and participate in more cellular pathways (protein hubs), and also with proteins that are central to many pathways and, therefore, occupy a more central position within these networks (protein bottlenecks)

(Dyer et al., 2008; Pichlmair et al., 2012; Shapira et al., 2009). For example, in a TAP–MS experiment were mapped 3787 complex associations between 54 viral proteins from different viruses and 1079 host proteins (Rozenblatt-Rosen et al., 2012), highlighting the high degree of connectivity of the interacting proteins.

As discussed above, some of the host factors predicted, by the combined transcriptional profiling and in silico analyses, to interact with host proteins implicated in direct binary contacts with influenza proteins (Y2H interactors) were demonstrated to influence viral replication in functional screenings (Shapira et al., 2009). These findings demonstrate that influenza, and potentially other viruses, hijacks cellular networks by combining physical and regulatory (transcriptional level) interactions with these pathways. This multilayer regulation offers further alternatives of blocking physical or regulatory interactions aiming to impair viral replication. Interestingly some of the host factors found in the networks of protein interacting with viral proteins are induced at the transcriptional level by infection in a type I IFN-independent manner (Shapira et al., 2009). This indicates that viruses modulate, at the transcriptional levels, the abundance of proteins that will engage in physical binary or multiprotein complex interactions. Therefore, computational analysis of the pathways discovered by functional genomics could lead to the identification of protein networks implicated in physical interactions with viral proteins.

Protein complexes established by viruses seem to determine the disease associated to the infection. For example, analysis of the interactome of E6 proteins from HPV types differing in their oncogenic potential associates with different subset of host proteins (Rozenblatt-Rosen et al., 2012). Therefore, disruption of these complexes could also prevent viral pathogenesis.

Almost 78% of the human-pathogen PPIs reported belong to HIV-1 (Dyer et al., 2008; Fahey et al., 2011; Ruepp et al., 2010). Generally host factors interacting with viral proteins tend to preserve the host protein interactions mapped in noninfected cells (Rozenblatt-Rosen et al., 2012; Yu et al., 2011), indicating that the normal host protein complexes rather than new virus-induced protein complexes are implicated in viral replication. This opens the opportunity to utilize the current body of knowledge on the human interactome to define the host complex hijacked by viruses. However, examples of virus-specific complex variants also occur. Among them, one of the best characterized is the complex between HIV-1 Vif and the transcriptional and protein degradation cellular machineries.

Vif removes APOBEC3 family restriction factors by targeting the ubiquitin ligase complex CUL5 to this restriction factor, and inducing its proteasome-mediated degradation. Other HIV-1 proteins, Vpu and Vpr, also exploit cullin-RING E3 ligase to degrade other restriction factors. Vif directly interacts with the substrate, the transcription factor CBF-β, the N-terminus of CUL5 (cullin-RING E3 ligase), and the heterodimeric substrate adaptor EloB/EloC. In turn, the C-terminus of CUL5 binds Rbx2 and recruits an E2 ubiquitin conjugation enzyme that mediates the transfer of poly-ubiquitin to the substrate proteins (Jager, Kim, et al., 2011; Zhang, Du, Evans, Yu, & Yu, 2011). CBF-β is required in this complex for the assembly of the Vif–CUL5 E3-ubiquitin-ligase complex, but not for the binding of Vif to the substrate (Zhang et al., 2011). CBF-β normally heterodimerizes with the RUNX family of transcription factors preventing its degradation and Vif competes with RUNX1 for binding to CBF-β (Guo et al., 2014).

The surface area buried at the interface of the interaction of Vif and CBF-β is large (~4800 A^2) suggesting that is undruggable (Salter, Morales, & Smith, 2014), as we will discuss later for other surfaces of PPIs. However, alanine scanning indicated that aa 5–11 of Vif are a hot spot (defined in a later section of this review) in the interaction with CBF-β. Furthermore, residues Phe68 and Ile55 of CBF-β establish important interactions with Trp5 in Vif through hydrophobic interactions (Desimmie, Smith, Matsuo, Hu, & Pathak, 2017), highlighting further opportunities for disruption of this complex with small molecules. Therefore, these structural characteristics of the Vif-CBF-β surface of interaction do not exclude its amenability to small-molecule interference, as we will show later for similar PPIs.

5. DEVELOPMENT OF SMALL MOLECULES DISRUPTING PPIs

Interfaces of PPI were considered undruggable for sometime, mainly because these surfaces of interactions are bigger than classical druggable protein surfaces, such as allosteric and catalytic sites in enzymes. The total area buried by the interactors in the binding site in PPIs is in average 1600 (±400) A^2, with some proteins extending to 2000–4660 A^2. In contrast, surfaces of interaction between proteins and small molecules are approximately from 300 to 1000 A^2 (Arkin, Tang, & Wells, 2014; Lo Conte, Chothia, & Janin, 1999). In addition, the surface of protein–protein interaction, although very variable in characteristics, was generally considered large patches of segmented complementary surfaces lacking small grooves or pockets (Hopkins & Groom, 2002; Jones & Thornton, 1996, 1997; Lo Conte et al., 1999).

Furthermore, attempts to find small molecules interfering with PPIs showed a very high failure rate (~60%) (Brown & Superti-Furga, 2003), therefore reinforcing the concept that surface of PPIs is not druggable. In correlation with this, estimates using very stringent concepts for drugs indicate that only approximately 10% of the human proteome could be targeted by oral, drug-like small molecule. These predictions also indicate that of the druggable gene products only between 5% and 50% were implicated in diseases (Hopkins & Groom, 2002).

Our understanding of the rules driving the PPIs importantly changed with the discovery that the different residues buried in the large surfaces of interactions contribute differentially to the binding strength of the interacting proteins (Clackson & Wells, 1995). This characteristic reduced considerably the area required to be targeted by small molecules. Interrogation of the surfaces of binding in PPIs by mutagenesis of individual residues (alanine scanning) indicated that mutation of a few residues implicated in these binding surfaces was sufficient to disrupt the PPIs. Residues importantly contributing to the binding free energy were defined as hot spots (Clackson & Wells, 1995). These residues tend to cluster in tightly packed regions in the center of the surfaces of interactions and are called hot regions (Clackson & Wells, 1995; Keskin, Ma, & Nussinov, 2005). Hot spots can be amenable to drug targeting due to their small surface area and important contribution to the total binding energy of the complex. However, not always hot spots are vulnerable to drugs, as additional topological constraints for drug binding limit the number of druggable hot spots (Zerbe, Hall, Vajda, Whitty, & Kozakov, 2012). In addition, in silico analysis indicated that interfaces could contain more than one hot region, and in some PPIs the establishment of the complex depends on the cooperative interaction of different hot spots within the region (Keskin et al., 2005).

By 2011, more than 40 PPIs have been reported to be targeted by small molecules (Morelli, Bourgeas, & Roche, 2011). Twenty-seven protein–protein complexes were included in the 2016 release of the small-molecule orthosteric modulators of PPIs database 2P2Idb. In this site are listed only those PPIs in which there is structural information of the protein–protein and protein–inhibitor complexes (Basse, Betzi, Morelli, & Roche, 2016).

The majority of the most successful PPIs inhibitors target hot-spot residues that cluster in small binding pockets (250–900 A^2), and the surface of interaction of the binding proteins has short primary sequences (Arkin et al., 2014).

Structural analysis of protein–inhibitor complexes indicated that in some cases small molecules interrupting PPIs bind to cavities not present in the surface of interaction of the protein–protein complex or in the interactor proteins when they are in isolation. These findings indicate some degree of adaptability in the surfaces of interaction initially considered lacking of druggable groves or pockets. According to in silico simulations the opening of some of these binding pockets is transient (Wells & McClendon, 2007).

Most of the small molecules interfering with PPIs bind directly to the implicated surfaces of interactions (orthosteric modulators) by targeting hot spot residues or by molecular mimicry of elements of secondary structures (Arkin et al., 2014; Basse et al., 2016; Fry, 2006; Wells & McClendon, 2007; Yin & Hamilton, 2005). However, several successful examples of small molecules disrupting PPIs act through their binding to allosteric sites (Crump et al., 2004; McMillan et al., 2000; Oswald et al., 2016; Roche et al., 2016; Szilagyi, Nussinov, & Csermely, 2013).

A well-characterized model of small-molecule disruption of a host PPI, deposited in the 2P2Idb database, is the MDM2/p53 interaction. MDM2 binds to p53, impairing its transcriptional activity and stability. The interaction surfaces are formed by a hydrophobic pocket (aa 25–109) in the N-terminus of MDM2 that is occupied by the hydrophobic side of an amphipathic α-helix in p53 (aa 19–26). Mutations at any of the four residues within the MDM2-binding pocket or of three residues within the p53 α-helix block this PPI (Moll & Petrenko, 2003). Small molecules interrupting this interaction bind to the p53-binding pocket in MDM2 (Vassilev et al., 2004).

Most of the data on small molecules targeting PPIs in the 2P2Idb database correspond to host–host PPIs and only two are relevant to viruses. That is the case of the interaction between the HPV proteins E2 and E1, and of allosteric inhibitors of HIV-1 integrase activity (ALLINIs) (Engelman, Kessl, & Kvaratskhelia, 2013). HPV E2 and E1 bind cooperatively to the viral origin of DNA replication and are required for initiation of DNA replication (Berg & Stenlund, 1997). Indandione derivatives that bind to the E2 transactivation domain block this interaction. One of these small molecules was shown to contact 7 of the 20 residues implicated in this interaction.

ALLINIs bind to the dimer interface of integrase to the same residues of the host protein LEDGF/p75 (Engelman et al., 2013). LEDGF/p75 is a chromatin-bound protein required for efficient HIV-1 cDNA integration (Llano et al., 2006; Shun et al., 2007), an essential step in the viral life cycle. LEDGF/p75 binds to the dimer interface of the catalytic core domain of integrase. Two residues in the integrase-binding domain of LEDGF/p75 are essential in this interaction. Asp366 in an interhelical loop of the host protein contacts, via hydrogen bonds, with residues Glu170 and His171 in one integrase monomer, whereas LEDGF/p75 residue Ile365 interacts with a hydrophobic pocket (Leu102, Ala128, Trp132) in the other integrase subunit (Cherepanov, Ambrosio, Rahman, Ellenberger, & Engelman, 2005; Cherepanov, Sun, et al., 2005). ALLINIs bind to the viral integrase at the LEDGF/p75-binding site, preventing the binding of LEDGF/p75 and affecting integrase catalytic core domain dimerization. This disrupts integrase assembly with viral DNA and allosterically inhibits its activity (Engelman et al., 2013). 2-(Quinolin-3-yl) acetic acid derivatives (LEDGINs), and in particular compound 6, showed potent inhibitory activity of HIV-1 replication (Christ et al., 2010). This compound establishes hydrogen bonds with integrase residues Glu170, His171, and Thr173 that are hot spots in the LEDGF/p75-binding site in integrase (Christ et al., 2010). As ALLINIs were more potent against HIV-1 in cells lacking LEDGF/p75, the contribution of LEDGF/p75-binding inhibition to their mechanism of action is not clear (Wang et al., 2012).

Another successful strategy in the design of small molecules interfering viral replication has been the targeting of multimeric viral RNA polymerases. Influenza Virus and Vaccinia Virus encode RNA polymerases formed by the essential interaction of several subunits. In Influenza Virus each of the three viral subunits of the polymerase, PB1, PB2, and PA, is required to assemble in a complex for the polymerase activity. Therefore, small molecules that interrupt the binding of PB1 into a hydrophobic groove in the C-terminus of PA (He et al., 2008) block replication of influenza A and B viruses (Muratore et al., 2012).

Similarly, Vaccinia Virus depends for replication on the formation of a trimeric complex between the DNA polymerase E9, the uracil DNA glycosylase D4, and the viral protein A20 (Stanitsa, Arps, & Traktman, 2006). Small molecules disrupting this complex by impairing the interaction D4-A20 inhibit replication of Vaccinia Virus and Cowpox Virus (Schormann et al., 2011).

6. TARGETING VIRUS–HOST PPIs WITH SMALL MOLECULES

There are two types of physical interactions that can be targeted to interrupt the utilization of cellular complexes by the virus: the virus–host interface or the host–host interface of complexes hijacked by the virus. Targeting protein interaction surfaces containing viral proteins is always associated with the selection of mutant viruses that escape control of the inhibitor (De Clercq, 2007; Strasfeld & Chou, 2010). This is the result of the highly mutagenic rate of viruses. However, targeting host complexes utilized by viruses at the host–host interface is expected to do not exhibit mechanisms of resistance, since host proteins are genetically more stable than their viral counterparts. Therefore, disruption of the host complexes exploited by viruses rather than interference of the virus–host protein interface is an attractive alternative to avoid the selection of escape mutants resistant to the inhibitors.

A major disadvantage of this strategy is the potential toxicity associated to the disruption of cellular complexes. Potentially, this could be minimized if the interruption of the complex is transient. Viral replication is a fast process that occurs in a few hours, and disruption of these complexes for a few hours could greatly affect the ability of the virus to replicate without affecting cellular viability. We will later discuss experimental findings supporting this view.

In addition, the high degree of functional overlap in the human proteome, that is absent in viruses because of their relative smaller genomes, could also ameliorate the toxicity caused by the disruption of host–host PPIs implicated in viral replication. The differences in functional redundancy suggest that the disruption of cellular complex will be more tolerable for the host than the pathogen. In support of this view, shRNA-based studies have showed that many host factors required for viral replication do not carry essential cellular functions. For example, out of 54,509 transcripts permanently targeted with lentivirus-encoded shRNAs, 17% were dispensable for the cell and deficient stable cell lines were developed but a fraction of them were required for efficient HIV-1 replication (Yeung, Houzet, Yedavalli, & Jeang, 2009).

Finally, targeting host complexes relevant for different viruses could allow the development of broad-spectrum antivirals.

7. SMALL MOLECULES DISRUPTING HOST COMPLEXES IMPLICATED IN VIRAL REPLICATION

Protein translation is essential for the virus and the cell. However, transient interference with protein synthesis at the initiation step has been demonstrated to be more detrimental for the pathogen than the host at a cellular level.

Both cellular mRNA and mRNA from different viruses carry a 7-methylguanosine cap (cap) on their 5′ terminal nucleotide. Therefore, cap-dependent cellular translation is exploited by many different viruses. Critical to this translation mechanism is the eukaryotic initiation factor 4F complex (eIF4F). This complex is required for the efficient recruitment of ribosomes to capped mRNAs. The eIF4F complex includes a cap-binding protein (eIF4E), an RNA helicase (eIF4A), and a large scaffolding subunit (eIF4G). eIF4E protein binds to cap-mRNA and then recruits eIF4G that binds eIF4A. Therefore, inhibitors of eIF4E–eIF4G interaction (i.e., 4E2RCat (Cencic, Hall, et al., 2011)) or the helicase activity (hippuristanol) drastically impair translation.

4E2RCat completely blocks the replication of the human Alphacoronavirus 229E (Cencic, Desforges, et al., 2011) or Murine Norovirus 1 (Royall et al., 2015) at doses that impair approximately 40% of the cellular protein synthesis (Cencic, Desforges, et al., 2011) or do not alter at all cell viability. At these doses 4E2RCat affects eIF4E–eIF4G interaction as determined by coimmunoprecipitation (Royall et al., 2015). These results indicate that Coronaviruses are more dependable on eIF4F for ribosome recruitment mRNAs than the host.

Hippuristanol is a natural product that inhibits the activity of eIF4A by binding to its C-terminal domain. This compound impairs eIF4A RNA-binding, ATPase, and helicase activities. Hippuristanol specifically impaired replication of poliovirus (Bordeleau et al., 2006), Caliciviruses (Chaudhry et al., 2006), Human Cytomegalovirus (Lenarcic, Ziehr, De Leon, Mitchell, & Moorman, 2014), and Junin Virus (Linero, Thomas, Boccaccio, & Scolaro, 2011). Despite the cell toxicity of hippuristanol (Bordeleau et al., 2006) at the doses and length of the treatment used in these reports, viral replication was affected specifically since this compound did not alter cellular viability. For example, 6-h treatment of RAW 264.7 cells with hippuristanol reduced cell viability by 10%, but Murine Norovirus production was reduced by over 3000-fold (Chaudhry et al., 2006).

As described earlier, eIF4E interacts with eIF4G and assembles into the eIF4F complex together with eIF4A on the capped mRNA. eIF4F complex formation leads to eIF4E phosphorylation at Ser209 by the eIF4G-associated kinase Mnk1. Inhibition of Mnk1 by CGP57380 specifically impairs protein synthesis in a variety of large DNA viruses (Walsh, Mathews, & Mohr, 2013). For example, this compound reduced by 10^2-fold the replication of HSV-1 in quiescent primary fibroblasts without affecting cell viability (Walsh & Mohr, 2004). In addition, CGP57380 inhibits reactivation of Kaposi's sarcoma-associated Herpesvirus (Walsh et al., 2013). Because Mnk1 is not essential for cell growth or development, it is a potential therapeutic target (Ueda, Watanabe-Fukunaga, Fukuyama, Nagata, & Fukunaga, 2004).

There are other examples of small molecules inhibiting PPIs in host complex that are not essential for the cell. Some compounds targeting host–virus PPIs implicated in the earlier steps of the HIV-1 life cycle are in this category. For example, the HIV-1 entry inhibitor Maraviroc, that reached clinical use, acts through an allosteric mechanism. This drug binds to a hydrophobic pocket located in the transmembrane domain of CCR5, altering the conformation of the extracellular domain of CCR5 and preventing in this manner the binding of Env (Roche et al., 2016).

In addition, the small molecule PF74, that inhibits HIV-1 replication by triggering early uncoating (Shi, Zhou, Shah, Aiken, & Whitby, 2011), binds to a pocket encompassing the N-terminal domain–C-terminal domain interface of CA in the assembled capsid (Bhattacharya et al., 2014). Cleavage and polyadenylation-specific factor 6 (CPSF6) also binds to the same pocket in the capsid during HIV trafficking to the nucleus. This PPI delays uncoating, allowing HIV-1 to evade the innate immune activation through cytoplasmic nucleic acid sensor signaling (Rasaiyaah et al., 2013).

8. CONCLUDING REMARKS

We have presented literature evidences indicating that viruses interact for replication with cellular complexes. Within these complexes, viral proteins bind to cellular proteins that are central (bottleneck proteins) in the interacting networks and establish multiple interactions in the human interactome (hub proteins). These PPIs offer different opportunities for the development of small molecules to block viral replication. Virus–virus, virus–host, and host–host PPIs are amenable for pharmacological disruption. Different viruses utilize similar host protein complexes during infection, showing overlap in the cellular processes hijacked. Targeting these shared cellular processes opens the possibility to generate broad-spectrum antivirals and general biomarkers of viral infection. Potentially, the innate immune system recognizes these shared pathogen-induced processes to counteract infection. A better understanding of this mechanism could greatly impact the development of stronger vaccine adjuvants.

ACKNOWLEDGMENTS

This work was supported by Grant Number SC1GM115240 to M.L. from the National Institute of Allergy and Infectious Diseases (NIAID), National Institutes of Health (NIH). M.A.P.-H. was funded by Facultad de Ciencias Biológicas, Universidad Autónoma de Nuevo León, Mexico.

We thank Zachary S. Martinez and Angelica Lopez, University of Texas at El Paso, for reviewing the manuscript.

REFERENCES

Arkin, M. R., Tang, Y., & Wells, J. A. (2014). Small-molecule inhibitors of protein-protein interactions: Progressing toward the reality. *Chemistry & Biology*, *21*(9), 1102–1114. https://doi.org/10.1016/j.chembiol.2014.09.001.

Basse, M. J., Betzi, S., Morelli, X., & Roche, P. (2016). 2P2Idb v2: Update of a structural database dedicated to orthosteric modulation of protein-protein interactions. *Database (Oxford), 2016*. https://doi.org/10.1093/database/baw007.

Berg, M., & Stenlund, A. (1997). Functional interactions between papillomavirus E1 and E2 proteins. *Journal of Virology*, *71*(5), 3853–3863.

Bhattacharya, A., Alam, S. L., Fricke, T., Zadrozny, K., Sedzicki, J., Taylor, A. B., et al. (2014). Structural basis of HIV-1 capsid recognition by PF74 and CPSF6. *Proceedings of the National Academy of Sciences of the United States of America*, *111*(52), 18625–18630. https://doi.org/10.1073/pnas.1419945112.

Bordeleau, M. E., Mori, A., Oberer, M., Lindqvist, L., Chard, L. S., Higa, T., et al. (2006). Functional characterization of IRESes by an inhibitor of the RNA helicase eIF4A. *Nature Chemical Biology*, *2*(4), 213–220. https://doi.org/10.1038/nchembio776.

Brander, C., & Walker, B. D. (2000). Modulation of host immune responses by clinically relevant human DNA and RNA viruses. *Current Opinion in Microbiology*, *3*(4), 379–386.

Brown, D., & Superti-Furga, G. (2003). Rediscovering the sweet spot in drug discovery. *Drug Discovery Today*, *8*(23), 1067–1077.

Calderwood, M. A., Venkatesan, K., Xing, L., Chase, M. R., Vazquez, A., Holthaus, A. M., et al. (2007). Epstein-Barr virus and virus human protein interaction maps. *Proceedings of the National Academy of Sciences of the United States of America*, *104*(18), 7606–7611. https://doi.org/10.1073/pnas.0702332104.

Cencic, R., Desforges, M., Hall, D. R., Kozakov, D., Du, Y., Min, J., et al. (2011). Blocking eIF4E-eIF4G interaction as a strategy to impair coronavirus replication. *Journal of Virology*, *85*(13), 6381–6389. https://doi.org/10.1128/JVI.00078-11.

Cencic, R., Hall, D. R., Robert, F., Du, Y., Min, J., Li, L., et al. (2011). Reversing chemoresistance by small molecule inhibition of the translation initiation complex eIF4F. *Proceedings of the National Academy of Sciences of the United States of America*, *108*(3), 1046–1051. https://doi.org/10.1073/pnas.1011477108.

Chaudhry, Y., Nayak, A., Bordeleau, M. E., Tanaka, J., Pelletier, J., Belsham, G. J., et al. (2006). Caliciviruses differ in their functional requirements for eIF4F components.

The Journal of Biological Chemistry, *281*(35), 25315–25325. https://doi.org/10.1074/jbc.M602230200.

Cherepanov, P., Ambrosio, A. L., Rahman, S., Ellenberger, T., & Engelman, A. (2005). Structural basis for the recognition between HIV-1 integrase and transcriptional coactivator p75. *Proceedings of the National Academy of Sciences of the United States of America*, *102*(48), 17308–17313.

Cherepanov, P., Sun, Z. Y., Rahman, S., Maertens, G., Wagner, G., & Engelman, A. (2005). Solution structure of the HIV-1 integrase-binding domain in LEDGF/p75. *Nature Structural & Molecular Biology*, *12*(6), 526–532.

Christ, F., Voet, A., Marchand, A., Nicolet, S., Desimmie, B. A., Marchand, D., et al. (2010). Rational design of small-molecule inhibitors of the LEDGF/p75-integrase interaction and HIV replication. *Nature Chemical Biology*, *6*(6), 442–448. https://doi.org/10.1038/nchembio.370.

Chukwurah, E., Handy, I., & Patel, R. C. (2017). ADAR1 and PACT contribute to efficient translation of transcripts containing HIV-1 trans-activating response (TAR) element. *The Biochemical Journal*, *474*(7), 1241–1257. https://doi.org/10.1042/BCJ20160964.

Clackson, T., & Wells, J. A. (1995). A hot spot of binding energy in a hormone-receptor interface. *Science*, *267*(5196), 383–386.

Clerzius, G., Shaw, E., Daher, A., Burugu, S., Gelinas, J. F., Ear, T., et al. (2013). The PKR activator, PACT, becomes a PKR inhibitor during HIV-1 replication. *Retrovirology*, *10*, 96. https://doi.org/10.1186/1742-4690-10-96.

Crump, M. P., Ceska, T. A., Spyracopoulos, L., Henry, A., Archibald, S. C., Alexander, R., et al. (2004). Structure of an allosteric inhibitor of LFA-1 bound to the I-domain studied by crystallography, NMR, and calorimetry. *Biochemistry*, *43*(9), 2394–2404. https://doi.org/10.1021/bi035422a.

Davis, Z. H., Verschueren, E., Jang, G. M., Kleffman, K., Johnson, J. R., Park, J., et al. (2015). Global mapping of herpesvirus-host protein complexes reveals a transcription strategy for late genes. *Molecular Cell*, *57*(2), 349–360. https://doi.org/10.1016/j.molcel.2014.11.026.

de Chassey, B., Navratil, V., Tafforeau, L., Hiet, M. S., Aublin-Gex, A., Agaugue, S., et al. (2008). Hepatitis C virus infection protein network. *Molecular Systems Biology*, *4*, 230. https://doi.org/10.1038/msb.2008.66.

De Clercq, E. (2007). The design of drugs for HIV and HCV. *Nature Reviews. Drug Discovery*, *6*(12), 1001–1018. https://doi.org/10.1038/nrd2424.

Desimmie, B. A., Smith, J. L., Matsuo, H., Hu, W. S., & Pathak, V. K. (2017). Identification of a tripartite interaction between the N-terminus of HIV-1 Vif and CBFbeta that is critical for Vif function. *Retrovirology*, *14*(1), 19. https://doi.org/10.1186/s12977-017-0346-5.

Dyer, M. D., Murali, T. M., & Sobral, B. W. (2008). The landscape of human proteins interacting with viruses and other pathogens. *PLoS Pathogens*, *4*(2), e32. https://doi.org/10.1371/journal.ppat.0040032.

Engelman, A., Kessl, J. J., & Kvaratskhelia, M. (2013). Allosteric inhibition of HIV-1 integrase activity. *Current Opinion in Chemical Biology*, *17*(3), 339–345. https://doi.org/10.1016/j.cbpa.2013.04.010.

Ewing, R. M., Chu, P., Elisma, F., Li, H., Taylor, P., Climie, S., et al. (2007). Large-scale mapping of human protein-protein interactions by mass spectrometry. *Molecular Systems Biology*, *3*, 89. https://doi.org/10.1038/msb4100134.

Fahey, M. E., Bennett, M. J., Mahon, C., Jager, S., Pache, L., Kumar, D., et al. (2011). GPS-Prot: A web-based visualization platform for integrating host-pathogen interaction data. *BMC Bioinformatics*, *12*, 298. https://doi.org/10.1186/1471-2105-12-298.

Fossum, E., Friedel, C. C., Rajagopala, S. V., Titz, B., Baiker, A., Schmidt, T., et al. (2009). Evolutionarily conserved herpesviral protein interaction networks. *PLoS Pathogens*, *5*(9), e1000570. https://doi.org/10.1371/journal.ppat.1000570.

Fry, D. C. (2006). Protein-protein interactions as targets for small molecule drug discovery. *Biopolymers*, *84*(6), 535–552. https://doi.org/10.1002/bip.20608.

Germain, M. A., Chatel-Chaix, L., Gagne, B., Bonneil, E., Thibault, P., Pradezynski, F., et al. (2014). Elucidating novel hepatitis C virus-host interactions using combined mass spectrometry and functional genomics approaches. *Molecular & Cellular Proteomics*, *13*(1), 184–203. https://doi.org/10.1074/mcp.M113.030155.

Guo, Y., Dong, L., Qiu, X., Wang, Y., Zhang, B., Liu, H., et al. (2014). Structural basis for hijacking CBF-beta and CUL5 E3 ligase complex by HIV-1 Vif. *Nature*, *505*(7482), 229–233. https://doi.org/10.1038/nature12884.

Havugimana, P. C., Hart, G. T., Nepusz, T., Yang, H., Turinsky, A. L., Li, Z., et al. (2012). A census of human soluble protein complexes. *Cell*, *150*(5), 1068–1081. https://doi.org/10.1016/j.cell.2012.08.011.

He, X., Zhou, J., Bartlam, M., Zhang, R., Ma, J., Lou, Z., et al. (2008). Crystal structure of the polymerase PA(C)-PB1(N) complex from an avian influenza H5N1 virus. *Nature*, *454*(7208), 1123–1126. https://doi.org/10.1038/nature07120.

Hirsch, A. J. (2010). The use of RNAi-based screens to identify host proteins involved in viral replication. *Future Microbiology*, *5*(2), 303–311. https://doi.org/10.2217/fmb.09.121.

Hiscott, J., Nguyen, T. L., Arguello, M., Nakhaei, P., & Paz, S. (2006). Manipulation of the nuclear factor-kappaB pathway and the innate immune response by viruses. *Oncogene*, *25*(51), 6844–6867. https://doi.org/10.1038/sj.onc.1209941.

Hopkins, A. L., & Groom, C. R. (2002). The druggable genome. *Nature Reviews. Drug Discovery*, *1*(9), 727–730. https://doi.org/10.1038/nrd892.

Jager, S., Cimermancic, P., Gulbahce, N., Johnson, J. R., McGovern, K. E., Clarke, S. C., et al. (2011). Global landscape of HIV-human protein complexes. *Nature*, *481*(7381), 365–370. https://doi.org/10.1038/nature10719.

Jager, S., Gulbahce, N., Cimermancic, P., Kane, J., He, N., Chou, S., et al. (2011). Purification and characterization of HIV-human protein complexes. *Methods*, *53*(1), 13–19. https://doi.org/10.1016/j.ymeth.2010.08.007.

Jager, S., Kim, D. Y., Hultquist, J. F., Shindo, K., LaRue, R. S., Kwon, E., et al. (2011). Vif hijacks CBF-beta to degrade APOBEC3G and promote HIV-1 infection. *Nature*, *481*(7381), 371–375. https://doi.org/10.1038/nature10693.

Jones, S., & Thornton, J. M. (1996). Principles of protein-protein interactions. *Proceedings of the National Academy of Sciences of the United States of America*, *93*(1), 13–20.

Jones, S., & Thornton, J. M. (1997). Analysis of protein-protein interaction sites using surface patches. *Journal of Molecular Biology*, *272*(1), 121–132. https://doi.org/10.1006/jmbi.1997.1234.

Keskin, O., Ma, B., & Nussinov, R. (2005). Hot regions in protein–protein interactions: The organization and contribution of structurally conserved hot spot residues. *Journal of Molecular Biology*, *345*(5), 1281–1294. https://doi.org/10.1016/j.jmb.2004.10.077.

Kitano, H. (2002a). Computational systems biology. *Nature*, *420*(6912), 206–210. https://doi.org/10.1038/nature01254.

Kitano, H. (2002b). Systems biology: A brief overview. *Science*, *295*(5560), 1662–1664. https://doi.org/10.1126/science.1069492.

Lee, S., Salwinski, L., Zhang, C., Chu, D., Sampankanpanich, C., Reyes, N. A., et al. (2011). An integrated approach to elucidate the intra-viral and viral-cellular protein interaction networks of a gamma-herpesvirus. *PLoS Pathogens*, 7(10), e1002297. https://doi.org/10.1371/journal.ppat.1002297.

Lenarcic, E. M., Ziehr, B., De Leon, G., Mitchell, D., & Moorman, N. J. (2014). Differential role for host translation factors in host and viral protein synthesis during human cytomegalovirus infection. *Journal of Virology*, *88*(3), 1473–1483. https://doi.org/10.1128/JVI.02321-13.

Li, S., Peters, G. A., Ding, K., Zhang, X., Qin, J., & Sen, G. C. (2006). Molecular basis for PKR activation by PACT or dsRNA. *Proceedings of the National Academy of Sciences of*

the United States of America, *103*(26), 10005–10010. https://doi.org/10.1073/pnas.0602317103.
Linero, F. N., Thomas, M. G., Boccaccio, G. L., & Scolaro, L. A. (2011). Junin virus infection impairs stress-granule formation in Vero cells treated with arsenite via inhibition of eIF2alpha phosphorylation. *The Journal of General Virology*, *92*(Pt. 12), 2889–2899. https://doi.org/10.1099/vir.0.033407-0.
Llano, M., Saenz, D. T., Meehan, A., Wongthida, P., Peretz, M., Walker, W. H., et al. (2006). An essential role for LEDGF/p75 in HIV integration. *Science*, *314*(5798), 461–464.
Lo Conte, L., Chothia, C., & Janin, J. (1999). The atomic structure of protein-protein recognition sites. *Journal of Molecular Biology*, *285*(5), 2177–2198.
McMillan, K., Adler, M., Auld, D. S., Baldwin, J. J., Blasko, E., Browne, L. J., et al. (2000). Allosteric inhibitors of inducible nitric oxide synthase dimerization discovered via combinatorial chemistry. *Proceedings of the National Academy of Sciences of the United States of America*, *97*(4), 1506–1511.
Moll, U. M., & Petrenko, O. (2003). The MDM2-p53 interaction. *Molecular Cancer Research*, *1*(14), 1001–1008.
Morelli, X., Bourgeas, R., & Roche, P. (2011). Chemical and structural lessons from recent successes in protein-protein interaction inhibition (2P2I). *Current Opinion in Chemical Biology*, *15*(4), 475–481. https://doi.org/10.1016/j.cbpa.2011.05.024.
Muratore, G., Goracci, L., Mercorelli, B., Foeglein, A., Digard, P., Cruciani, G., et al. (2012). Small molecule inhibitors of influenza A and B viruses that act by disrupting subunit interactions of the viral polymerase. *Proceedings of the National Academy of Sciences of the United States of America*, *109*(16), 6247–6252. https://doi.org/10.1073/pnas.1119817109.
Navratil, V., de Chassey, B., Combe, C. R., & Lotteau, V. (2011). When the human viral infectome and diseasome networks collide: Towards a systems biology platform for the aetiology of human diseases. *BMC Systems Biology*, *5*, 13. https://doi.org/10.1186/1752-0509-5-13.
Oswald, C., Rappas, M., Kean, J., Dore, A. S., Errey, J. C., Bennett, K., et al. (2016). Intracellular allosteric antagonism of the CCR9 receptor. *Nature*, *540*(7633), 462–465. https://doi.org/10.1038/nature20606.
Patel, R. C., & Sen, G. C. (1998). PACT, a protein activator of the interferon-induced protein kinase, PKR. *The EMBO Journal*, *17*(15), 4379–4390. https://doi.org/10.1093/emboj/17.15.4379.
Pichlmair, A., Kandasamy, K., Alvisi, G., Mulhern, O., Sacco, R., Habjan, M., et al. (2012). Viral immune modulators perturb the human molecular network by common and unique strategies. *Nature*, *487*(7408), 486–490. https://doi.org/10.1038/nature11289.
Rajagopala, S. V., Casjens, S., & Uetz, P. (2011). The protein interaction map of bacteriophage lambda. *BMC Microbiology*, *11*, 213. https://doi.org/10.1186/1471-2180-11-213.
Ramage, H. R., Kumar, G. R., Verschueren, E., Johnson, J. R., Von Dollen, J., Johnson, T., et al. (2015). A combined proteomics/genomics approach links hepatitis C virus infection with nonsense-mediated mRNA decay. *Molecular Cell*, *57*(2), 329–340. https://doi.org/10.1016/j.molcel.2014.12.028.
Rasaiyaah, J., Tan, C. P., Fletcher, A. J., Price, A. J., Blondeau, C., Hilditch, L., et al. (2013). HIV-1 evades innate immune recognition through specific cofactor recruitment. *Nature*, *503*(7476), 402–405. https://doi.org/10.1038/nature12769.
Roche, M., Borm, K., Flynn, J. K., Lewin, S. R., Churchill, M. J., & Gorry, P. R. (2016). Molecular gymnastics: Mechanisms of HIV-1 resistance to CCR5 antagonists and impact on virus phenotypes. *Current Topics in Medicinal Chemistry*, *16*(10), 1091–1106.
Royall, E., Doyle, N., Abdul-Wahab, A., Emmott, E., Morley, S. J., Goodfellow, I., et al. (2015). Murine norovirus 1 (MNV1) replication induces translational control of the host

by regulating eIF4E activity during infection. *The Journal of Biological Chemistry*, *290*(8), 4748–4758. https://doi.org/10.1074/jbc.M114.602649.

Rozenblatt-Rosen, O., Deo, R. C., Padi, M., Adelmant, G., Calderwood, M. A., Rolland, T., et al. (2012). Interpreting cancer genomes using systematic host network perturbations by tumour virus proteins. *Nature*, *487*(7408), 491–495. https://doi.org/10.1038/nature11288.

Ruepp, A., Waegele, B., Lechner, M., Brauner, B., Dunger-Kaltenbach, I., Fobo, G., et al. (2010). CORUM: The comprehensive resource of mammalian protein complexes—2009. *Nucleic Acids Research*, *38*(Database issue), D497–501. https://doi.org/10.1093/nar/gkp914.

Salter, J. D., Morales, G. A., & Smith, H. C. (2014). Structural insights for HIV-1 therapeutic strategies targeting Vif. *Trends in Biochemical Sciences*, *39*(9), 373–380. https://doi.org/10.1016/j.tibs.2014.07.001.

Schormann, N., Sommers, C. I., Prichard, M. N., Keith, K. A., Noah, J. W., Nuth, M., et al. (2011). Identification of protein-protein interaction inhibitors targeting vaccinia virus processivity factor for development of antiviral agents. *Antimicrobial Agents and Chemotherapy*, *55*(11), 5054–5062. https://doi.org/10.1128/AAC.00278-11.

Shapira, S. D., Gat-Viks, I., Shum, B. O., Dricot, A., de Grace, M. M., Wu, L., et al. (2009). A physical and regulatory map of host-influenza interactions reveals pathways in H1N1 infection. *Cell*, *139*(7), 1255–1267. https://doi.org/10.1016/j.cell.2009.12.018.

Shi, J., Zhou, J., Shah, V. B., Aiken, C., & Whitby, K. (2011). Small-molecule inhibition of human immunodeficiency virus type 1 infection by virus capsid destabilization. *Journal of Virology*, *85*(1), 542–549. https://doi.org/10.1128/JVI.01406-10.

Shun, M. C., Raghavendra, N. K., Vandegraaff, N., Daigle, J. E., Hughes, S., Kellam, P., et al. (2007). LEDGF/p75 functions downstream from preintegration complex formation to effect gene-specific HIV-1 integration. *Genes & Development*, *21*(14), 1767–1778.

Stanitsa, E. S., Arps, L., & Traktman, P. (2006). Vaccinia virus uracil DNA glycosylase interacts with the A20 protein to form a heterodimeric processivity factor for the viral DNA polymerase. *The Journal of Biological Chemistry*, *281*(6), 3439–3451. https://doi.org/10.1074/jbc.M511239200.

Strasfeld, L., & Chou, S. (2010). Antiviral drug resistance: Mechanisms and clinical implications. *Infectious Disease Clinics of North America*, *24*(3), 809–833. https://doi.org/10.1016/j.idc.2010.07.001.

Szilagyi, A., Nussinov, R., & Csermely, P. (2013). Allo-network drugs: Extension of the allosteric drug concept to protein-protein interaction and signaling networks. *Current Topics in Medicinal Chemistry*, *13*(1), 64–77.

Trigg, S. A., Garza, R. M., MacWilliams, A., Nery, J. R., Bartlett, A., Castanon, R., et al. (2017). CrY2H-seq: A massively multiplexed assay for deep-coverage interactome mapping. *Nature Methods*, *14*(8), 819–825. https://doi.org/10.1038/nmeth.4343.

Ueda, T., Watanabe-Fukunaga, R., Fukuyama, H., Nagata, S., & Fukunaga, R. (2004). Mnk2 and Mnk1 are essential for constitutive and inducible phosphorylation of eukaryotic initiation factor 4E but not for cell growth or development. *Molecular and Cellular Biology*, *24*(15), 6539–6549. https://doi.org/10.1128/MCB.24.15.6539-6549.2004.

Uetz, P., Dong, Y. A., Zeretzke, C., Atzler, C., Baiker, A., Berger, B., et al. (2006). Herpesviral protein networks and their interaction with the human proteome. *Science*, *311*(5758), 239–242. https://doi.org/10.1126/science.1116804.

Vance, R. E., Isberg, R. R., & Portnoy, D. A. (2009). Patterns of pathogenesis: Discrimination of pathogenic and nonpathogenic microbes by the innate immune system. *Cell Host & Microbe*, *6*(1), 10–21. https://doi.org/10.1016/j.chom.2009.06.007.

Vassilev, L. T., Vu, B. T., Graves, B., Carvajal, D., Podlaski, F., Filipovic, Z., et al. (2004). In vivo activation of the p53 pathway by small-molecule antagonists of MDM2. *Science*, *303*(5659), 844–848. https://doi.org/10.1126/science.1092472.

Walsh, D., Mathews, M. B., & Mohr, I. (2013). Tinkering with translation: Protein synthesis in virus-infected cells. *Cold Spring Harbor Perspectives in Biology*, *5*(1), a012351. https://doi.org/10.1101/cshperspect.a012351.

Walsh, D., & Mohr, I. (2004). Phosphorylation of eIF4E by Mnk-1 enhances HSV-1 translation and replication in quiescent cells. *Genes & Development*, *18*(6), 660–672. https://doi.org/10.1101/gad.1185304.

Wang, H., Jurado, K. A., Wu, X., Shun, M. C., Li, X., Ferris, A. L., et al. (2012). HRP2 determines the efficiency and specificity of HIV-1 integration in LEDGF/p75 knockout cells but does not contribute to the antiviral activity of a potent LEDGF/p75-binding site integrase inhibitor. *Nucleic Acids Research*, *40*(22), 11518–11530. https://doi.org/10.1093/nar/gks913.

Wells, J. A., & McClendon, C. L. (2007). Reaching for high-hanging fruit in drug discovery at protein-protein interfaces. *Nature*, *450*(7172), 1001–1009. https://doi.org/10.1038/nature06526.

Wong, P., Althammer, S., Hildebrand, A., Kirschner, A., Pagel, P., Geissler, B., et al. (2008). An evolutionary and structural characterization of mammalian protein complex organization. *BMC Genomics*, *9*, 629. https://doi.org/10.1186/1471-2164-9-629.

Yeung, M. L., Houzet, L., Yedavalli, V. S., & Jeang, K. T. (2009). A genome-wide short hairpin RNA screening of jurkat T-cells for human proteins contributing to productive HIV-1 replication. *The Journal of Biological Chemistry*, *284*(29), 19463–19473. https://doi.org/10.1074/jbc.M109.010033.

Yin, H., & Hamilton, A. D. (2005). Strategies for targeting protein-protein interactions with synthetic agents. *Angewandte Chemie (International Edition in English)*, *44*(27), 4130–4163. https://doi.org/10.1002/anie.200461786.

Yu, H., Tardivo, L., Tam, S., Weiner, E., Gebreab, F., Fan, C., et al. (2011). Next-generation sequencing to generate interactome datasets. *Nature Methods*, *8*(6), 478–480. https://doi.org/10.1038/nmeth.1597.

Zak, D. E., Tam, V. C., & Aderem, A. (2014). Systems-level analysis of innate immunity. *Annual Review of Immunology*, *32*, 547–577. https://doi.org/10.1146/annurev-immunol-032713-120254.

Zerbe, B. S., Hall, D. R., Vajda, S., Whitty, A., & Kozakov, D. (2012). Relationship between hot spot residues and ligand binding hot spots in protein-protein interfaces. *Journal of Chemical Information and Modeling*, *52*(8), 2236–2244. https://doi.org/10.1021/ci300175u.

Zhang, W., Du, J., Evans, S. L., Yu, Y., & Yu, X. F. (2011). T-cell differentiation factor CBF-beta regulates HIV-1 Vif-mediated evasion of host restriction. *Nature*, *481*(7381), 376–379. https://doi.org/10.1038/nature10718.

CHAPTER EIGHT

Investigating the Influence of Hotspot Mutations in Protein–Protein Interaction of IDH1 Homodimer Protein: A Computational Approach

D. Thirumal Kumar*, P. Sneha*, Jennifer Uppin*, S. Usha†, C. George Priya Doss*,[1]

*School of Biosciences and Technology, VIT University, Vellore, Tamil Nadu, India
†School of Social Sciences and Languages, VIT University, Vellore, Tamil Nadu, India
[1]Corresponding author: e-mail addresses: georgecp77@yahoo.co.in; georgepriyadoss@vit.ac.in

Contents

Abstract

Protein–protein interaction (PPI) helps in maintaining the cellular homeostasis. In particular, the homodimeric proteins play a crucial role as cell regulators. Studying the critical functions of each PPI on the living system is very challenging. The mutations in the PPIs have given birth to various diseases including many types of cancers and it has soon become the target for drug discovery. The mutations in IDH1, an asymmetric homodimer in the cytoplasm, leads to various diseases including gliomas. In this study,

Advances in Protein Chemistry and Structural Biology, Volume 111
ISSN 1876-1623
https://doi.org/10.1016/bs.apcsb.2017.08.002
© 2018 Elsevier Inc.
All rights reserved.

we have used extensive computational approaches to identify the impact of missense mutations (R132C, R132G, R132H, R132L, R132S, and V178I) occurring in the interacting region of the IDH1 homodimer. By in silico pathogenicity analysis, all the mutations occurring at the positions 132 and 178 were found to be pathogenic and neutral respectively. Furthermore, the mutants R132C and R132G were found to be responsible for increasing the stability, whereas the mutants R132H, R132L, and R132S were found to be responsible for the decrease in stability by stability analysis. R132H, R132L, and R132S mutants exhibited higher destabilization when compared to the structures of R132C and R132G mutants by molecular docking and molecular dynamics analysis.

1. INTRODUCTION

Protein–protein interactions (PPIs) serve as an essential element for many biological functions (Jones & Thornton, 1995). PPIs are becoming a promising target for drug discovery as they are more accurate to disease-related pathways (Murakami, Tripathi, Prathipati, & Mizuguchi, 2017). Functionally important proteins such as tyrosine kinase receptors, transcriptional factors, G-protein-coupled receptors, ion channels, and enzyme complexes occur as homodimers or heterodimers (Marianayagam, Sunde, & Matthews, 2004; Milligan, Ramsay, Pascal, & Carrillo, 2003; Rönnstrand, 2004; Simon & Goodenough, 1998). The PPIs within the proteins could be a homodimer or heterodimer. In PPIs, homodimers are formed by the combination of symmetric monomers, whereas heterodimers are formed by the combination of asymmetric monomers (Zhanhua, Gan, Lei, Sakharkar, & Kangueane, 2005). Isocitrate dehydrogenase (IDH) is a crucial enzyme that catalyzes a two-step process involved in the conversion of oxidative decarboxylation of isocitrate, producing alpha-ketoglutarate (α-ketoglutarate) and CO_2. Abundantly, IDH exists in three isoforms; IDH1 and IDH2 catalyze the same reaction of the TCA cycle and use $NADP^+$ as a cofactor. Meanwhile, IDH3 catalyzes the conversion of NAD^+ to NADH in the citric acid cycle (Corpas et al., 1999).

Isocitrate dehydrogenase 1 (*IDH1*) gene delivers instructions to synthesize the enzyme IDH which is an essential enzyme in cellular and lipid metabolism. *IDH1* gene is located on the 2q33.3 chromosome, encoding 414 amino acid homodimer protein (Tefferi, 2010). The IDH1 is mainly found in the cytoplasm (43%) and lesser in peroxisomes (21%). *IDH1* gene is soluble (46.7 kDa) and involved in catalyzing the oxidative decarboxylation of isocitrate into alpha-ketoglutarate using $NADP^+$ and MG^{2+} as cofactors (Shechter, Dai, Huo, & Guan, 2003). The wild-type IDH1 holds its normal function of oxidative decarboxylation of isocitrate to alpha-ketoglutarate, whereas the mutants gain a new neomorphic catalytic activity in the conversion of alpha-ketoglutarate to 2-hydroxyglutarate (2-HG). During the catalyzation, the $NADP^+$ is reduced to NADPH along with the release of CO_2 (Yang, Ye, Guan, & Xiong, 2012). The alpha-ketoglutarate plays an intermediate step in citric acid cycle producing energy. The alpha-ketoglutarate controls the histone methylation, DNA methylation, hypoxic conditions, as well as many other cellular processes. The 2-HG hinders the deoxygenases of prolyl hydroxylase and histone demethylases, leading to angiogenesis and unique gene expression (Rotili & Mai, 2011). The 2-HG acts as an oncometabolite in causing cancer in humans by changing the chromatin structure. The recent studies have disclosed that mutated form of IDH mutations causes a block in DNA and histone methylation (Dang, Yen, & Attar, 2016). The mutation in IDH1 shows high histone methylation in comparison to wild-type IDH1 (Lu, Ward, Kapoor, & Rohle, 2012). The glutamine pathway is activated under hypoxia condition as cancer cells require constant energy to proliferate; the glutamate is converted to alpha-ketoglutarate in the TCA cycle by glutamate dehydrogenases and aminotransferases supporting the growth of cancer cells (Altman, Stine, & Dang, 2016). Excess acquisition of the oncometabolite D2HG hinders the alpha-ketoglutarate-dependent dehydrogenase and further alters the gene expression by histone and DNA methylases (Nakajima & Kunimoto, 2014). This hindrance leads to tumorigenesis blocking the cell proliferation, and hypermethylation and aberrant histone deacetylation. Further, the alpha-ketoglutarate-dependent enzyme ten-eleven-translocation 2 (TET2), the alteration in the function of TET2, leads to the increased DNA methylation (Figueroa et al., 2010).

In 2008, *IDH1* encoding genes were first reported in low-grade gliomas and secondary glioblastomas (GBM) in cancer genome project. The study involved sequencing of 20,661 genes samples obtained from GBM (Yang et al., 2012). They confine to the cytosol, mitochondrion, and peroxisome. The IDH1 was observed largely in WHO graded 2 and 3 gliomas (75%), acute myeloid leukemia (20%), chondrosarcomas (tumor of bones) (75%), cholangiocarcinomas (23%), in CN-AML a rare blood cancer (16%), and 90% of the gliomas at R132H point mutation (Amary et al., 2011; Borger et al., 2012; Marcucci, Haferlach, & Döhner, 2011; Yang et al., 2012). The conserved arginine amino acid at position 132 of the IDH1 is the main site for amino acid substitution, usually causing missense mutations leading to cancers (Yang et al., 2012). There was only one case reported to have hormone receptor-positive breast carcinoma with increased levels of

2-HG in the patient's lymph node metastasis, bearing the IDH1 mutant R132L which was somatically acquired. The mutations in R132C were also reported with infiltrating mucinous and ductal carcinoma. The IDH1 mutant R132H is most commonly found in gliomas, but very recently Ang et al. (2012) reported the presence of a mutation in breast carcinoma (Ang et al., 2012). In this study, we have used an extensive computational analysis to understand the behavior of mutation present in the protein–protein interaction region of IDH1. A series of analysis such as conservation analysis, pathogenicity analysis, stability analysis, molecular docking, and molecular dynamics analysis were employed to study the change in the behavior of protein due to mutations.

2. MATERIALS AND METHODS

2.1 Dataset and Protein Modeling

The amino acids involved in the PPI of IDH1 homodimer protein were retrieved from PDBsum (Laskowski, 2001). The FASTA sequence of the protein and the natural variants (mutations) of IDH1 were obtained from the UniProt database with UniProtKB—O75874. The 3D structure of IDH1 with the resolution of 2.70 Å was retrieved from the Protein Data Bank with PDB ID 1T09. The mutations were introduced in the native molecule using SwissPDB Viewer and energy minimized using the same (Guex & Peitsch, 1997).

2.2 Conservation and Pathogenicity Analysis

The conservation pattern of the each amino acids of IDH1 protein was evaluated using the ConSurf Server (Glaser et al., 2003). The FASTA sequence obtained from the UniProt database was given as the input. The server generated multiple sequence alignment for the given sequence and results were predicted in the range of scale 1–9, where scale 1 is highly variable and scale 9 is highly conserved. In addition to the conservation scale, the tool also predicts whether the amino acid is in the exposed or buried and functional or structural.

The pathogenicity of the mutations was evaluated using the PredictSNP server (Bendl et al., 2014). In addition to its prediction, the PredictSNP server comprises the pathogenicity prediction of other standard tools such as MAPP, PhD-SNP, PolyPhen-1, PolyPhen-2, SIFT, SNAP, nsSNPAnalyzer, and PANTHER. The FASTA sequence obtained from the UniProt and the mutations were given as the input for prediction. The results were categorized as deleterious or neutral.

2.3 Stability Analysis

Further to the pathogenicity prediction analysis, the change in stability due to mutation was evaluated using iStable server (Chen, Lin, & Chu, 2013). Similar to PredictSNP server, the iStable server comprises the stability prediction of other standard tools such as i-Mutant2.0 and MUpro servers. The protein sequence obtained from the UniProt and the mutations were given as the input for prediction. The results have been achieved as increase or decrease in stability.

2.4 Molecular Docking and Molecular Dynamics Simulations

Protein–protein docking was initiated using ZDOCK online server to understand the change in the binding pattern of homodimer due to the mutation (Pierce et al., 2014). Further to the docking, molecular dynamics simulations (MDSs) were carried out for the native and mutant complexes using Gromacs 4.5.6 (Pronk et al., 2013). GROMOS 53a6 force field was employed for the topology building. The system was neutralized by replacing six water molecules with six Na^+ ions. Energy minimization and equilibrations (NVT and NPT) were carried out for 100 and 1000 ps, respectively. As a final step, MDSs were performed for 50 ns and trajectories were obtained. g_rms, r_gyrate, and g_hbond plugins of Gromacs were used to analyze the root mean square deviation (RMSD), radius of gyration (R_g), and intramolecular hydrogen bond formation, respectively.

3. RESULTS

3.1 Protein–Protein Interaction

The amino acid involved in the PPI of IDH1 homodimer was obtained using the PDBsum of the 3D structure with PDB ID 1T09. From the analysis, we observed that 52 residues of IDH1 A chain and 53 residues of IDH1 B chain were involved in the interaction of homodimer (Fig. 1A). From the UniProt database, six mutations were obtained (R132C, R132G, R132H, R132L, R132S, and V178I). Both the positions 132 and 178 of the mutation were found to be in the interacting A chain and B chain of IDH1 (Fig. 1B). These mutations were subjected to series of analysis such as pathogenicity prediction, stability prediction, conservation analysis, physicochemical property changes, molecular docking, and molecular dynamics.

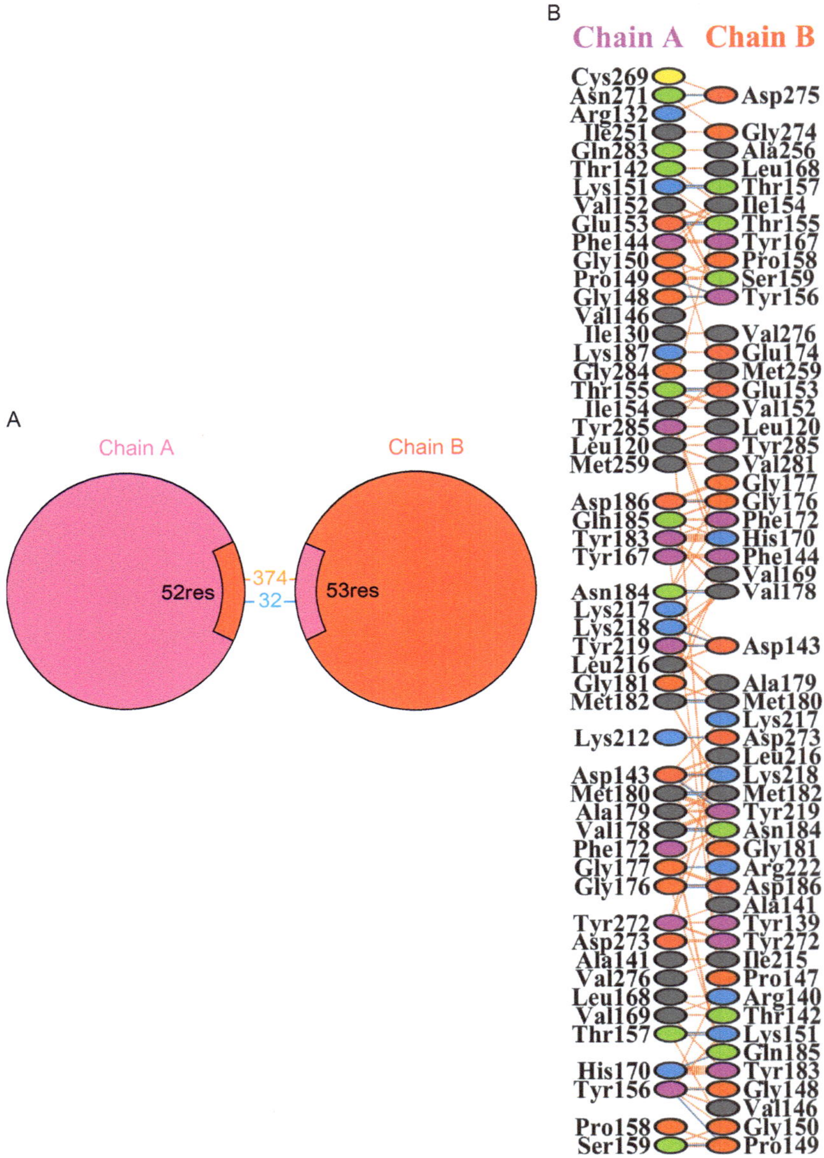

Fig. 1 (A) PDBsum illustration showing the number of interacting amino acids from A chain and B chain of the IDH1 homodimer. (B) PDBsum illustration showing the list of amino acids of A chain and B chain of IDH1 involved in protein–protein interaction.

3.2 Conservation and Pathogenicity Analysis

The evolutionary conservation of the amino acids along with their properties (buried or exposed and structural or functional) was analyzed using the ConSurf server. From the conservation analysis, we found that the position 132 of IDH1 which has arginine in the native form is highly conserved with the scale of 9 and also located in the exposed region serving as functional residue. On the other hand, the position 178 which has valine in the native form is in the buried region with medium conservation scale of 6 (Fig. 2).

Pathogenicity of the mutations obtained from the UniProt was evaluated using PredictSNP server. All the five mutations at the position 132 (R132C, R132G, R132H, R132L, and R132S) were predicted to be deleterious by all the nine tools (PredictSNP, MAPP, PhD-SNP, PolyPhen-1, PolyPhen-2, SIFT, SNAP, nsSNPAnalyzer, and PANTHER), whereas the mutation V178I was predicted to be nonpathogenic by all the nine tools (Table 1). The mutations found to be conserved and pathogenic are further taken to the stability analysis.

3.3 Stability Analysis

In addition to the pathogenicity analysis, the change in stability due to mutation was analyzed for the five mutants (R132C, R132G, R132H, R132L, and R132S) using the iStable server. The mutants R132H, R132L, and R132S were predicted to decrease the stability by all the three tools (i-Mutant2.0, MUpro, and iStable), whereas the mutants R132C and R132G were predicted to be decreasing the stability only by MUpro server, and other two servers (i-Mutant2.0 and iStable) predicted them to be increasing the stability (Table 2).

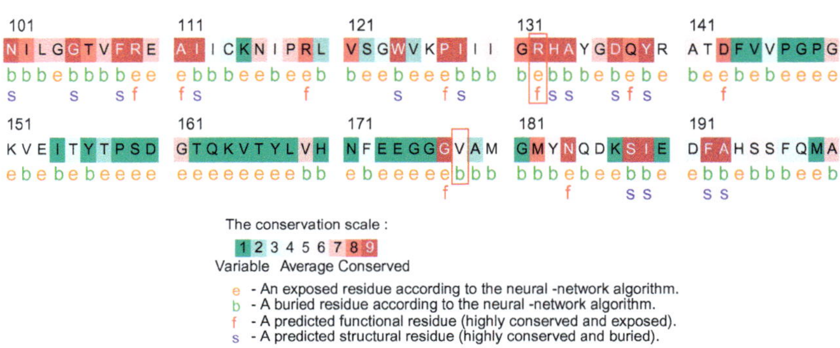

Fig. 2 Prediction of amino acid conservation using the online ConSurf server.

Table 1 Pathogenicity Prediction Analysis of IDH1 Protein

	R132C	R132G	R132H	R132L	R132S	V178I
PredictSNP	Deleterious	Deleterious	Deleterious	Deleterious	Deleterious	Neutral
MAPP	Deleterious	Deleterious	Deleterious	Deleterious	Deleterious	Neutral
PhD-SNP	Deleterious	Deleterious	Deleterious	Deleterious	Deleterious	Neutral
PolyPhen-1	Deleterious	Deleterious	Deleterious	Deleterious	Deleterious	Neutral
PolyPhen-2	Deleterious	Deleterious	Deleterious	Deleterious	Deleterious	Neutral
SIFT	Deleterious	Deleterious	Deleterious	Deleterious	Deleterious	Neutral
SNAP	Deleterious	Deleterious	Deleterious	Deleterious	Deleterious	Neutral
nsSNPAnalyzer	Deleterious	Deleterious	Deleterious	Deleterious	Deleterious	Neutral
PANTHER	Deleterious	Deleterious	Deleterious	Deleterious	Deleterious	Neutral

Table 2 Stability Prediction Analysis of IDH1 Protein

	R132C	R132G	R132H	R132L	R132S
i-Mutant2.0	Increase	Increase	Decrease	Decrease	Decrease
MUpro	Decrease	Decrease	Decrease	Decrease	Decrease
iStable	Increase	Increase	Decrease	Decrease	Decrease

3.4 Molecular Docking Analysis

The molecular docking analysis was initiated to understand the change in binding modes of the IDH1 homodimer. Molecular docking analysis was carried using the ZDOCK online server. The server works based on the rigid-body docking program. From the analysis, we obtained a highest ZDOCK score of 3249.195 for the native and least ZDOCK score for the mutant R132H (3214.93). It was followed with least ZDOCK score of 3215.891 and 3217.277 for R132L and R132S, respectively. However, the mutants R132C and R132G which were found to increase the stability by iStable and i-Mutant2.0 showed a similar ZDOCK score of 3223.289 and 3223.121, respectively. The graphical representation of native docked complex figure is shown in Fig. 3.

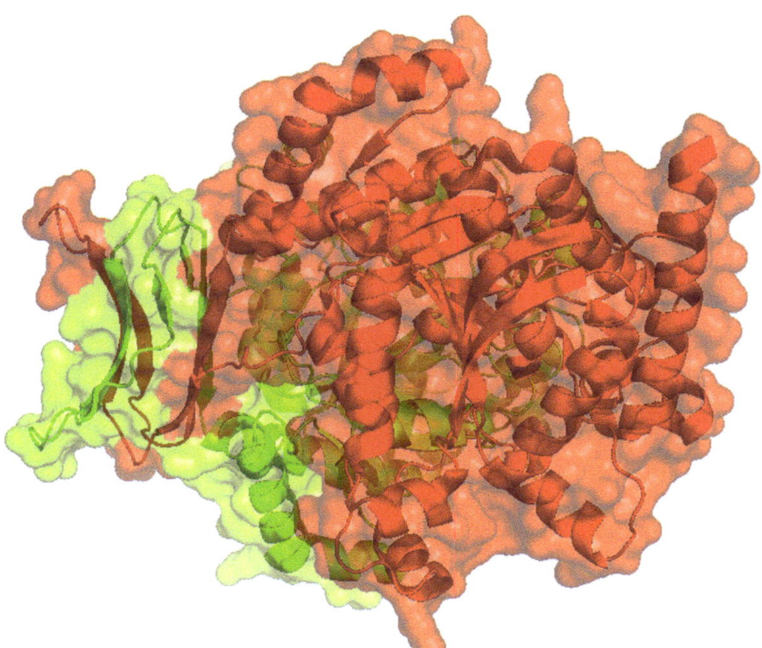

Fig. 3 The graphical representation of native homodimer complex visualized using PyMol. Color scheme: *green*—A chain, *red*—B chain.

3.5 Molecular Dynamics Simulations (MDS)

MDS was carried out for the docked complexes to understand the impact of mutations at the atomic level of the homodimer protein. The MDSs were carried out for 50 ns followed by employing g_rms, g_gyrate, and g_hbond plugins of Gromacs to analyze the RMSD, change in compactness of the proteins, and change in intramolecular hydrogen bond formation. From the RMSD analysis, we observed that the protein complexes attained convergence at the end of the 50 ns period. The mutant complexes R132C and R132G showed lesser deviation pattern up to ~0.15 nm than that of the native protein. Meanwhile, the mutant complexes R132H, R132L, and R132S exhibited higher deviation pattern up to ~0.45 nm when compared to the native complex throughout the simulation period of 50 ns (Fig. 4). Further, the change in compactness of the protein due to mutation was analyzed using a radius of gyration (R_g). From the radius of gyration analysis, we found that the native complex showed least radius of gyration value, whereas R132H and R132L complexes showed higher radius of gyration values. Mutant complex R132S showed radius of gyration value and ranked in third after the R132H and R132L complexes. However, R132C and R132G

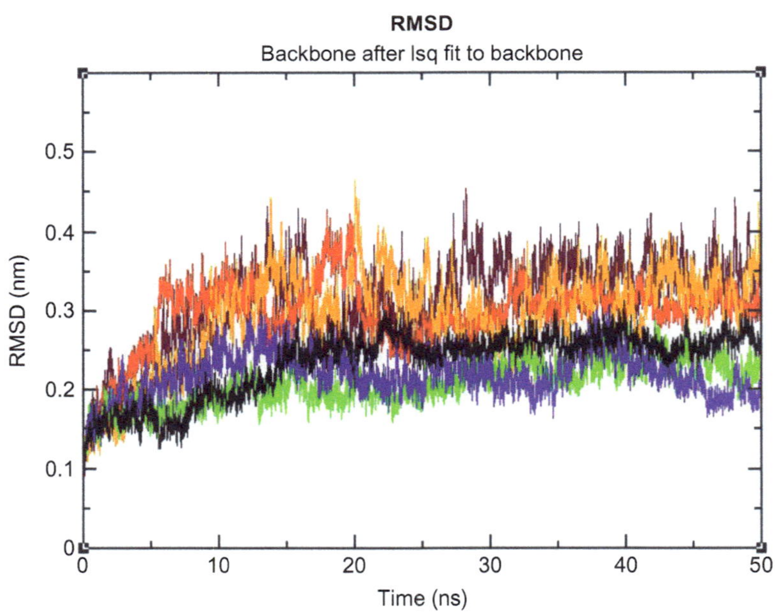

Fig. 4 Comparative backbone RMSD graph of 50 ns simulation of native and mutant complexes. Color scheme native (*black*), mutant R132H (*maroon*), mutant R132L (*orange*), mutant R132S (*red*), mutant R132C (*green*), and mutant R132G (*blue*).

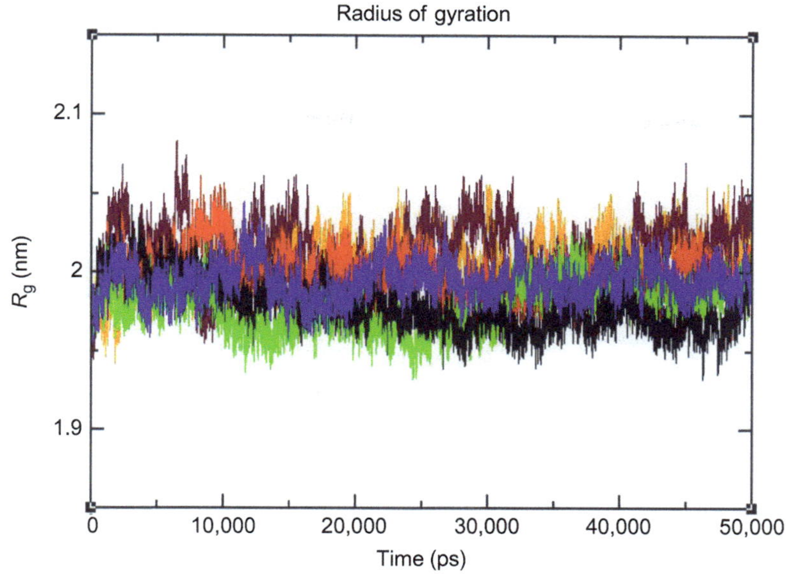

Fig. 5 Comparative radius of gyration graph of 50 ns simulation of native and mutant complexes. Color scheme native (*black*), mutant R132H (*maroon*), mutant R132L (*orange*), mutant R132S (*red*), mutant R132C (*green*), and mutant R132G (*blue*).

complexes maintained radius of gyration pattern similar to that of the native complex (Fig. 5). To address the reason for the change in compactness, change in intramolecular hydrogen bond formation was analyzed. From the analysis, we found that the native complex possessed substantiate number of intramolecular hydrogen bonds (~200 to ~250 number) throughout the analysis. The mutant complexes R132H and R132L exhibited minimal participation in intramolecular hydrogen bond formation (up to 180 number). On the other hand, R132G and R132C complexes possessed a higher number of intramolecular hydrogen bonds when compared to the native complex and other mutant complexes (up to ~250 numbers). Mutant R132S complex possessed a similar number of intramolecular hydrogen bonds to that of the native complex (Fig. 6).

4. DISCUSSION

Protein–protein interaction helps in the normal biological functions such as cell regulation and cell signaling. The interactions within the protein help in forming stable dimers (heterodimers or homodimers) (Zhanhua et al., 2005). The mutations occurring in the PPI interfaces result in the

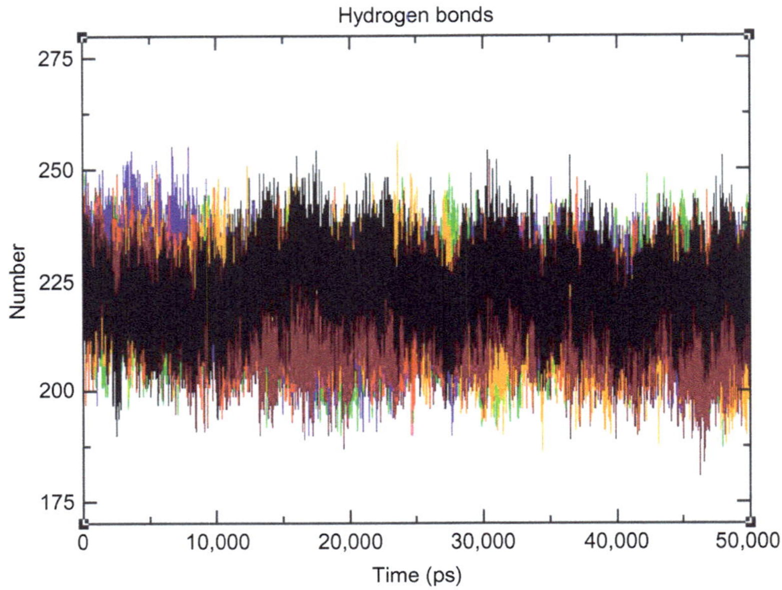

Fig. 6 Comparative intramolecular hydrogen bonds graph of 50 ns simulation of native and mutant complexes. Color scheme native (*black*), mutant R132H (*maroon*), mutant R132L (*orange*), mutant R132S (*red*), mutant R132C (*green*), and mutant R132G (*blue*).

interruption of networks within and between the cells (Jubb et al., 2017). The mutations occurring in the highly conserved amino acids possess higher impact than the mutations occurring in the lesser conserved amino acids (Shastry, 2007; Thirumal Kumar et al., 2017). To analyze this factor, ConSurf server was used to predict the scale of conservation of the amino acids positioned at 132 and 178 where the natural variants were reported in UniProt. From this analysis, we found that the position 132 which has arginine in the native form is highly conserved at the scale of 9 and the position 178 which has a valine in the native form has average conservation at the scale of 5 (Fig. 2). Also, the arginine at the position 132 was found to be functional and located in the exposed region. On the other hand, the valine at 178 was found in the buried region and does not possess any functional impact. Further, to address the deleterious nature of the mutations, the pathogenicity analysis was initiated. It is not necessary that all the mutants reported need to be deleterious. The mutants might be deleterious or nondeleterious, but the deleterious mutants might lead to the dysfunction of the protein. Online PredictSNP server was used to analyze the pathogenicity of the mutants. In addition to its algorithm for predicting the pathogenicity, it also uses the algorithm of other pathogenicity predictors such as MAPP, PhD-SNP, PolyPhen-1, PolyPhen-2, SIFT, SNAP, nsSNPAnalyzer, and PANTHER. The integration of multiple algorithms for the pathogenicity prediction makes the results more accurate (Bendl et al., 2014; George Priya Doss, Chakraborty, Narayan, & Thirumal Kumar, 2014; Sneha, Thirumal, & Tanwar, et al., 2017). By pathogenicity analysis all the mutations at hotspot 132nd position (R132C, R132G, R132H, R132L, and R132S) were predicted to be deleterious, whereas the mutation at 178th position (V178I) was predicted to be neutral (Table 1). Mutant V178I which was found to be located in the medium-conserved region and predicted to be neutral by pathogenicity predictors was ignored for the further analysis.

The pathogenic mutants may or may not influence the stability of the protein (George Priya Doss et al., 2014). The stability changes due to mutation were analyzed using the iStable server. In addition to its prediction, iStable predicts the stability changes using the other predictors such as i-Mutant2.0 and MUpro. From the iStable server results, mutants R132H, R132L, and R132S were predicted to decrease the stability by all the three tools (i-Mutant2.0, MUpro, and iStable), whereas the mutants R132C and R132G were predicted to be decreasing the stability only by MUpro server, and other two servers (i-Mutant2.0 and iStable) predicted them to be increasing the stability (Table 2).

The amino acids in the core region of dimer help in maintaining the stability of the protein, whereas the amino acids in the interface region contribute in maintaining the interaction of dimer (Thirumal Kumar et al., 2017; Tina, Bhadra, & Srinivasan, 2007). The mutations occurring at the interface of the protein–protein interaction might lead to weakening of the interactions (destabilization) between the proteins, thereby resulting in diseased conditions (Petukh, Kucukkal, & Alexov, 2015). In other words, it can be stated that protein with higher interactions possess higher binding efficiency (Hridya et al., 2016; Thirumal Kumar & George Priya Doss, 2016a, 2016b). Protein–protein docking with the native and mutants was initiated using ZDOCK server to study the change in the interaction between the A chain and B chain of the IDH1 homodimer. From the analysis, we obtained a highest ZDOCK score for the native complex (3249.195) and least ZDOCK score for the mutant R132H complex (3214.93). It was followed by least ZDOCK score of 3215.891 and 3217.277 for R132L and R132S complexes, respectively. However, the mutant complexes R132C and R132G which were found to increase the stability by iStable and i-Mutant2.0 showed a similar ZDOCK score of

3223.289 and 3223.121, respectively. This elucidates that the interaction between the two chains of IDH1 is highly disturbed by the mutation R132H which was followed by the mutation R132L and R132S. Though R132C and R132G showed comparatively higher ZDOCK score than other mutants, still the score was comparatively lower than that of the native which indicates that there is a disturbance in the interaction due to R132C and R132G.

The docked complexes were further taken for the molecular dynamics analysis. Molecular dynamics has become one of the best studies to understand the proteins at the atomic level. Several earlier studies have shown the best correlation between the experimental results and the results obtained from molecular dynamics (Anantharaman et al., 2016; Naine, Devi, Mohanasrinivasan, Doss, & Kumar, 2015; Sudhakar et al., 2016; Sujitha et al., 2016). Simulations were carried out for 50 ns to study the variation in the deviation pattern, compactness, and intramolecular hydrogen bonds using g_rms, g_gyrate, and g_hbond, respectively (Sneha & George Priya Doss, 2016). RMSD was analyzed to understand the convergence of the proteins at the end simulation. Protein with higher RMSD value indicates decrease in stability (Sneha, Thirumal, & Tanwar, et al., 2017; Yun & Guy, 2011). From the RMSD analysis, we observed that the protein complexes attained convergence at the end of the 50 ns period. The mutant complexes R132C and R132G which were predicted to be increasing the stability by iStable and i-Mutant2.0 showed a lesser deviation than the native complex. On the other hand, the mutant complexes R132H, R132L, and R132S showed higher deviation when compared to the native complex throughout the simulation period of 50 ns (Fig. 4). This indicates that the stability is highly influenced by the mutants R132H and R132L when compared to the other mutations. Further, the change in compactness of the protein due to mutation was analyzed using a radius of gyration (R_g). Protein with higher radius of gyration elucidates that the protein possesses lower compactness and vice versa (Lobanov, Bogatyreva, & Galzitskaia, 2008). From the radius of gyration analysis, we found that the native complex possessed higher compactness, whereas mutant complexes R132H and R132L possessed least compactness. However, the compactness of mutant complexes R132S, R132H, and R132L was found to be in between the R132H and R132L mutant complexes. On the other hand, R132C and R132G mutant complexes exhibited similar compactness to that of native complex (Fig. 5). To address the reason for the change in compactness, change in a number of intramolecular hydrogen bonds was analyzed. Hydrogen bonds play a crucial role in maintaining the stability of the protein. Protein with higher number of hydrogen bonds results in higher stability. Meanwhile, a decrease in a number of hydrogen bonds can lead to the loss of stability of the protein (Pace et al., 2014; Sneha, Thirumal, George, Siva, & Zayed, 2017). From the analysis, we found that the native complex possessed a substantial number of intramolecular hydrogen bonds throughout the analysis. The mutant complexes R132H and R132L exhibiting higher radius of gyration values were also found to have a minimal number of hydrogen bonds. On the other hand, on few occasions, R132G and R132C complexes exhibited higher number of intramolecular hydrogen bond formation when compared to the native and other mutant complexes. Mutant R132S complex possessed a similar number of intramolecular hydrogen bonds to that of the native complex (Fig. 6). This overall higher deviation with the loss of compactness and intramolecular hydrogen bonds makes the mutations R132H and R132L more deleterious than the other mutations. However, the other mutations (R132S, R132C, and R132G) have also found to deviate with the loss of compactness from the native and possess minimal effect over the protein.

Though the amino acids in the core region of protein–protein interaction play a major role in maintaining the stability of the protein, the amino acids in the interface of PPI serve responsibly for both structural and functional modifications (Guharoy & Chakrabarti, 2005; Thirumal Kumar et al., 2017). This phenomenon makes our work interesting to understand the mutation at the interface of the protein–protein. The analysis of such a protein–protein interaction through experimental approaches is very tedious (Jordán, Nguyen, & Liu, 2012). The usage of computers in biological research has become a promising alternative to the traditional techniques. The tool such as molecular dynamics helps in understanding the complex biological interaction at an atomic level. The results obtained from this not only assist in understanding the IDH1 mutations but also significantly express the role of conserved amino acid mutation positioned at the protein–protein interface.

5. CONCLUSION

Protein–protein interaction plays a crucial role in the biological system. In some cases, the disruption of PPI due to the mutation in conserved amino acids positioned at the interface might lead to both structural and functional loss. In the case of homodimer IDH1, the position 132 which is found to be conserved amino acid at the interface of PPI is prone to five mutations (R132C, R132G, R132H, R132L, and R132S). In this study, we conclude that mutants R132H and R132L disrupt the PPI in higher range when compared to the other three (R132S, R132G and R132C) mutations and contribute to the destabilization of the protein. We believe that this type of computational approach could highly help in understanding the mutations in the complex PPI of the biological system.

ACKNOWLEDGMENTS

The authors take this opportunity to thank the management of VIT University and (BRAF) at C-DAC for providing the facilities to carry out this work. The authors have no conflict of interest to declare.

REFERENCES

Altman, B. J., Stine, Z. E., & Dang, C. V. (2016). From Krebs to clinic: Glutamine metabolism to cancer therapy. *Nature Reviews Cancer, 16*(10), 619–634.

Amary, M. F., Bacsi, K., Maggiani, F., Damato, S., Halai, D., Berisha, F., et al. (2011). IDH1 and IDH2 mutations are frequent events in central chondrosarcoma and central and periosteal chondromas but not in other mesenchymal tumours. *Journal of Pathology, 224*(3), 334–343.

Anantharaman, A., Hemachandran, H., Mohan, S., Manikoth Ayyathan, D., Kumar, D. T., George Priya Doss, C., et al. (2016). Induction of apoptosis by apocarotenoids in B16 melanoma cells through ROS-mediated mitochondrial-dependent pathway. *Journal of Functional Foods, 20*, 346–357.

Ang, D., VanSandt, A. M., Beadling, C., Warrick, A., West, R. B., Corless, C., et al. (2012). Biphasic papillary and lobular breast carcinoma with PIK3CA and IDH1 mutations. *Diagnostic Molecular Pathology, 21*(4), 221–224.

Bendl, J., Stourac, J., Salanda, O., Pavelka, A., Wieben, E. D., Zendulka, J., et al. (2014). Predict SNP: Robust and accurate consensus classifier for prediction of disease-related mutations. *PLoS Computational Biology, 10*(1), e1003440.

Borger, D. R., Tanabe, K. K., Fan, K. C., Lopez, H. U., Fantin, V. R., Straley, K. S., et al. (2012). Frequent mutation of isocitrate dehydrogenase (IDH)1 and IDH2 in cholangiocarcinoma identified through broad-based tumor genotyping. *The Oncologist, 17*(1), 72–79.

Chen, C.-W., Lin, J., & Chu, Y.-W. (2013). iStable: Off-the-shelf predictor integration for predicting protein stability changes. *BMC Bioinformatics, 14*(Suppl. 2), S5.

Corpas, F. J., Barroso, J. B., Sandalio, L. M., Palma, J. M., Lupiáñez, J. A., & del Río, L. A. (1999). Peroxisomal NADP-dependent isocitrate dehydrogenase. Characterization and activity regulation during natural senescence. *Plant Physiology, 121*(3), 921–928.

Dang, L., Yen, K., & Attar, E. C. (2016). IDH mutations in cancer and progress toward development of targeted therapeutics. *Annals of Oncology, 27*(4), 599–608.

Figueroa, M. E., Abdel-Wahab, O., Lu, C., Ward, P. S., Patel, J., Shih, A., et al. (2010). Leukemic IDH1 and IDH2 mutations result in a hypermethylation phenotype, disrupt TET2 function, and impair hematopoietic differentiation. *Cancer Cell, 18*(6), 553–567.

George Priya Doss, C., Chakraborty, C., Narayan, V., & Thirumal Kumar, D. (2014). Computational approaches and resources in single amino acid substitutions analysis toward clinical research. *Advances in Protein Chemistry and Structural Biology, 94*, 365–423.

Glaser, F., Pupko, T., Paz, I., Bell, R. E., Bechor-Shental, D., Martz, E., et al. (2003). ConSurf: Identification of functional regions in proteins by surface-mapping of phylogenetic information. *Bioinformatics (Oxford, England)*, *19*(1), 163–164.

Guex, N., & Peitsch, M. C. (1997). SWISS-MODEL and the Swiss-PdbViewer: An environment for comparative protein modeling. *Electrophoresis*, *18*(15), 2714–2723.

Guharoy, M., & Chakrabarti, P. (2005). Conservation and relative importance of residues across protein-protein interfaces. *Proceedings of the National Academy of Sciences of the United States of America*, *102*(43), 15447–15452.

Hridya, H., Amrita, A., Mohan, S., Gopalakrishnan, M., Dakshinamurthy, T. K., Doss, G. P., et al. (2016). Functionality study of santalin as tyrosinase inhibitor: A potential depigmentation agent. *International Journal of Biological Macromolecules*, *86*, 383–389.

Jones, S., & Thornton, J. M. (1995). Protein-protein interactions: A review of protein dimer structures. *Progress in Biophysics and Molecular Biology*, *63*(1), 31–65.

Jordán, F., Nguyen, T.-P., & Liu, W.-C. (2012). Studying protein-protein interaction networks: A systems view on diseases. *Briefings in Functional Genomics*, *11*(6), 497–504.

Jubb, H. C., Pandurangan, A. P., Turner, M. A., Ochoa-Montaño, B., Blundell, T. L., & Ascher, D. B. (2017). Mutations at protein-protein interfaces: Small changes over big surfaces have large impacts on human health. *Progress in Biophysics and Molecular Biology*, *128*, 3–13.

Laskowski, R. A. (2001). PDBsum: Summaries and analyses of PDB structures. *Nucleic Acids Research*, *29*(1), 221–222.

Lobanov, M. I., Bogatyreva, N. S., & Galzitskaia, O. V. (2008). Radius of gyration is indicator of compactness of protein structure. *Molekuliarnaia Biologiia*, *42*(4), 701–706.

Lu, C., Ward, P., Kapoor, G., & Rohle, D. (2012). IDH mutation impairs histone demethylation and results in a block to cell differentiation. *Nature*, *483*(7390), 474–478.

Marcucci, G., Haferlach, T., & Döhner, H. (2011). Molecular genetics of adult acute myeloid leukemia: Prognostic and therapeutic implications. *Journal of Clinical Oncology: Official Journal of the American Society of Clinical Oncology*, *29*(5), 475–486.

Marianayagam, N. J., Sunde, M., & Matthews, J. M. (2004). The power of two: Protein dimerization in biology. *Trends in Biochemical Sciences*, *29*(11), 618–625.

Milligan, G., Ramsay, D., Pascal, G., & Carrillo, J. J. (2003). GPCR dimerisation. *Life Sciences*, *74*, 181–188.

Murakami, Y., Tripathi, L. P., Prathipati, P., & Mizuguchi, K. (2017). Network analysis and in silico prediction of protein-protein interactions with applications in drug discovery. *Current Opinion in Structural Biology*, *44*, 134–142.

Naine, S. J., Devi, C. S., Mohanasrinivasan, V., Doss, C. G. P., & Kumar, D. T. (2015). Binding and molecular dynamic studies of sesquiterpenes (2R-acetoxymethyl-1,3,3-trimethyl-4t-(3-methyl-2-buten-1-yl)-1t-cyclohexanol) derived from marine Streptomyces sp. VITJS8 as potential anticancer agent. *Applied Microbiology and Biotechnology*, *100*(6), 2869–2882.

Nakajima, H., & Kunimoto, H. (2014). TET2 as an epigenetic master regulator for normal and malignant hematopoiesis. *Cancer Science*, *105*(9), 1093–1099.

Pace, C. N., Fu, H., Lee Fryar, K., Landua, J., Trevino, S. R., Schell, D., et al. (2014). Contribution of hydrogen bonds to protein stability. *Protein Science: A Publication of the Protein Society*, *23*(5), 652–661.

Petukh, M., Kucukkal, T. G., & Alexov, E. (2015). On human disease-causing amino acid variants: Statistical study of sequence and structural patterns. *Human Mutation*, *36*(5), 524–534.

Pierce, B. G., Wiehe, K., Hwang, H., Kim, B. H., Vreven, T., & Weng, Z. (2014). ZDOCK server: Interactive docking prediction of protein-protein complexes and symmetric multimers. *Bioinformatics*, *30*(12), 1771–1773.

Pronk, S., Páll, S., Schulz, R., Larsson, P., Bjelkmar, P., Apostolov, R., et al. (2013). GROMACS 4.5: A high-throughput and highly parallel open source molecular simulation toolkit. *Bioinformatics (Oxford, England)*, *29*(7), 845–854.

Rönnstrand, L. (2004). Signal transduction via the stem cell factor receptor/c-Kit. *Cellular and Molecular Life Sciences: CMLS*, *61*(19–20), 2535–2548.

Rotili, D., & Mai, A. (2011). Targeting histone demethylases: A new avenue for the fight against cancer. *Genes & Cancer*, *2*(6), 663–679.

Shastry, B. S. (2007). SNPs in disease gene mapping, medicinal drug development and evolution. *Journal of Human Genetics*, *52*(11), 871–880.

Shechter, I., Dai, P., Huo, L., & Guan, G. (2003). IDH1 gene transcription is sterol regulated and activated by SREBP-1a and SREBP-2 in human hepatoma HepG2 cells: Evidence that IDH1 may regulate lipogenesis in hepatic cells. *Journal of Lipid Research*, *44*(11), 2169–2180.

Simon, A. M., & Goodenough, D. A. (1998). Diverse functions of vertebrate gap junctions. *Trends in Cell Biology*, *8*(12), 477–483.

Sneha, P., & George Priya Doss, C. (2016). Molecular dynamics: New frontier in personalized medicine. *Advances in Protein Chemistry and Structural Biology*, *102*, 181–224.

Sneha, P., Thirumal, K. D., George, P. D. C., Siva, R., & Zayed, H. (2017). Determining the role of missense mutations in the POU domain of HNF1A that reduce the DNA-binding affinity: A computational approach. *PLoS One*, *12*(4), e0174953.

Sneha, P., Thirumal, K. D., Tanwar, H., Siva, R., George, P. D. C., & Zayed, H. (2017). Structural analysis of G1691S variant in the human *filamin B* gene responsible for Larsen syndrome: A comparative computational approach. *Journal of Cellular Biochemistry*, *118*(7), 1900–1910.

Sudhakar, N., George Priya Doss, C., Thirumal Kumar, D., Chakraborty, C., Anand, K., & Suresh, M. (2016). Deciphering the impact of somatic mutations in exon 20 and exon 9 of PIK3CA gene in breast tumors among Indian women through molecular dynamics approach. *Journal of Biomolecular Structure & Dynamics*, *34*(1), 29–41.

Sujitha, S. P., Kumar, D. T., Doss, C. G. P., Aavula, K., Ramesh, R., Lakshmanan, S., et al. (2016). DNA repair gene (XRCC1) polymorphism (Arg399Gln) associated with schizophrenia in south Indian population: A genotypic and molecular dynamics study. *PLoS One*, *11*(1), e0147348.

Tefferi, A. (2010). Novel mutations and their functional and clinical relevance in myeloproliferative neoplasms: JAK2, MPL, TET2, ASXL1, CBL, IDH and IKZF1. *Leukemia*, *24*(6), 1128–1138.

Thirumal Kumar, D., & George Priya Doss, C. (2016a). Investigating the inhibitory effect of wortmannin in the hotspot mutation at codon 1047 of PIK3CA kinase domain: A molecular docking and molecular dynamics approach. *Advances in Protein Chemistry and Structural Biology*, *102*, 267–297.

Thirumal Kumar, D., & George Priya Doss, C. (2016b). Role of E542 and E545 missense mutations of PIK3CA in breast cancer: A comparative computational approach. *Journal of Biomolecular Structure & Dynamics*, *27*, 1–13.

Thirumal Kumar, D., George Priya Doss, C., Sneha, P., Tayubi, I. A., Siva, R., Chakraborty, C., et al. (2017). Influence of V54M mutation in giant muscle protein titin: A computational screening and molecular dynamics approach. *Journal of Biomolecular Structure & Dynamics*, *35*(5), 917–928.

Tina, K. G., Bhadra, R., & Srinivasan, N. (2007). PIC: Protein interactions calculator. *Nucleic Acids Research*, *35*(Suppl. 2), 473–476.

Yang, H., Ye, D., Guan, K. L., & Xiong, Y. (2012). IDH1 and IDH2 mutations in tumorigenesis: Mechanistic insights and clinical perspectives. *Clinical Cancer Research*, *18*(20), 5562–5571.

Yun, S., & Guy, H. R. (2011). Stability tests on known and misfolded structures with discrete and all atom molecular dynamics simulations. *Journal of Molecular Graphics & Modelling*, *29*(5), 663–675.

Zhanhua, C., Gan, J. G.-K., Lei, L., Sakharkar, M. K., & Kangueane, P. (2005). Protein subunit interfaces: Heterodimers versus homodimers. *Bioinformation*, *1*(2), 28–39.

CHAPTER NINE

Human Interactomics: Comparative Analysis of Different Protein Interaction Resources and Construction of a Cancer Protein–Drug Bipartite Network

Javier De Las Rivas*^,1, Diego Alonso-López*, Mónica M. Arroyo*^,†

*Bioinformatics and Functional Genomics Group, Cancer Research Center (CiC-IMBCC, CSIC/USAL/IBSAL), Consejo Superior de Investigaciones Científicas (CSIC) and University of Salamanca (USAL), Salamanca, Spain

†Pontifical Catholic University of Puerto Rico (PCUPR), Ponce, Puerto Rico

[1]Corresponding author: e-mail address: jrivas@usal.es

Contents

Abstract

Unraveling the protein interaction wiring that occurs in human cells as a scaffold of biological processes requires the identification of all elements that constitute such molecular interaction networks. Proteome-wide experimental studies and bioinformatic comprehensive efforts have provided reliable and updated compendiums of the human protein interactome. In this work, we present a current view of available *databases of human protein–protein interactions* (PPIs) that allow building protein interaction networks. We also investigate human proteins as targets of specific drugs to analyze how chemicals interact with different target proteins, placing also the study in a network relational space. Hence, we undertake a description of several major *drug–target resources* to provide a present perspective of the associations between human proteins and specific chemicals. The identification of molecular targets for specific drugs is a critical step to improve disease therapy. As different diseases have different biomolecular

Advances in Protein Chemistry and Structural Biology, Volume 111
ISSN 1876-1623
https://doi.org/10.1016/bs.apcsb.2017.09.002

© 2018 Elsevier Inc.
All rights reserved.

scenarios, we addressed the identification of drug-targeted genes focusing our investigations on cancer and cancer genes. So, a description of *resources* that provide curated compendiums of human *cancer genes* is presented. Cancer is a complex disease where multiple genetic changes rewire cellular networks during carcinogenesis. This indicates that cancer drug therapy needs the implementation of network-driven studies to reveal multiplex interactions between cancer genes and drugs. To make progress in this direction, in the last part of this work we provide a *bipartite network* of cancer genes and their drugs shown in a graph landscape that disclose the existence of specific drug–target modules.

1. INTRODUCTION

The analysis at systems level of the protein interaction wiring that occurs in human cells along the different biological processes and functions that they perform requires a previous identification of the elements that constitute the molecular interaction network, i.e., the *interactome*. Recent large-scale proteome-wide experimental studies (Rolland et al., 2014) and bioinformatic comprehensive efforts (Alonso-López et al., 2016) have provided reliable and updated compendiums of the human protein–protein interactome. We present in this work a revision of several available *databases and resources of human protein–protein interactions* (PPIs), explaining the scope of each one and evaluating their coverage. We also investigate in this study the human proteins as targets of specific drugs in a therapeutic and pharmacoproteomic context, since it is very interesting to analyze how different drugs interact with different target proteins, not in a single drug–target binary approach (which can be quite reductionist) but in a more holistic scenario where the multiple interactions and cross talks between proteins and drugs can be explored and examined. To achieve this, we need collections of drug-targeted genes placed in a network relational space that can be combined with specific disease PPI networks. This type of disease-associated maps can be obtained generating subsets of the human interactome with the genes/proteins that are most affected, altered, or involved in a given disease (or disease family) of interest (Ellinghaus et al., 2016). In this work, as a second objective, we undertake a description of several major *human drug–target resources* to provide a current view of our knowledge of the associations and interactions between human proteins and specific chemicals used as drugs. The identification of molecular targets of specific drugs is a critical step to improve disease therapy and to advance toward a molecular-driven precision medicine. Since very different diseases have divergent cellular and molecular scenarios, in this work we undertake the identification of drug-targeted genes in a disease-related PPI network focusing our investigations on cancer and cancer genes. Therefore, in the third part of this study, we present a description of *cancer gene/protein resources* that provide curated compendiums of human genes or gene products directly involved in cancer disease (i.e., in tumorigenesis and tumor evolution). Cancer is a complex disease that affects the cells at systems level. This complexity is based on the observation that multiple possible genetic changes and environmental influence rewire cellular networks during carcinogenesis and can provide different types of tumors and malignant states. This scenario indicates that cancer drug therapy cannot be addressed with single or simple drug–target binary approaches and we need a better implementation of network-driven studies to show multiple interactions between cancer genes and drugs and to reveal the existence of complex molecular modules that may be critical during carcinogenesis rewiring. To make progress in this direction, in the last part of this work we provide an elaborated network-based view of multiple cancer genes and their drugs (FDA approved or on-trial) showed in a modular graph landscape. This *cancer protein–drug bipartite network* includes 92 proteins, 83 drugs, and 176 interactions and reveals the existence of multiple drug–target modules that enclose groups of human genes correlated by their association to specific drugs.

2. PRIMARY DATABASES OF PPIs

The dissemination of new knowledge among the scientific community is based on the publication of experimental results in the research literature. Some of these results are also stored and classified in well-organized databases to allow better access, use, and computational management. This is the case of protein–protein interactions (PPIs) where the international community started to work more than two decades ago (HUPO PSI-MI, http://www.psidev.info/) in the development of accessible compendiums of all known (i.e., experimentally demonstrated) molecular interactions between proteins pairs in different biological systems and conditions. Working in this direction, several research groups constructed PPI databases based on the reading and curation of hundreds of original scientific articles. These bioinformatics resources were called primary databases (De Las Rivas & Fontanillo, 2010) because they obtain their information directly from the scientific literature thanks to the work of compilation and interpretation carried out by experts called curators. These experts follow common annotation protocols such as those set by Orchard et al. (2007).

The most important primary databases of protein interactions that include traceable information, derived from the scientific literature, about the experimental methods that demonstrate each interaction are BioGRID, DIP, HPRD, IntAct, and MINT. IntAct (Kerrien et al., 2012) is a database created and maintained by the Molecular Interaction group of the European Institute of Bioinformatics (EBI-EMBL, Cambridge, UK). It is a comprehensive information system that includes detailed data on thousands of interactions, as well as a powerful web-based query system. It stores data on PPIs and also proteins and other small molecules. The development team is part of HUPO PSI-MI (http://www.psidev.info/), mentioned earlier, which is responsible for establishing standards on molecular interactions and encouraging their use. The protein interaction data produced by IntAct is integrated into the main universal database for proteins, UniProt (http://www.uniprot.org/). MINT (Licata et al., 2012) is also a database dedicated to the extraction of experimental data on protein interactions from the scientific literature. Since the end of 2013, this database uses the technological infrastructure of IntAct to make its data available to researchers (Orchard et al., 2014). Hence, the data on protein interactions cured by the MINT group of experts appear integrated and registered by IntAct. The Database of Interacting Proteins (DIP) (Salwinski et al., 2004) also stores data on PPIs extracted from the scientific literature, but it is much smaller than IntAct and MINT. These three primary databases (DIP, IntAct, and MINT) formed the core founders of the International Molecular Exchange consortium (IMEx, http://www.imexconsortium.org/) and they follow all standards and protocols determined by PSI-MI for PPIs, including the use of open biological/biomedical ontologies (OBO, http://www.obofoundry.org/) and controlled vocabularies. The Human Protein Reference Database (HPRD) (Keshava Prasad et al., 2009) is specialized only in human proteins. Its objective is to offer a web platform with general information on structural domains, posttranslational modifications, interactions, and association with pathologies for the different proteins that form the human proteome. HPRD, therefore, is not a database only focused on protein interactions; rather, it contains this information as one annotation among many. Thus, the specific PPIs reported are poorly annotated, often with little details about the experimental methods that prove the interactions, without UniProt identifiers for proteins and without the use of biological ontologies to allow cross-referencing with other biological databases. Finally, BioGRID (Biological General Repository for Interaction Datasets) (Chatr-Aryamontri et al., 2015) is another large primary database that provides data on experimentally proven PPIs, but also on genetic interactions (Mani, St. Onge, Hartman, Giaever, & Roth, 2008) that do not reflect physical molecular interactions between proteins (De Las Rivas & de Luis, 2004).

3. META-DATABASES INTEGRATING AND UNIFYING MULTIPLE PPI RESOURCES

There are also other databases that compile and integrate the information on protein interactions available from several primary databases, combining and unifying the original data to avoid redundancies and also to increase coverage. These databases are known as meta-databases since they process and cluster the data available in multiple primary PPI resources. One of the most complete is APID (Agile Protein Interactomes DataServer) (Alonso-López et al., 2016), a newly updated biological resource that provides a comprehensive collection of protein interactomes for more than 400 organisms based on the integration of known experimentally validated protein–protein physical interactions (PPIs). Construction of the interactomes is done with a methodological approach to report quality levels and coverage over the proteomes for each organism included. Moreover, APID provides interactomes from specific organisms that in 27 cases have more than 500 proteins. At present, APID includes a comprehensive compendium of more than 94,000 distinct proteins and more than 767,000 singular interactions. The analytical and integrative effort done in APID unifies PPIs from six primary databases of mole cular interactions (BIND, BioGRID, DIP, HPRD, IntAct, and MINT), from other original resources (e.g., BioPlex, http://bioplex.hms.harvard.edu/), and also from experimentally resolved 3D structures (PDB) where more than two distinct proteins have been identified. More than 8000 structures have been analyzed to find specific PPIs reported with details of their molecular interfaces. APID also includes a new data visualization web tool that allows the construction of subinteractomes using query lists of proteins of interest and the visual exploration of the corresponding networks, including an interactive selection of the properties of the interactions (i.e., the reliability of the edges on the network) and an interactive mapping of the functional environment of the proteins (i.e., the functional annotations of the nodes in the network). APID also complies with PSI-MI standards.

Mentha (Calderone, Castagnoli, & Cesareni, 2013) is another meta-database that contains only protein interactions described experimentally. Mentha includes data of several species and offers detailed information on each interaction while maintaining the link to the original source. It also establishes a *score* for each interaction based on the number of experimental evidence that describe it. In terms of data integration, the creators of Mentha designed an automated process to get their database updated. In this procedure, they use the PSICQUIC platform (http://www.ebi.ac.uk/Tools/webservices/psicquic/view/main.xhtml) (Aranda et al., 2011), processing five primary databases included in it: BioGRID, DIP, IntAct, MINT, and MatrixDB (Chautard, Fatoux-Ardore, Ballut, Thierry-Mieg, & Ricard-Blum, 2011). This method for updating is fast, but it is not always very accurate and provides only the data available in PSICQUIC. Finally, Mentha is oriented to work with particular sets or subsets of interactions and, therefore, is not designed to offer complete interactomes from specific organisms. Another integrated meta-database is iRefWeb (Turner et al., 2010) which contains a broad compendium of PPIs coming from multiple public resources: BIND, BioGRID, CORUM, DIP, IntAct, HPRD, MINT, MPact, MPPI, and OPHID. These sources are not all primary databases, but collections of PPIs produced with different scopes. Moreover, the iRefWeb meta-database integrates several types of protein associations (including genetic interactions) and, in the case of PPIs, it combines experimental and predicted data. The iRefWeb website offers a powerful search and filtering interface, providing outputs with detailed information about the interaction and the original records of the integrated resources. In addition, for many interactions, iRefWeb also provides a molecular interaction *score* (*MI score*) based on the scoring system for interaction confidence originally used by the MINT and described by Ceol et al. (2010). The use of a single *score* (as the ones described earlier provided by Mentha and iRefWeb) can sometimes be misleading or equivocal about the value of specific interactions since a score weighs the evidence based on some fixed criteria that may not have the same value for different types of PPIs. To avoid this problem, some resources, like APID (http://apid.dep.usal.es/), provide for each PPI the number of experimental proofs organized in categories so that the user can decide in a transparent way the ranking criteria based on the type of PPIs that they want to analyze (Alonso-López et al., 2016).

4. HUMAN INTERACTOME: PROTEOME-WIDE HUMAN PPI COMPENDIUMS

Despite that, all the described protein interaction resources, either primary databases or meta-databases, include many human PPIs; there are not many providing the complete compendiums of all known human PPIs (i.e., the human interactome), including the experimental evidence with the references to the corresponding original scientific literature that demonstrate each interaction. One resource that provides one of the better-characterized compendiums of the human interactome is APID (Alonso-López et al., 2016) since it allows the selection of the interactome at different quality levels and different coverage over the corresponding human reference proteome (http://www.uniprot.org/proteomes/UP000005640). Other meta-databases, like Mentha and iRefWeb, also allow downloading complete interactomes though not showing a clear mapping to the human proteome. Another interesting meta-database that has done a major effort to provide good quality, well-demonstrated human PPIs is HINT (High-quality INTeractomes) (Das & Yu, 2012). HINT includes a curated compilation of high-quality PPIs from several primary interactome resources, where interactions are filtered both systematically and manually to remove erroneous and low-quality interactions. The repository of high-quality interactions in HINT contains only manually validated high-throughput experiments and interactions from small-scale studies that have been reported at least twice in the literature. Another interesting characteristic of this resource is that it separates the interactions in two different types according to the experimental methodology that had been used in the determination: binary and cocomplex interactions (De Las Rivas & Fontanillo, 2010; Yu et al., 2008). The separation and accurate distinction between binary physical interactions vs cocomplex associations can be critical to achieving an accurate mapping of a protein interactome. Binary interactions represent a direct biophysical interaction between two proteins. On the other hand, cocomplex associations provide information about comembership in a complex, and in most cases, these associations actually represent indirect physical interactions. The biological information conveyed by these two kinds of interactions is different and, for many applications, it is necessary to have a clear distinction between these two (Yu et al., 2008).

Two recent resource publications based on specific large-scale experimental procedures (one based on a binary method and another on a cocomplex approach) attempt to provide a comprehensive proteome level map of the human interactome. These two resources are: HI (Human Interactome) (http://interactome.dfci.harvard.edu/H_sapiens/) that is based on a binary two-hybrid high-throughput methodology (Rolland et al., 2014); and BioPlex (Biophysical interactions of ORFeome-based complexes) (http://bioplex.hms.harvard.edu/) that is based on a cocomplex high-throughput methodology of tandem affinity purification coupled with mass spectrometry (TAP-MS) (Huttlin et al., 2015). Both resources set up accessible repositories to allow users to download the data and generate the corresponding protein interactome networks.

Finally, a more recent bioinformatic comprehensive effort has created the Human Reference Protein Interactome project (HuRI, http://interactome.baderlab.org/) to generate a first reference map of the human protein–protein interactome network. To reach this target, this project is identifying binary PPIs by systematically interrogating all pairwise combinations of human protein-coding genes using a proteome-scale yeast two-hybrid technology (Y2H) as the primary screening method followed by validation of subsets of PPIs in multiple orthogonal assays for binary PPI detection. This effort is at present providing the most reliable and updated compendium of the human protein–protein interactome.

5. HUMAN DRUG–TARGET INTERACTION RESOURCES

After describing different types of PPI resources that allow the construction of human protein interaction networks, we present a current overview of the bioinformatics databases that provide reliable accessible information about drug–target associations focusing on human proteins as druggable biomolecular preys. This study does not pretend to present an exhaustive review of the numerous pharmacogenomic databases and resources available because they are quite heterogeneous and many times out of the specific focus of our investigation. Therefore, the databases featured herein correspond to a selection done using the following criteria: (i) they should report human protein–drug associations mostly derived or based on experimental data (e.g., not databases focused just on proteins and reporting ligands or databases focused only chemicals and reporting targets derived from other sources); (ii) they should be repositories actively maintained and updated in the last 3 years; and (iii) they should be accessible via the web and public (i.e., available for download). Considering these criteria we found the following drug–target repositories that we describe and analyze later: the Binding Database (BindingDB); the Drug Gene Interaction Database (DGIdb); the Drug SIGnatures DataBase (DSigDB); the IUPHAR/BPS Guide to PHARMACOLOGY (GtoPdb); the Pharmacogenomics Knowledgebase (PharmGKB); and the Therapeutic Target Database (TTD).

The Binding Database (BindingDB, http://www.bindingdb.org) is a public, web-accessible database that contains information for about 7000 protein targets and over 0.6 million small molecules (providing over 1.3 million binding data). This database can be of interest to molecular chemists and pharmacologists given its in-depth collection of experimental, affinity, and binding data from chemical biology and biochemistry journals that are not included in other databases (Gilson et al., 2016). Furthermore, BindingDB includes web-based tools to perform virtual compound screening, validated datasets for computer-aided drug design, and to allow finding compounds for targets of interest or vice versa. In addition to being well supported by documentation, BindingDB allows all target–drug data to be downloaded as TSV files and also as a SQL dump file. As a whole, the drug–target information is displayed in a clear binary way, easy to query and access.

The Drug Gene Interaction Database (DGIdb, http://dgidb.genome.wustl.edu/) is a manually curated meta-database that brings together the information available on drug–target interactions from several resources (for example, PharmGKB and TTD, also described here). The database is now in its version 3.0 and it integrates information from 15 source databases. According to the report published in January 2016, DGIdb contained about 26,300 unique drug–gene interaction claims, involving over 2600 genes and over 7500 drugs. The authors also indicated that about 10,100 drug–gene interactions were potentially relevant to clinical actionability (Wagner et al., 2016). Queries to this database may be performed by gene or drug of interest with the optional application of filters, such as selecting for FDA-approved drugs, as well as browse the gene categories. Results are returned in a clear binary way showing drug–target pair interactions, easy to query and access, and they can be downloaded in TSV format. Results also include other parameters such as potentially druggable gene categories (41 categories) and gene–drug interaction types (33 types). An additional resource for DGIdb is the Bioconductor R package *rDGIdb* (Thurnherr, Singer, Stekhoven, & Beerenwinkel, 2016) that allows queries as in the web but with the ability to integrate them into other bioinformatic analyses. It should be noted that the meta-databases, both in the package and the web interface, do not necessarily include most updated information, as they do not use the latest version available from the source databases.

The Drug SIGnatures DataBase (DSigDB, http://tanlab.ucdenver.edu/DSigDB) is both a manually and automatically curated resource of drugs and small molecules and their target genes (Yoo et al., 2015). The database is organized in four collections: approved drugs, kinase inhibitors, perturbed gene signatures (gene expression profiles perturbed by compounds), and computational drug signatures. Upon performing a search, the user may determine the data collection from which to see the results. For example, from a compound search it is possible to obtain the approval status in several countries, whether the molecule fulfills Lipinski's rules, known activity values (e.g., IC_{50} and K_i), gene targets, and other valuable information for further studies. A gene search, on the other hand, will return its interactions with compounds and the source to discriminate the study of the results. Thus, the results of the queries are returned in a clear binary way showing drug–target pair interactions. The current version (release 1.0) comprises about 22,500 gene sets (i.e., 19,531 unique genes) with over 17,000 compounds, of which 1200 are FDA approved. For researchers who wish to do an in-depth analysis of the data, the entire database is available for download both in TXT and tab-delimited gene set GMT format. Through the DSigDB Collections page, it is also possible to download the individual databases that contribute to the overall collection in GMT format.

The IUPHAR/BPS Guide to PHARMACOLOGY (GtoPdb, http://www.guidetopharmacology.org) provides a searchable database with quantitative information on human drug targets and the prescription medicines and experimental drugs that interact with them. It is also defined as an expert-driven guide to pharmacological targets and the substances that act on them, including both approved drugs or candidate on trial chemicals (Southan et al., 2016). This database is of interest for chemical, biological, and clinical researchers given its comprehensive content. Updated on a quarterly basis, the most current database (version 2017.4) contains information for about 2800 targets, 8900 ligands (including 247 natural products, 1325 approved drugs, and 67 withdrawn drugs), and over 1600 human targets with interactions. Queries may be performed by protein target search or by chemical ligand search, returning clear drug–target binary interactions with an assortment of options to narrow the search. For example, target searches can be performed using a wide variety of database identifiers, limited to transporters, kinases, ion channels, and even as a literature search by PubMed ID. The results provide the identifiers with links to external databases (like experimentally determined 3D structures, extra information on inhibitors, modulators and ligands, or GO annotations). Ligand searches, on the other hand, can be performed using a string search, different chemical identifiers (such as DrugBank IDs, PubChem CID, UniProt IDs), or even drawing the structure on a JavaScript application (Marvin JS App from ChemAxon, https://marvinjs-demo.chemaxon.com/latest/). In addition to the structure and physicochemical properties of the ligand, search results include biological activity, clinical data (including ADME, pharmacokinetics, and organ function impairment), and structurally similar ligands. GtoPdb offers CSV data files for download or a copy of the database in PostgreSQL. The download page also offers a set of generic slides and a poster for use in teaching and presentations.

The Pharmacogenomics Knowledgebase (PharmGKB, https://www.pharmgkb.org/) is a publicly available, online knowledge base resource that provides consistent gene variant–drug response interactions (i.e., interactions of drugs with proteins and protein variants) derived from manually curated pharmacogenomics or pharmacogenetics information from the primary literature (Whirl-Carrillo et al., 2012). Accordingly, it offers pharmacogenomic drug properties, information on the genes and their associations with drugs, variant and clinical annotations, and drug label annotations obtained from: the US Food and Drug Administration (FDA), the European Medicines Agency (EMA), the Japanese Pharmaceuticals and Medical Devices Agency (PMDA), or the Health Canada Santé Canada (HCSC). A useful feature of PharmGKB is the ability to search by gene, chemical, disease, pathway, dosing guidelines, and drug label annotations of several regulating bodies. Upon a drug search, for example, the user is shown several tabs with data that include useful information such as mechanism of action, pharmacology, contraindications, ADMET details, and basic chemical properties. It is also possible to retrieve pathways based on literature evidence and gene targets. PharmGKB has information on about 27,000 unique genes, about 3200 drugs, and over 3500 diseases. It should be noted that some of the compound entries represent groups of drugs, such as alkylating agents and topoisomerase I inhibitors; and also that due to the complexity of the information the data are not displayed in an easy drug–target binary way. The database is frequently updated, with literature and variant annotations revised. Data are available for download, but to do so and to explore the drug label annotations, the variants information and the gene and drug relationship information data require registration and sign-in.

The Therapeutic Target Database (TTD, http://bidd.nus.edu.sg/group/ttd/ttd.asp) provides information about the known and explored therapeutic protein and nucleic acid targets, the targeted disease, and the corresponding drugs directed at each of these targets linked to the correlated diseases. The database includes about 2500 protein targets and about 30,000 drugs (including about 2000 approved, about 9500 on clinical trials, and about 18,000 experimental). The database cross-links to biological pathways resources such as KEGG, Reactome, PID, and PANTHER

(Yang et al., 2016). A biomarker search will return, for example, the disease(s) in which it is involved, known treatments, FDA approval status of a drug, and external links to various other resources. An advantage of the TTD is the ability to perform various types of searches, such as disease, target, drug, biomarker, and pathway, thus being accessible to researchers and facilitating the association between drugs, targets, and diseases. The TTD was last updated in late 2015 and its raw data can be downloaded from the website.

6. CANCER GENES/PROTEINS COMPENDIUM

As indicated earlier, in this work we undertake the identification of drug-targeted genes focusing our investigations on cancer and cancer genes. Before delving into the construction and analysis of specific protein–drug interaction networks corresponding to cancer genes, we present a description of the main resources that provide curated compendiums of human genes (gene products/proteins) directly involved in cancer (i.e., genes/proteins that are conductors of the malignant transformation of normal cells into tumoral, by tumorigenesis, progression, and tumor evolution).

Probably the best-known and most referenced resource on human cancer genes is the Cancer Gene Census (CGC, http://cancer.sanger.ac.uk/census). The CGC is an ongoing effort to catalog those genes for which mutations have been causally implicated in cancer. The first original census was published in 2004 in *Nature Reviews Cancer* (Futreal et al., 2004), and since then, it has been regularly updated. A quick perusal of the census table provides the gene symbol with its name, mutation type, tumor type, tissue type, and whether the gene is dominant or recessive (if known), and other useful molecular features. Upon choosing a gene of interest, the user is provided with a wealth of information, including tissue and mutation type distribution, variants, known drug resistance, and a list of references associated with the gene. The latest available version of the Cancer Gene Census available for download (v.82 from 31 August 2017) includes 567 human genes. Within this set, the census distinguishes four main subtypes of cancer genes: (i) the oncogenes (OCG) that are 209 genes and are causal of cancer by a gain-of-function behavior; (ii) the tumor-suppressor genes (TSG) that are 198 genes and are causal of cancer by a loss-of-function behavior; (iii) a group 67 genes that have an unclear dual or mix behavior (i.e., sometimes as oncogenes but others as tumor-suppressor genes, depending on the type or subtype of cancer and the tumor circumstances);

and (iv) and a final group of 93 genes that correspond to genes suffering genomic fusion or rearrangement events, usually forming chimeric genes in tumoral malignant cells.

The Cancer Gene Census is included in the web portal of the Catalogue of Somatic Mutations In Cancer (COSMIC) developed by the Sanger Institute (http://www.sanger.ac.uk/) that is the world's largest and most comprehensive resource for exploring the impact of somatic mutations in human cancer. COSMIC is an expert-curated database of cancer somatic mutations with almost 4.5 million coding mutations from all types of human cancer, including rare types, from over 1.3 million samples in over 24,800 publications. COSMIC is manually curated and provides details on the gene variants (including mutations, fusions, copy number variation, expression, and methylation), tissue distribution, and many other attributes (Forbes et al., 2017). It should be noted that the academic user has to register into the COSMIC portal to receive free access to all data inside and receive email announcements, such as when a new release is available.

As indicated earlier, the collection of cancer genes included in CGC is quite large (over 500 genes) since they are associated with the information on mutations registered in COSMIC for millions of cancer samples. A more restrictive and well-curated compendium of cancer genes based both on detailed genomic profiling and clinical information was published by Bert Vogelstein, who is one of the world's experts in cancer (Vogelstein et al., 2013). In this study, Vogelstein and collaborators indicate that comprehensive analyses of the genomic landscapes of common forms of human cancer have revealed the existence of 138 genes that, when altered by intragenic mutations, can promote or drive tumorigenesis. Therefore, the set of real causal cancer genes conferring selective growth advantage is smaller; and in the CGC collection, most of the genes may be considered passengers rather than driver cancer genes. The set of 138 driver genes reported by Vogelstein et al. (2013) includes 64 oncogenes (OCG) and 74 tumor-suppressor genes (TSG) associated to three major biological processes: cell fate, cell survival, and genome maintenance. Further analysis of the mutation profile of these genes allows reducing the number of the set to include only those that have a significant mutation rate in tested samples, therefore selecting around 90 cancer genes that are most frequently mutated. This final subset is the one that we use in the construction of a cancer protein–drug bipartite network at the end of this study.

7. CONSTRUCTION OF A PPI NETWORK WITH CANCER GENES

Based on the set of 567 human genes included in the Cancer Gene Census, described earlier, we map such genes to human proteins. This mapping provided 582 proteins that were used to build a protein interaction network retrieving all the known physical interactions between any of them. This PPI network, produced using APID interactomes server (Alonso-López et al., 2016), is presented in Fig. 1 and displays in the first part (Fig. 1A) a view of the network corresponding to the 582 cancer proteins and including 4968 interactions. The network is laid out in four sections according to the degree of the proteins. The first section includes, in a circle with red and orange colors, the 33 most connected proteins selected by having the highest degree. Within these proteins, the top-10 are 4 transcription factors (TFs) and 6 other proteins. The six most connected proteins are: NTRK1 (143 interactions), TP53 (123 interactions), EP300 (106 interactions), XPO1 (106 interactions), EGFR (103 interactions), and CTNNB1 (100 interactions). This group includes some well-known key cancer proteins, such as TP53 and EGFR, which appear altered and involved in many types and subtypes of cancer. The TFs found most connected to other cancer proteins are: JUN (136 interactions), ESR1 (113 interactions), CREBBP (95 interactions), and MYC (94 interactions). The identification of these four TFs with high connectivity can indicate the importance of these regulatory genes as drivers of the common biological processes that occur in a cancer cell. In fact, CREBBP, JUN, and MYC belong to a large family of basic leucine zipper (bZIP)-containing transcription factors that frequently show correlated expression profiles.

In order to investigate the transcriptomic coregulation of the genes involved in cancer, we performed a second analysis of the set of 582 human proteins. Fig. 1B presents the results of the search for common transcription factor binding sites (TFBS) in the promoter regions of the cancer genes corresponding to the set of 582 proteins studied. To achieve this, we performed a genome-wide search for TFBS applying an enrichment analysis provided by the iRegulon tool (Janky et al., 2014). This bioinformatic tool allows the identification of the TFs that are most significantly enriched as coregulators of the genes included in our search. The six TFs that showed the best enrichment, called master regulators, are presented in Fig. 1B: CTBP2, FOXA2, FOXJ3, TFDP3, EGR2, and OTUD4. These TFs are

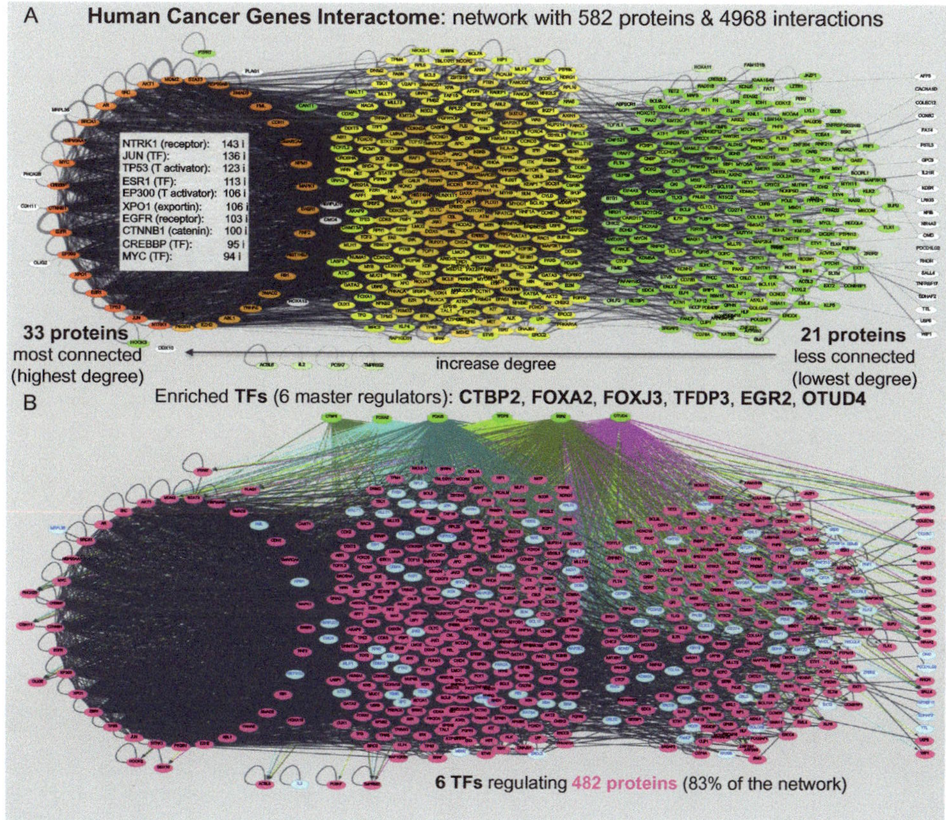

Fig. 1 See figure legend on next page.

putative regulators of the expression of 482 proteins of the total set analyzed, showing that 83% of the network can follow a common coexpression program coordinated by a small set of master genes. For example, CTBP2 is a gene regulated by HIPK2 that is at the same time regulated by CREBBP, PML, and TP53, all TFs very much implicated in cancer and malignant progression. Likewise, FOXA2 (together with its close homolog FOXA1) and EGR2 (together with its close homolog EGR1) are TFs that have been closely associated with carcinogenesis.

8. CANCER PROTEIN–DRUG BIPARTITE NETWORK REVEALS DRUGGABLE MODULES

As a final part of this study, we present the construction of a cancer protein–drug bipartite network derived from the database of genomic biomarkers for cancer drugs and clinical targetability in solid tumors, developed by Dienstmann, Jang, Bot, Friend, and Guinney (2015). We select a subset of 92 cancer driver genes, based on the work by Vogelstein et al. (2013) as reported earlier, and we map the chemicals that are known to target specific proteins within this set. Thus, we build a relational map including 92 proteins and 83 drugs that included 176 directed interactions. This map is presented in Fig. 2A that shows a protein–drug bipartite network. A bipartite network (or bigraph) is a graph whose vertices can be divided into two disjoint and independent sets of nodes, that we called D (drugs) and T (protein targets), such that every edge connects a vertex in D to one in T. The edges between a drug and a protein in this network are different according to the

Fig. 1 Human Cancer Genes Interactome. Protein–protein interaction network including 582 human proteins from the Cancer Gene Census (CGS, http://cancer.sanger.ac.uk/census/) plus all the known physical interactions between them obtained from the APID interactomes server (Alonso-López et al., 2016). (A) Visualization of such PPI network (with 582 proteins and 4968 interactions) using a four-section layout (*red*, *yellow*, *green*, and *white*) according to the degree of the proteins. The first section includes (in a *circle* with *red* to *orange colors*) 33 proteins selected by having the highest degree (i.e., they are the most connected proteins). The top-10 proteins with the highest degree are also included in a *white box* indicating the number of interactions they have. (B) Visualization of the same PPI network marking in *purple* 482 proteins (out of 582 total) that are candidate regulatory targets of six transcriptions factors (TFs: CTBP2, FOXA2, FOXJ3, TFDP3, EGR2, OTUD4), since they are the most enriched TFs according to the TFBS (Transcription Factor Binding Sites) present in the promoter regions of the genes corresponding to such proteins of the network.

A **Cancer gene drug targets**: drug–target bipartite network including 92 proteins, 83 drugs, & 176 interactions

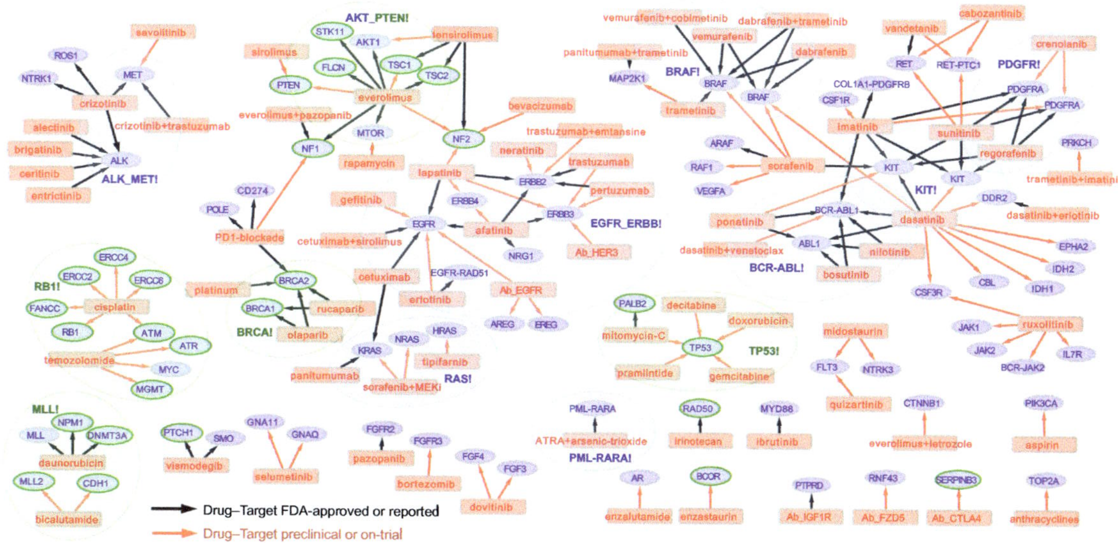

13 major cancer druggable modules: AKT_PTEN, ALK_MET, BCR-ABL, BRAF, BRCA, EGFR_ERBB, KIT, MLL, PDGFR, PML-RARA, RAS, RB1, TP53

B

Drugs	Degree	Genes	Degree
dasatinib	11	EGFR	7
everolimus	9	ERBB2	7
imatinib	7	BCR-ABL1	6
sorafenib	6	BRAF	6
sunitinib	6	KIT	6
cisplatin	6	ALK	5
afatinib	5	BRAF	5
lapatinib	5	TP53	5
ruxolitinib	5	Drugs & genes with ≥ 5 interactions	

C

Tumor supressor cancer genes (loss of function): 27 genes		
ATM	ERCC6	PTCH1
ATR	FANCC	PTEN
BCOR	FLCN	RAD50
BRCA1	MGMT	RB1
BRCA2	MLL2	SERPINB3
CDH1	NF1	STK11
DNMT3A	NF2	TP53
ERCC2	NPM1	TSC1
ERCC4	PALB2	TSC2

Fig. 2 See figure legend on next page.

therapeutic status of the connection, since there are drug–target links approved by FDA and used in the clinic (marked as black arrows in Fig. 2A) and links that correspond to drug–target associations still in preclinical or trial phases (marked as pale red arrows in Fig. 2A). Thanks to this labeling we can distinguish in the network the existence of some modules that reveal well-established cancer targets, such as: (i) the module of EGFR (i.e., ERBB1) and ERBB2 with their paralogous proteins (ERBB3, 4) interacting with drugs like afatinib, lapatinib, and trastuzumab which play a central therapeutic role in breast and lung cancer; (ii) the double module that includes BCR-ABL1 and KIT associated with the action of imatinib and dasatinib as very efficient cancer drugs (in fact imatinib under the brand name *Gleevec*, FDA approved in 2001, was the first bullet drug efficiently used against cancer and derived from a rational molecular drug design); or (iii) the module of BRAF targeted by vemurafenib and dabrafenib that are powerful inhibitors of this oncogene, very relevant in skin cancer, melanoma.

The analysis of the data included in the cancer protein–drug bipartite network (Fig. 2) also reveals the existence of other cancer druggable modules enclosing groups of human tumor-suppressor genes (TSG) correlated by their association to specific drugs. Within these modules we highlight: the BRCA module, including BRCA1 and BRCA2, targeted both by olaparib and rucaparib; the RB1 module that brings together important tumor suppressors like ATM, ATR, and the ERCC family targeted by cisplatin, an anticancer (antineoplastic or cytotoxic) chemotherapy drug used to treat a large number of cancers; or the module formed by multiple tumor suppressors (NF1, PTEN, FLCN, TSC1, and TSC2) indirectly targeted by the drug everolimus, an inhibitor of the mammalian target of rapamycin (mTOR,

Fig. 2 Bipartite network of cancer gene–drug targets. Cancer protein–drug bipartite network including 92 cancer driver genes/proteins and 83 selected drugs that interact with such proteins. (A) Visualization of the network that includes 176 directed interactions, showing with *black arrows* the links corresponding to drugs approved by FDA and with *pale red arrows* the links corresponding to drugs that are still on-trial or preclinical use. The nodes correspond to cancer proteins, with *blue ellipses* corresponding to oncogenes (OCG) and *green line ellipses* corresponding to tumor-suppressor genes (TSG). The network includes a series of colored areas (*blue* for OCGs and *green* for TSGs) to show 13 regions where the connections bring together certain proteins and drugs. These regions allow defining what we called cancer druggable modules and are labeled: AKT_PTEN, ALK_MET, BCR-ABL, BRAF, BRCA, EGFR_ERBB, KIT, MLL, PDGFR, PML-RARA, RAS, RB1. (B) Table showing the degree of the proteins and the drugs that have the highest number of interactions ($\geq$5). (C) Table showing the list of 27 proteins that are tumor suppressors.

oncogene linked to the inactivation of such set of tumor-suppressor genes). Finally, we measured the degree of the proteins and the drugs in the bipartite graph (Fig. 2B) to evaluate their centrality or weight in the network. Accordingly, we observed that many of the key cancer genes (like EGFR, ERBB2, BCR-ABL1, and BRAF) show a high number of interactions.

REFERENCES

Alonso-López, D., Gutiérrez, M. A., Lopes, K. P., Prieto, C., Santamaría, R., & De Las Rivas, J. (2016). APID interactomes: Providing proteome-based interactomes with controlled quality for multiple species and derived networks. *Nucleic Acids Research*, *44*(W1), W529–35.

Aranda, B., Blankenburg, H., Kerrien, S., Brinkman, F. S. L., Ceol, A., Chautard, E., et al. (2011). PSICQUIC and PSISCORE: Accessing and scoring molecular interactions. *Nature Methods*, *8*(7), 528–529.

Calderone, A., Castagnoli, L., & Cesareni, G. (2013). Mentha: A resource for browsing integrated protein-interaction networks. *Nature Methods*, *10*(8), 690–691.

Ceol, A., Chatr-Aryamontri, A., Licata, L., Peluso, D., Briganti, L., Perfetto, L., et al. (2010). MINT, the molecular interaction database: 2009 update. *Nucleic Acids Research*, *38*(Database issue), D532–9.

Chatr-Aryamontri, A., Breitkreutz, B.-J., Oughtred, R., Boucher, L., Heinicke, S., Chen, D., et al. (2015). The BioGRID interaction database: 2015 update. *Nucleic Acids Research*, *43*(Database issue), D470–8.

Chautard, E., Fatoux-Ardore, M., Ballut, L., Thierry-Mieg, N., & Ricard-Blum, S. (2011). MatrixDB, the extracellular matrix interaction database. *Nucleic Acids Research*, *39*(Database issue), D235–40.

Das, J., & Yu, H. (2012). HINT: High-quality protein interactomes and their applications in understanding human disease. *BMC Systems Biology*, *6*, 92.

De Las Rivas, J., & de Luis, A. (2004). Interactome data and databases: Different types of protein interaction. *Comparative and Functional Genomics*, *5*(2), 173–178.

De Las Rivas, J., & Fontanillo, C. (2010). Protein-protein interactions essentials: Key concepts to building and analyzing interactome networks. *PLoS Computational Biology*, *6*(6), e1000807.

Dienstmann, R., Jang, I. S., Bot, B., Friend, S., & Guinney, J. (2015). Database of genomic biomarkers for cancer drugs and clinical targetability in solid tumors. *Cancer Discovery*, *5*(2), 118–123.

Ellinghaus, D., Jostins, L., Spain, S. L., Cortes, A., Bethune, J., Han, B., et al. (2016). Analysis of five chronic inflammatory diseases identifies 27 new associations and highlights disease-specific patterns at shared loci. *Nature Genetics*, *48*(5), 510–518.

Forbes, S. A., Beare, D., Boutselakis, H., Bamford, S., Bindal, N., Tate, J., et al. (2017). COSMIC: Somatic cancer genetics at high-resolution. *Nucleic Acids Research*, *45*(D1), D777–D783.

Futreal, P. A., Coin, L., Marshall, M., Down, T., Hubbard, T., Wooster, R., et al. (2004). A census of human cancer genes. *Nature Reviews Cancer*, *4*(3), 177–183.

Gilson, M. K., Liu, T., Baitaluk, M., Nicola, G., Hwang, L., & Chong, J. (2016). BindingDB in 2015: A public database for medicinal chemistry, computational chemistry and systems pharmacology. *Nucleic Acids Research*, *44*(D1), D1045–53.

Huttlin, E. L., Ting, L., Bruckner, R. J., Gebreab, F., Gygi, M. P., Szpyt, J., et al. (2015). The BioPlex network: A systematic exploration of the human interactome. *Cell*, *162*(2), 425–440.

Janky, R., Verfaillie, A., Imrichová, H., Van de Sande, B., Standaert, L., Christiaens, V., et al. (2014). iRegulon: From a gene list to a gene regulatory network using large motif and track collections. *PLoS Computational Biology*, *10*(7), e1003731.

Kerrien, S., Aranda, B., Breuza, L., Bridge, A., Broackes-Carter, F., Chen, C., et al. (2012). The IntAct molecular interaction database in 2012. *Nucleic Acids Research*, *40*(Database issue), D841–6.

Keshava Prasad, T. S., Goel, R., Kandasamy, K., Keerthikumar, S., Kumar, S., Mathivanan, S., et al. (2009). Human Protein Reference Database—2009 update. *Nucleic Acids Research*, *37*(Database issue), D767–72.

Licata, L., Briganti, L., Peluso, D., Perfetto, L., Iannuccelli, M., Galeota, E., et al. (2012). MINT, the molecular interaction database: 2012 update. *Nucleic Acids Research*, *40*(Database issue), D857–61.

Mani, R., St. Onge, R. P., Hartman, J. L., 4th, Giaever, G., & Roth, F. P. (2008). Defining genetic interaction. *Proceedings of the National Academy of Sciences of the United States of America*, *105*(9), 3461–3466.

Orchard, S., Ammari, M., Aranda, B., Breuza, L., Briganti, L., Broackes-Carter, F., et al. (2014). The MIntAct project—IntAct as a common curation platform for 11 molecular interaction databases. *Nucleic Acids Research*, *42*(Database issue), D358–63.

Orchard, S., Salwinski, L., Kerrien, S., Montecchi-Palazzi, L., Oesterheld, M., Stümpflen, V., et al. (2007). The minimum information required for reporting a molecular interaction experiment (MIMIx). *Nature Biotechnology*, *25*(8), 894–898.

Rolland, T., Taşan, M., Charloteaux, B., Pevzner, S. J., Zhong, Q., Sahni, N., et al. (2014). A proteome-scale map of the human interactome network. *Cell*, *159*(5), 1212–1226.

Salwinski, L., Miller, C. S., Smith, A. J., Pettit, F. K., Bowie, J. U., & Eisenberg, D. (2004). The database of interacting proteins: 2004 update. *Nucleic Acids Research*, *32*(Database issue), D449–51.

Southan, C., Sharman, J. L., Benson, H. E., Faccenda, E., Pawson, A. J., Alexander, S. P. H., et al. (2016). The IUPHAR/BPS Guide to PHARMACOLOGY in 2016: Towards curated quantitative interactions between 1300 protein targets and 6000 ligands. *Nucleic Acids Research*, *44*(D1), D1054–68.

Thurnherr, T., Singer, F., Stekhoven, D. J., & Beerenwinkel, N. (2016). Genomic variant annotation workflow for clinical applications. *F1000Research*, *5*, 1963.

Turner, B., Razick, S., Turinsky, A. L., Vlasblom, J., Crowdy, E. K., Cho, E., et al. (2010). iRefWeb: Interactive analysis of consolidated protein interaction data and their supporting evidence. *Database: The Journal of Biological Databases and Curation*, *2010*, baq023.

Vogelstein, B., Papadopoulos, N., Velculescu, V. E., Zhou, S., Diaz, L. A., Jr., & Kinzler, K. W. (2013). Cancer genome landscapes. *Science*, *339*(6127), 1546–1558.

Wagner, A. H., Coffman, A. C., Ainscough, B. J., Spies, N. C., Skidmore, Z. L., Campbell, K. M., et al. (2016). DGIdb 2.0: Mining clinically relevant drug-gene interactions. *Nucleic Acids Research*, *44*(D1), D1036–44.

Whirl-Carrillo, M., McDonagh, E. M., Hebert, J. M., Gong, L., Sangkuhl, K., Thorn, C. F., et al. (2012). Pharmacogenomics knowledge for personalized medicine. *Clinical Pharmacology and Therapeutics*, *92*(4), 414–417.

Yang, H., Qin, C., Li, Y. H., Tao, L., Zhou, J., Yu, C. Y., et al. (2016). Therapeutic target database update 2016: Enriched resource for bench to clinical drug target and targeted pathway information. *Nucleic Acids Research*, *44*(D1), D1069–74.

Yoo, M., Shin, J., Kim, J., Ryall, K. A., Lee, K., Lee, S., et al. (2015). DSigDB: Drug signatures database for gene set analysis. *Bioinformatics*, *31*(18), 3069–3071.

Yu, H., Braun, P., Yildirim, M. A., Lemmens, I., Venkatesan, K., Sahalie, J., et al. (2008). High-quality binary protein interaction map of the yeast interactome network. *Science*, *322*(5898), 104–110.